$z = z_0$

z_0	0.00	0.01	0.02	0.03	0.04	0.05	0.06	0.07	0.08	0.09
0.0	0.5000	0.5040	0.5080	0.5120	0.5160	0.5199	0.5239	0.5279	0.5319	0.5359
0.1	0.5398	0.5438	0.5478	0.5517	0.5557	0.5596	0.5636	0.5675	0.5714	0.5753
0.2	0.5793	0.5832	0.5871	0.5910	0.5948	0.5987	0.6026	0.6064	0.6103	0.6141
0.3	0.6179	0.6217	0.6255	0.6293	0.6331	0.6368	0.6406	0.6443	0.6480	0.6517
0.4	0.6554	0.6591	0.6628	0.6664	0.6700	0.6736	0.6772	0.6808	0.6844	0.6879
0.5	0.6915	0.6950	0.6985	0.7019	0.7054	0.7088	0.7123	0.7157	0.7190	0.7224
0.6	0.7257	0.7291	0.7324	0.7357	0.7389	0.7422	0.7454	0.7486	0.7517	0.7549
0.7	0.7580	0.7611	0.7642	0.7673	0.7704	0.7734	0.7764	0.7794	0.7823	0.7852
0.8	0.7881	0.7910	0.7939	0.7967	0.7995	0.8023	0.8051	0.8078	0.8106	0.8133
0.9	0.8159	0.8186	0.8212	0.8238	0.8264	0.8289	0.8315	0.8340	0.8365	0.8389
1.0	0.8413	0.8438	0.8461	0.8485	0.8508	0.8531	0.8554	0.8577	0.8599	0.8621
1.1	0.8643	0.8665	0.8686	0.8708	0.8729	0.8749	0.8770	0.8790	0.8810	0.8830
1.2	0.8849	0.8869	0.8888	0.8907	0.8925	0.8944	0.8962	0.8980	0.8997	0.9015
1.3	0.9032	0.9049	0.9066	0.9082	0.9099	0.9115	0.9131	0.9147	0.9162	0.9177
1.4	0.9192	0.9207	0.9222	0.9236	0.9251	0.9265	0.9279	0.9292	0.9306	0.9319
1.5	0.9332	0.9345	0.9357	0.9370	0.9382	0.9394	0.9406	0.9418	0.9429	0.9441
1.6	0.9452	0.9463	0.9474	0.9484	0.9495	0.9505	0.9515	0.9525	0.9535	0.9545
1.7	0.9554	0.9564	0.9573	0.9582	0.9591	0.9599	0.9608	0.9616	0.9625	0.9633
1.8	0.9641	0.9649	0.9656	0.9664	0.9671	0.9678	0.9686	0.9693	0.9699	0.9706
1.9	0.9713	0.9719	0.9726	0.9732	0.9738	0.9744	0.9750	0.9756	0.9761	0.9767
2.0	0.9772	0.9778	0.9783	0.9788	0.9793	0.9798	0.9803	0.9808	0.9812	0.9817
2.1	0.9821	0.9826	0.9830	0.9834	0.9838	0.9842	0.9846	0.9850	0.9854	0.9857
2.2	0.9861	0.9864	0.9868	0.9871	0.9875	0.9878	0.9881	0.9884	0.9887	0.9890
2.3	0.9893	0.9896	0.9898	0.9901	0.9904	0.9906	0.9909	0.9911	0.9913	0.9916
2.4	0.9918	0.9920	0.9922	0.9925	0.9927	0.9929	0.9931	0.9932	0.9934	0.9936
2.5	0.9938	0.9940	0.9941	0.9943	0.9945	0.9946	0.9948	0.9949	0.9951	0.9952
2.6	0.9953	0.9955	0.9956	0.9957	0.9959	0.9960	0.9961	0.9962	0.9963	0.9964
2.7	0.9965	0.9966	0.9967	0.9968	0.9969	0.9970	0.9971	0.9972	0.9973	0.9974
2.8	0.9974	0.9975	0.9976	0.9977	0.9977	0.9978	0.9979	0.9979	0.9980	0.9981
2.9	0.9981	0.9982	0.9982	0.9983	0.9984	0.9984	0.9985	0.9985	0.9986	0.9986
3.0	0.9987	0.9987	0.9987	0.9988	0.9988	0.9989	0.9989	0.9989	0.9990	0.9990
3.1	0.9990	0.9991	0.9991	0.9991	0.9992	0.9992	0.9992	0.9992	0.9993	0.9993
3.2	0.9993	0.9993	0.9994	0.9994	0.9994	0.9994	0.9994	0.9995	0.9995	0.9995
3.3	0.9995	0.9995	0.9995	0.9996	0.9996	0.9996	0.9996	0.9996	0.9996	0.9997
3.4	0.9997	0.9997	0.9997	0.9997	0.9997	0.9997	0.9997	0.9997	0.9997	0.9998
3.5	0.9998	0.9998	0.9998	0.9998	0.9998	0.9998	0.9998	0.9998	0.9998	0.9998
3.6	0.9998	0.9998	0.9999	0.9999	0.9999	0.9999	0.9999	0.9999	0.9999	0.9999
3.7	0.9999	0.9999	0.9999	0.9999	0.9999	0.9999	0.9999	0.9999	0.9999	0.9999
3.8	0.9999	0.9999	0.9999	0.9999	0.9999	0.9999	0.9999	0.9999	0.9999	0.9999
3.9	1.0000	1.0000	1.0000	1.0000	1.0000	1.0000	1.0000	1.0000	1.0000	1.0000

PRENTICE HALL SERIES IN MATHEMATICS FOR MIDDLE SCHOOL TEACHERS

JOHN BEEM *Geometry Connections*
ASMA HARCHARRAS and DORINA MITREA *Calculus Connections*
IRA J. PAPICK *Algebra Connections*
DEBRA A. PERKOWSKI and MICHAEL PERKOWSKI *Data and Probability Connections*

Data and Probability Connections

Debra A. Perkowski
Mathematics Department
Westminster College

Michael Perkowski
Learning Center
University of Missouri-Columbia

Upper Saddle River, New Jersey 07458

Library of Congress Cataloging-in-Publication Data

Perkowski, Debra A.
  Data and probability connections / Debra A. Perkowski, Michael Perkowski.
    p. cm.—(Prentice Hall series in mathematics for middle school teachers)
  Includes bibliographical references and index.
  ISBN 0-13-144922-2
  1. Statistics–Study and teaching (Middle school) 2. Probabilities–Study and teaching (Middle school) I. Perkowski, Michael. II. Title. III. Series.

QA276.18.P467 2007
519.5071′2—dc22

2006035817

Editor in Chief: *Sally Yagan*
Executive Acquisitions Editor: *Petra Recter*
Project Manager: *Michael Bell*
Editorial Assistant/Supplement Editor: *Joanne Wendelken*
Production Management: *Progressive Publishing Alternatives*
Assistant Managing Editor: *Bayani Mendoza de Leon*
Senior Managing Editor: *Linda Mihatov Behrens*
Executive Managing Editor: *Kathleen Schiaparelli*
Manufacturing Manager: *Alexis Heydt-Long*
Manufacturing Buyer: *Maura Zaldivar*
Director of Marketing: *Patrice Jones*
Art Director: *Jayne Conte*
Cover Designer: *Bruce Kenselaar*
Art Studio/Formatter: *Laserwords*
Cover Image: *The Freedom Bridge (Szabadsag hid) over the Danube River at Gellert Hill, Budapest, Hungary*—© *Getty Images, Inc., Lonely Planet Images*

©2007 Pearson Education, Inc.
Pearson Prentice Hall
Pearson Education, Inc.
Upper Saddle River, New Jersey 07458

All rights reserved. No part of this book may be reproduced, in any form or by any other means, without permission in writing from the publisher.

Pearson Prentice Hall™ is a trademark of Pearson Education, Inc.
Development of these materials was supported by a grant from the National Science Foundation (ESI 0101822).

Printed in the United States of America
10 9 8 7 6 5 4 3 2 1

ISBN: 0-13-144922-2

Pearson Education LTD., *London*
Pearson Education Australia PTY, Limited, *Sydney*
Pearson Education Singapore, Pte. Ltd.
Pearson Education North Asia Ltd., *Hong Kong*
Pearson Education Canada, Ltd., *Toronto*
Pearson Educación de Mexico, S.A. de C.V.
Pearson Education—Japan, *Tokyo*
Pearson Education Malaysia, Pte. Ltd.

Contents

Preface .... **vii**

List of Figures .... **xi**

1 Elements of Statistics **1**
   1.1   Getting Started 1
   1.2   Types of Data 10

2 Organizing and Displaying Data **20**
   2.1   Displaying Categorical Data 20
   2.2   Displaying Quantitative Data 25
   2.3   Misleading Graphs 37

3 Describing Data with Numbers **48**
   3.1   Measures of Center 48
   3.2   Measures of Spread 57
   3.3   Measures of Position 70
   3.4   Box-and-Whisker Plots 74

4 Data with Two Variables **85**
   4.1   Scatter Plots and Correlation 85
   4.2   Pearson's Correlation Coefficient 90
   4.3   Slopes and Equations of Fitted Lines 96
   4.4   The Least Squares Line 100
   4.5   The Median-Median Line 106

5 Probability **118**
   5.1   What Is Probability? 119
   5.2   Outcomes and Events 120
   5.3   Basic Probability Rules 130
   5.4   Conditional Probability and Independence 133
   5.5   Multiplication Rules 137
   5.6   Geometric Probability 141

6 Counting Techniques **152**
   6.1   The Multiplication Principle or the Fundamental Counting Principle 154
   6.2   Permutations 160
   6.3   Combinations 168
   6.4   Mixed Counting Problems 173

Contents

7 Random Variables and Probability Distributions — **183**
- 7.1 What Is a Random Variable? — 183
- 7.2 The Mean of a Random Variable — 189
- 7.3 Variance and Standard Deviation — 193
- 7.4 Binomial Random Variables — 200
- 7.5 The Normal Curve — 207
- 7.6 Normal Approximations — 219

8 Distributions from Random Samples — **234**
- 8.1 Random Sampling — 234
- 8.2 The Distribution of Sample Means — 242
- 8.3 The Distribution of Sample Proportions — 253

9 Estimating with Confidence — **266**
- 9.1 Confidence Intervals for Proportions — 267
- 9.2 Confidence Intervals for Means — 280
- 9.3 Sample Size — 293

10 Testing Hypotheses — **301**
- 10.1 What Is a Hypothesis Test? — 301
- 10.2 Tests about Proportions — 309
- 10.3 The P-Value for a Test — 318
- 10.4 Tests about Means — 326

Some Final Thoughts — **342**

Glossary — **343**

References — **358**

Answers and Hints for Odd Numbered Exercises — **359**

Photo Credits — **380**

Index — **385**

Preface

Improving the quality of mathematics education for middle school students is of critical importance at this time, and creating new opportunities for students to learn important mathematical skills under the leadership of well-prepared and dedicated teachers is essential. New standards-based curricula and instructional models, coupled with on-going professional development and teacher preparation, are fundamental to this effort.

These sentiments are eloquently articulated in the Glenn Commission Report: *Before It's Too Late: A Report to the Nation from the National Commission on Mathematics and Science Teaching for the 21st Century* (U.S. Department of Education, 2000). In fact, the principal message of the Glenn Commission Report is that America's students must improve their mathematics and science performance if they are to be successful in our rapidly changing technological world. To this end, the report recommends that we greatly intensify our focus on improving the quality of mathematics and science teaching in grades K–12 by bettering the quality of teacher preparation, and it also stresses the necessity of developing creative plans to attract and retain substantial numbers of future mathematics and science teachers.

Some fifteen years ago, mathematics teachers, mathematics educators, and mathematicians collaborated to develop the architecture for standards-based reform, and their recommendations for the improvement of school mathematics, instruction, and assessment were articulated in three seminal documents published by the National Council of Teachers of Mathematics (*Curriculum and Evaluation Standards for School Mathematics* [1989], *Professional Standards for School Mathematics* [1995], *Assessment Standards in School Mathematics* [1995]; more recently, these three documents were updated and combined into a single book, *NCTM Principles and Standards for School Mathematics, a.k.a. PSSM* [2000]).

The vision of school mathematics laid out in these three foundational documents was outstanding in spirit and content, yet it has remained abstract in practice. Concrete exemplary models reflecting those standards have been lacking, and implementations of the "recommendations" could not be realized without significant commitment of resources. Recognizing the opportunity to stimulate improvement in student learning, the National Science Foundation (NSF) made a strong commitment to bring life to the documents' messages by supporting several K–12 mathematics curriculum development projects (standards-based curriculum) and a number of other related dissemination and implementation projects.

Standards-based middle school curricula are designed to engage students in a variety of mathematical experiences, including thoughtfully planned classroom explorations that provide and reinforce fundamental skills while illuminating the power and utility of mathematics in our world. These materials integrate central concepts in algebra, geometry, data analysis and probability, and mathematics of change and focus on important unifying ideas such as proportional reasoning.

The mathematical content of standards-based middle grade mathematics materials is challenging and relevant to our technological world. Its effective classroom implementation is dependent upon teachers having a strong and appropriate mathematical preparation. *The Connecting Middle School and College Mathematics Project* $(CM)^2$ is a three-year (2001–2004) National Science Foundation funded project addressing the need for improved teacher qualifications and viable recruitment plans for middle grade mathematics teachers through the development of four foundational mathematics courses with accompanying support materials and the creation and implementation of effective teacher recruitment models.

The $(CM)^2$ materials are built upon a framework laid out in the *CBMS Mathematical Education of Teachers Book* (MET) (2001). This report outlines recommendations for the mathematical preparation of middle grade teachers that differ significantly from those for the preparation of elementary teachers and provides guidance to those developing new programs. Our books are designed to provide middle grade mathematics teachers with a strong mathematical foundation and connect the mathematics they are learning with the mathematics they will be teaching. Their focus is on algebraic and geometric structures, data analysis and probability, and mathematics of change, and they employ standards-based middle grade mathematics curricular materials as a springboard to explore and learn mathematics in more depth. They have been extensively piloted in Summer Institutes, courses offered at school-based sites, a variety of professional development programs, and in courses offered at a number of universities throughout the nation.

Since the publication of the NCTM's original vision of school mathematics, *Curriculum and Evaluation Standards for School Mathematics* (1989), data analysis and probability have become integral parts of not only the middle school curricula but also the elementary and secondary school curricula as well. We live in a data driven society. There is no way to avoid the impact of statistics on our daily lives. Most colleges and universities strive as part of their educational goals to educate their students for mathematical literacy. Statistical literacy needs to be part of that goal. The GAISE College Report (6/11/2005), based on a project funded by the American Statistical Association, had six recommendations for the teaching of college-level statistics.

1. Emphasize statistical literacy and develop statistical thinking.
2. Use real data.
3. Stress conceptual understanding rather than mere knowledge of procedures.
4. Foster active learning in the classroom.
5. Use technology for developing conceptual understanding and analyzing data.
6. Use assessments to improve and evaluate student learning.

Though we started this book before the publication of the GAISE recommendations, many (if not all) of these recommendations are present throughout this book. Some—striving for conceptual development, active learning, and the use of technology—are very conspicuous in each chapter. These are recommendations that we have been integrating into our elementary statistics courses for years and would hope that others would as well.

The purpose of this book is to explicitly connect the mathematics of a typical elementary statistics course at the college level to the statistical concepts that middle school teachers will actually teach. To accomplish this goal, numerous illustrations and problems are pulled from four middle school mathematics curricula funded by the National Science Foundation. These four curricula we believe most fully embody the goals and vision set forth by the NCTM in *Principles and Standards for School Mathematics* (2002).

This is not a methods book, though it might seem so to some more used to a traditional statistics or mathematics text. In order to foster the active learning and development of conceptual ideas (now recommended by the GAISE report), many of the chapters in the book start with a classroom exploration problem for students to discuss. These exploration problems lay the foundation for developing key concepts and ideas throughout the chapter and provide a more active way for the instructor to engage the students in the class rather than simply lecturing over definitions and terms. Throughout the text, students are given opportunities to apply chapter concepts through what we have chosen to call "Focus on Understanding" activities. Each of these activities is directly connected to a particular excerpt from a middle school text and asks the reader to apply and extend statistical concepts under discussion in that section. The Focus on Understanding activities also give students the opportunity to work with their peers either in small groups or pairs to develop ideas and share understanding about the nature of statistics.

In keeping with the *Principles and Standards for School Mathematics* (2000), technology is an integral part of this text—both calculator technology and computer software. We have written explicit directions for using the TI-83 Plus calculator throughout the text. After much debate, we decided to leave the choice of computer software up to each individual instructor rather than to try to write directions in the text for using Minitab™, Excel™, Fathom™, SAS™, Tinkerplots™, and numerous other software packages currently available for use in the classroom.

Through our many years of experience in teaching elementary statistics at the college level, we are very aware of the anxiety statistics sometimes fosters in students and the common misconception that the subject is deadly dull. To counteract both of these, we have chosen to keep the tone of the text very informal, as if we were having a personal conversation with the reader. Consequently, we sometimes use phrases or make comments that might be considered "trite" or "unimportant," but we believe they help the student relax a bit when reading the text and create more of a rapport with the student. We have also purposefully tried to focus on developing concepts by relating them to students' prior experience or their future careers as middle school teachers and how they might introduce those concepts to their own students in the future.

We would like to thank Ira Papick for inviting us to be a part of this project. Our thanks also go to James Tarr for his help in getting us started. We hope that we stayed true to his initial vision. We appreciate the comments and contributions of Paul Rahmoeller and Mary Majerus. We would also like to thank the following reviewers for their helpful suggestions: Paul Ache, Kutztown University of Pennsylvania; Keith R. Kull, Kutztown University of Pennsylvania; Teresa Floyd, Mississippi College; Victor Cifarelli, UNC Charlotte; and C. E. Davis; North Carolina Department of

Public Instruction. Finally, we would like to thank Michael Bell and Petra Recter at Pearson/Prentice-Hall and Heather Meledin at Progressive Publishing Alternatives for their help in getting this book into print.

Debbie Perkowski
Mike Perkowski

List of Figures

Figure	Page Number	Title
1.1.1	8	Sampling strategies from "Random Samples" lesson from *Connected Mathematics Project*
1.1.2	9	Critiquing sampling strategies from *Connected Mathematics Project*
1.1.3	10	Examining bias in polling from *Connected Mathematics Project*
1.1.4	11	Letter to the editor critical of the use of non-scientific polls
1.2.1a	12	Pet data
1.2.1b	13	More pet data
1.2.1c	13	*Connected Mathematics* Problem 2.2
2.1.1	21	Frequency table for categorical data
2.1.2	22	Completed frequency table
2.1.3	22	Frequency table with relative frequency and percentages
2.1.4	24	Characteristics of different data displays
2.1.5	25	Three representations of the same data set
2.2.1	26	Homework from *MathThematics, Book 3*, on frequency tables
2.2.2	27	Frequency table with optional classes
2.2.3	28	Frequency table showing class boundaries
2.2.4	29	Frequency table with overlapping classes—not good
2.2.5	29	Frequency table to be completed
2.2.6	31	A Stem-and-leaf plot
2.2.7	32	Back-to-back stem-and-leaf plot
2.2.8	32	Frequency table and corresponding histogram
2.2.9	34	Frequency tables and histograms
2.2.10	34	Line plot from *MathThematics*
2.2.11	35	Line plot from *Mathscape*
2.2.12	36	Line Graph from *Connected Mathematics Project*
2.2.13a	36	Choosing a data display
2.2.13b	37	Choosing a data display (*Continued*)
2.3.1	38	"Projected mortality," adapted from *Newsweek*, April 22, 2002
2.3.2	38	Bar graph used for advertising Magical Fantastic Casino

xii   List of Figures

Figure	Page Number	Title
2.3.3	39	"Stiff competition," adapted from *Newsweek*, May 20, 2002
2.3.4a	39	Misleading graphs
2.3.4b	40	Misleading graphs (*Continued*)
2.3.5	41	"Church converts level off," adapted from *Newsweek*, March 4, 2002
2.3.6	41	Gross earnings of "*A Beautiful Mind*"
2.3.7	41	Gross earnings of "*A Beautiful Mind*" with a modified horizontal scale
2.3.8	42	Quickvote poll graph from www.cnn.com
2.3.9	43	"Affected groups by age and gender, 2000," adapted from *Newsweek*, April 22, 2002
2.3.10	43	Enlarging an image in two dimension
3.1.1	49	Modes from stem-and-leaf plots
3.1.2	51	The mean as a balancing point
3.1.3	52	Evening things out
3.1.4	53	Mean, median, mode, and range
3.2.1	65	Comparing soda consumption
3.4.1	74	Five-number summaries and their relation to box plots
3.4.2	75	Determining the five-number summary from a plot
3.4.3	77	Constructing box-and-whisker plots
3.4.4	78	Comparing distributions using box plots
3.4.5	78	Comparing box plots and line plots
4.1.1	86	Bat data with scatter plot
4.1.2	86	Lines with positive and negative slope
4.1.3	87	Different types of correlation
4.1.4	88	Describing correlation
4.1.5	89	Stock prices and manatees killed
4.2.1	91	Dividing the plane into four sections
4.2.2	92	The sign of $(x - \bar{x})(y - \bar{y})$
4.2.3	92	Rectangle areas represented by $(x - \bar{x})(y - \bar{y})$
4.2.4	94	Values of r and strength of correlation
4.2.5	94	Computations for Pearson's correlation coefficient
4.2.6	95	Data and computations for Murre Island bats
4.3.1	96	Comparing fitted lines
4.3.2	97	Slope of a line
4.4.1	101	Some data on tents
4.4.2	102	Scatter plot of the tent data
4.4.3	102	Illustrating the sum of squares
4.4.4	103	The least squares line
4.4.5	104	Data and computations for the least squares line
4.5.1	107	The tent data, sorted and grouped

Figure	Page Number	Title
4.5.2	108	Scatter plot showing groups
4.5.3	109	The median points and three lines
5.0.1	118	The Carnival Collection
5.1.1	121	Beginning concepts about probability
5.2.1	125	Frequency table for theoretical and experimental probabilities
5.2.2	126	Thinking about the Law of Large Numbers
5.2.3	127	Finding probabilities from an experiment
5.2.4	128–129	Predicting to win
5.3.1	132	Venn diagram
5.3.2	133	Contingency table for students and their lunch choices
5.4.1	134	Contingency table for students and their lunch choices
5.4.2	136	Venn diagram
5.5.1	139	Deep in the dungeon
5.5.2	140	Simulating multistage events
5.6.1	142	The Cover-Up Game
5.6.2	144	Lunch spinners
5.6.3	145	More geometric probabilities
6.0.1	153	The Unique Diner
6.1.1	154	Robert's clothes
6.1.2	155	Matching Robert's shirts and pants
6.1.3	155	Arrows matching shirts and pants
6.1.4	156	Tree diagram of shirt and pants outfits
6.1.5	158	Making faces
6.1.6	159	Lunch specials
6.1.7	160	Creating a new counting problem from a tree diagram
6.2.1	164	The Battle of the Bands
6.2.2	166	Opening locks
6.3.1	169	Combinations of horses
6.3.2	169	Combinations of horses (*Continued*)
6.3.3	173	Confused student
7.1.1	183	A one-and-one situation
7.1.2	184	A one-and-one situation (*Continued*)
7.1.3	186	Area model for one-and-one situation
7.1.4	187	A two-shot situation
7.1.5	188	The Special Sums Game
7.2.1	189	Expected value
7.2.2	191	The Prime Number Multiplication Game
7.2.3	192	Tasa's choice
7.3.1	193	Deviations and squared deviations

xiv List of Figures

Figure	Page Number	Title
7.3.2	194	Area model for option 2A
7.3.3	198	A probability distribution in the TI-83 plus
7.3.4	198	One-variable statistics from the TI-83 plus
7.4.1	204	Mindy's test
7.5.1	208	The number of heads in 100 tosses
7.5.2	208	The standard normal curve
7.5.3	210	The area represented by $P(z \leq 1.28)$
7.5.4	210	Using Table 1
7.5.5	211	The area for $P(z \geq -0.67)$
7.5.6	212	The area for $P(-1.25 < z < -0.50)$
7.5.7	212	Using a probability to find a z-value
7.5.8	212	Using Table 1 to find a z-value
7.5.9	214	The probability that a vehicle is speeding
7.5.10	215	The fastest 10% of vehicles
7.5.11	217	ShadeNorm($-1.25, -0.5$) on the TI-83 plus
7.6.1	220	Assorted binomial distributions
7.6.2	222	Estimating a binomial probability with a normal curve
7.6.3	225	Estimating $P(10 \leq x \leq 15)$
7.6.4	226	Estimating $P(x = 12)$
7.6.5	226	The y-values included with each x-value
7.6.6	227	Tossing tacks
8.1.1	235	Random samples
8.1.2	237	Selecting a random sample
8.1.3	238	Data from 100 eighth graders
8.1.4	240	Some results for the means of random samples
8.1.5	240	Values within half an hour of the population mean
8.2.1	243	Three possibilities for the distribution of an estimator
8.2.2	244	Selecting a random sample of size 2
8.2.3	245	The four possible samples of size 2
8.2.4	247	The eight possible samples of size 3
8.2.5	249	The distribution of $\bar{x}$ for samples of size 100
8.2.6	249	The distribution of coins in the population being sampled
8.2.7	251	Finding $P(0.86 \leq \bar{x} \leq 1.06)$
8.3.1	254	Theoretical vs. experimental probability
8.3.2	256	The two possible samples of size 1
8.3.3	259	The distribution of sample proportions
9.1.1	268	100 confidence intervals for p using $E = 0.10$
9.1.2	269	Using the distribution of $\hat{p}$ to find an error bound for a 95% confidence interval
9.1.3	271	100 confidence intervals for p using $E = 0.15$
9.1.4	273	100 confidence intervals for p using $E = 1.96\sqrt{\frac{\hat{p}\hat{q}}{n}}$

Figure	Page Number	Title
9.1.5	274	Tossing tacks
9.1.6	275	Using the distribution of $\hat{p}$ to find a general error bound
9.1.7	276	Finding $z_{\alpha/2}$
9.1.8	278	Using Table 2
9.1.9	279	Lifespan of a mayfly
9.2.1	281	Comparing the distributions of $\hat{p}$ and $\bar{x}$
9.2.2	281	Using the distribution of $\bar{x}$ to find a 95% confidence interval
9.2.3	283	Using the distribution of $\bar{x}$ to find a general confidence interval
9.2.4	284	Mayfly lifespan data
9.2.5	286	Comparing a T-distribution with the standard normal curve
9.2.6	288	Dot plot of sleep hours for the Grade 8 database
9.2.7	288	Finding a critical value from a T-distribution
9.2.8	289	Keisha's chocolate chip data
9.3.1	296	Drawing conclusions from samples
10.2.1	311	Using the distribution of $\hat{p}$ to determine the decision rule
10.2.2	312	Drawing conclusions from samples
10.2.3	315	Using the distribution of $\hat{p}$ for a lower-tailed test
10.2.4	316	Results from 50 random samples of 500 coin tosses each
10.2.5	317	Why do we choose $p = p_0$?
10.3.1	319	The P-value of a hypothesis test
10.3.2	320	Relating the P-value to α
10.3.3	323	Finding the P-value for a two-tailed test
10.3.4	325	A one-proportion z-test on the TI-83 plus
10.4.1	327	The P-value for a two-tailed test about μ
10.4.2	331	The rejection region in a T-distribution
10.4.3	331	Finding a critical value from a T-distribution
10.4.4	332	The P-value for a T-test
10.4.5	333	Mean annual snowfall

Elements of Statistics

CHAPTER 1

1.1 GETTING STARTED
1.2 TYPES OF DATA

Classroom Exploration 1.1

You have been hired as the newest member of a data collection corporation. Your first job is to collect information for the corporation on the workload and responsibilities of middle school teachers in the United States.

1. In very general terms, how would you conduct such a study? Would you contact every middle school teacher or every middle school in the United States? Why or why not?
2. What types or kinds of information would you want from each teacher/school in your study? Make a list of the kinds of information you think would be important in describing the workload and responsibilities of middle school teachers in the United States.
3. If possible, compare your list with that of someone else and discuss why you chose the items you did. Are there some types of information that might be harder to get than others? Why do you think that would be the case?

1.1 GETTING STARTED

The task set forth in Exploration 1.1 might seem rather overwhelming at first, but it is the type of problem that many information-gathering companies, government agencies, or interest groups face on a daily basis. Our society wants information ranging from what kind of dog food dogs prefer to the best shampoo for dry hair to the best mathematics book to use with sixth graders. Who is responsible for getting this information, how do they go about getting it, and what do they do with it after they get it? We hope to answer all of these questions as we move through this course together!

What Is Statistics?

Statistics is a branch of mathematical sciences that helps us to understand how to collect, organize, and interpret numbers or other information about some topic. The numbers or other pieces of information are referred to as **data**. Statistics, data analysis, and probability have emerged as integral components of school mathematics, and span the curriculum from kindergarten through grade 12.

Descriptive versus Inferential Statistics

As you might guess, **descriptive statistics** is the branch of statistical study that deals with ways of collecting, organizing, and describing data. For example, when an organization reports that the average age of its members is 35.5 years and the average gross annual income of those members is $56,342, they are reporting descriptive statistics. **Inferential statistics** is the branch of statistical study that uses information (usually descriptive statistics) obtained from samples (some subset of a group of people or objects that they are interested in) to make an estimate or to draw a conclusion about a population. Typically, most statistics textbooks will spend the first third or so of the book exploring descriptive statistics, the middle third on probability and probability distributions, and the last third on inferential statistics. Inferential statistics combines what we learned about describing data with probability distributions in order to test conjectures about a population or a population parameter. For example, suppose the publisher of the mathematics textbook you are using in your seventh grade class claims that students using their textbook increase their ACT scores by at least two points. How does a person go about supporting or even making such a claim? That is what the realm of inferential statistics is all about!

What Are Statistics?

Statistics themselves are essentially numerical values used to describe some attribute or attributes of a group of objects. Statistics are commonplace in American culture. It is virtually impossible to pick up a newspaper, listen to the radio, watch television, or surf the Internet without encountering statistics. Polls have become staples of Internet sites for news (e.g., www.cnn.com) and sports (e.g., espn.go.com). Each week, movie box office reports detail the weekend's top grossing motion pictures (e.g., movies.go.com/boxoffice), and A. C. Nielsen publishes reports on the nation's television viewing habits. For example, suppose it is reported that 19 million viewers tuned in to the season finale of *Law and Order*. Who counted these people? If you were one of the viewers, it is likely that you were not personally contacted to indicate whether or not you watched the show. In reality, only about 5,000 households are solicited each week to participate in the Nielsen ratings. The nation's viewing habits are inferred from this sample.

Exploration 1.1 question 2 asked you to decide what information you would want from each teacher or school. In formulating your responses, you probably generated a long list of information you would like to have about the workload and responsibilities of middle school teachers. Your list might include some of the following:

- Subjects typically taught in middle school.
- Amount of time spent on each subject in a typical day.
- Number of students in each class.
- Number of students in each grade level.
- Grade levels typically included in a middle school.
- Salaries and other benefits to teachers.
- Types of duties other than teaching (lunchroom supervisor, bus supervisor, etc.).
- Amount of time spent on school-related work outside of school hours.

Each of these characteristics of a teacher's workload and responsibilities is called a **variable**. In other words, the response would vary from teacher to teacher or from school to school. The responses of a particular teacher or a particular school are called **data** (data is plural, datum is singular) on that teacher or school. The collection of data on all members of a study or survey is called a **data set**.

Question 1 of Exploration 1.1 was to start you thinking about how many teachers or schools you really need to survey in order to form an accurate picture of the workload and responsibilities of middle school teachers. Do you think it would be possible to survey *every* middle school teacher in the United States? Would it be feasible? How many would be "enough?" How are you going to choose the teachers or schools to survey? Getting good information can be a lot trickier than it seems at first glance!

Populations versus Samples

The **population** in a statistical study is the complete set of people or objects being studied. Sometimes we say it is the set of all possible people or objects under consideration. In Exploration 1.1, the population would be *every* middle school teacher in the United States. **Population parameters** are specific characteristics of the population. For example, the average number of subjects taught per day would be a specific characteristic of our middle school teacher population. Population parameters are generally denoted using Greek letters such as μ (mu) or capital letters such as N (used to represent the population size).

A **sample** is a subset of the population from which data are obtained. **Sample statistics** are characteristics of the sample found by consolidating or summarizing the raw data. These statistics are generally used to estimate characteristics of a population. For example, you might gather data on the age upon marriage from a sample of 10,000 couples nationwide. The average age could then be computed from the sample data and this sample statistic would be used to estimate the average age upon marriage for all couples within the United States. Sample statistics are normally represented using lowercase letters. **Variables**, as we have mentioned earlier, are characteristics of individuals in the population that vary from individual to individual. For example, height is a characteristic of people that varies from person to person, so height would be considered a variable. **Raw data** are the bits of information on individuals in the population or sample like the eye color, height, weight, martial status, and years of teaching experience of a particular individual or individuals. The collection of all the raw data on all the individuals in the sample or population constitutes a data set.

Focus on Understanding

A local school district would like to conduct a survey to estimate the percentage of registered voters in their district who would support a school bond levy. To determine the level of support, the school board surveys 1,000 registered voters from their school district. From this scenario, describe the:

a. Population
b. Sample
c. Variable(s)
d. Raw data
e. Sample statistics
f. Population parameters

In this case, all the registered voters within the school district represent the population; 1,000 registered voters surveyed represent the sample; responses to the survey of 1,000 registered voters comprise the raw data; the percentage of 1,000 registered voters surveyed who indicate they support the school bond levy is a sample statistic; and the actual percentage of registered voters who support the levy is the population parameter being estimated by this survey of 1,000 registered voters.

Sampling Techniques

Sampling techniques are required to collect data. There are many different means by which elements of the population can be sampled. Here are some sampling techniques:

Simple Random Sampling: Every sample of the same size has an equal chance of being selected. Computers or calculators are often used to determine which people (or elements) are selected.

EXAMPLE Suppose that every single middle school mathematics teacher in the United States has been identified (not likely!) and given a number. The random number generator on a computer generates a list of 10,000 numbers. Teachers with these identification numbers are sent surveys on the workload and responsibilities of middle school teachers in the United States. ∎

Section 1.1 Getting Started 5

To generate a random sample using a TI-83 plus calculator, press the MATH button to get to the MATH menu. Use the left or right arrow keys to move the cursor to PRB at the top of the screen. The PRB menu should now appear.

1. rand
2. nPr
3. nCr
4. !
5. randInt(
6. randNorm(
7. randBin(

Notice that four out of the seven selections contain "rand." These four selections are used for generating various types of random numbers. We will discuss the other selections in later chapters of this book. Right now, we are interested in selection number 5, the random integer generator. If each student in your class is given a unique number (in other words, no repeated numbers) from 1 to 24, we can generate a random sample of seven students by doing the following:

1. Use the down arrow key to move the cursor to selection number 5 and press ENTER.
2. randInt(should now appear on your screen. We need to tell it where our numbers begin and end and, if you want, how many numbers you need to generate. To do this we enter after randInt(1,24,7). Our numbers started at 1, went to 24, and we wanted seven people to be selected. The calculator gave us the numbers {18 2 16 20 18 9 5}. Your calculator may give you a different set of seven numbers. Do not worry about having different numbers. Remember it is a *random* sample. We will talk more about what is going on with your calculator in another chapter.
3. Notice that our calculator-generated random sample has a slight problem. Person 18 appears to be selected twice! We need to generate one more number (that is not one of the six numbers we already have) so we need to go back to randInt(and enter randInt(1,24,1). This time the calculator shows the number 6, so the people selected to be in our sample are {18 2 16 20 9 5 6}.

Note: There are other ways to generate random numbers on your calculator. You can get your calculator to generate one integer at a time by either entering a "1" for the number of numbers to be generated, as we did in step 3, or you can simply enter randInt(1,24) and press ENTER. Each time you press the ENTER key, the calculator will generate another number within the range of 1 to 24.

EXAMPLE You would like to select 10 reading books at random to make a mini-reading corner in your mathematics classroom. You have 30 books from which to choose and you have given each one a number. Using your calculator you generate the following random numbers: 28, 5, 16, 13, 23, 2, 11, 30, 7, 24. These are the books you select for the reading corner.

EXAMPLE There are 12 couples at a dance contest. Since the dance floor is rather small, only four couples can compete at a time. Each couple has a competition number ranging from 1 to 12. The first four couples to compete are drawn at random from a large bowl. The following four numbers are drawn: 12, 3, 5, 1. The first four couples to compete would be couples 12, 3, 5, and 1. ∎

Systematic Sampling: In this case, every nth element of the population is selected (e.g., every 50th person is selected).

EXAMPLE Using our fictitious list in which every middle school teacher in the United States has an identification number, we select every 50th teacher on the list to receive a survey on the workload and responsibilities of middle school teachers in the United States. ∎

EXAMPLE Every twelfth student in the milk line is asked whether or not they prefer chocolate milk or regular milk. ∎

EXAMPLE Every twentieth parent listed in the school directory is called to serve as a chaperone at an upcoming school activity. ∎

Convenience Sampling: Selecting participants (or elements) that are available.

EXAMPLE Every middle school teacher who attends the annual meeting of the National Council of Teachers of Mathematics (NCTM) is given a survey on the workload and responsibilities of middle school teachers in the United States. (Note: Participants at NCTM conferences typically identify what grades levels they teach when they register for the conference.) ∎

EXAMPLE Your school is considering raising the price of admission to home games by 25¢. In order to get a feel for what students think about this idea you survey your homeroom class. ∎

EXAMPLE The principal would like to know what students think about the new discipline program he launched at the beginning of the school year. He decides to survey those students who come to the principal's office during the course of one week. (Do you see a problem with the principal's "convenient" sample?) ∎

Cluster Sampling: Divide the population areas into sections, randomly select a few of those sections, and then choose all members in them.

EXAMPLE There are 50 states (sections) in the United States. The states are put in alphabetical order and each assigned a number. Six states are selected at random. All the middle school teachers in these six states are sent surveys on the workload and responsibilities of middle school teachers in the United States. ∎

EXAMPLE There are 500 students in your school who normally ride the bus to school. There are 13 buses that carry these 500 students to and from school everyday. The school board would like to know the average number of miles traveled each day by the students who ride the buses, so they randomly select 5 of those 13 buses and survey each of the students on those 5 buses. ■

EXAMPLE The seats at a sporting arena are divided up into 30 sections with approximately 100 people per section. Six sections are randomly selected and all of the people within each of those sections are polled for their opinion about needed renovations to the arena. ■

Stratified Sampling: Partition the population into several strata, and then draw a sample from each.

EXAMPLE There are 50 states in the United States. A random sample of 100 middle school teachers is selected from each state. These teachers are sent surveys on the workload and responsibilities of middle school teachers in the United States. ■

EXAMPLE Your school has three grade levels: sixth, seventh, and eighth. From each of those grade levels, 15 students are selected at random to participate in a survey about drug use in schools. ■

EXAMPLE The eighth grade choir consists of 30 students. Eight are to be randomly selected to sing in an all-district choir. (Yes, we know this is not really how this works!) The choir is divided into four sections: sopranos, altos, tenors, and basses. From each of these four subgroups, two students are selected at random. ■

Ideally, we seek to generate a sample that is *representative* of the population. The likelihood of the sample being representative can vary widely. For example, suppose you hear that 37% of adults surveyed indicate they do not support a tax levy on an upcoming ballot. It could be that all persons in opposition to the tax levy were selected to participate in the poll; if another sample were selected, perhaps far fewer than 37% of respondents would indicate their opposition to the referendum. In general, larger samples are more likely to be representative of the population. In a related sense, we can estimate the actual population parameters using larger samples. How big a sample is "sufficient" depends upon quite a few things. You might have heard that a sample size of 30 is "good enough." In reality, it depends upon what kinds of questions you are trying to answer and how much error you are willing to accept in your estimate. We will explore the idea of sample size and error in detail in later chapters!

Classroom Connection

A portion of *Samples and Populations* from the *Connected Mathematics Project* (page 27) is shown in Figure 1.1.1. In this problem, students are asked to consider the advantages and disadvantages of four sampling techniques.

This investigation helps students to grapple with various means for selecting a sample. Answer questions A and B from the investigation in Figure 1.1.1. Is there a single "right" answer for determining which sample is most likely to be representative of the population? Why or why not? ◆

8 Chapter 1 Elements of Statistics

> **Problem 2.2** 
>
> Ms. Baker's class wants to find out how many students in their school wear braces on their teeth. The class divides into four groups. Each group devises a plan for sampling the school population.
>
> - Each member of group 1 will survey the students who ride on his or her school bus.
> - Group 2 will survey every fourth person in the cafeteria line.
> - Group 3 will read a notice on the school's morning announcements asking for volunteers for their survey.
> - Group 4 will randomly select 30 students for their survey from a list of three-digit student ID numbers. They will roll a 10-sided number cube three times to generate each number.
>
> **A.** What are the advantages and disadvantages of each sampling plan?
>
> **B.** Which plan do you think would most accurately predict the number of students in the school who wear braces? That is, which plan do you think will give the most *representative* sample? Explain your answer.

FIGURE 1.1.1 Sampling strategies from "Random Samples" lesson from *Connected Mathematics Project*

Focus on Understanding

1. Using vocabulary introduced on the types of sampling, identify the sampling technique used by each group.
2. How can stratified sampling be used to generate a sample? Explain.

Classroom Connection

Consider the sampling techniques offered in problems 7 and 8 from the same lesson in *Connected Mathematics Project* (page 32) in Figure 1.1.2.

For each of these, describe the advantages and disadvantages of the sampling technique. If potential problems are evident, propose an alternative sampling strategy that is more likely to avoid similar problems.

In some cases, results of a survey or the generation of a sample do not appear to be random at all. For example, if you frequently travel by air, you may feel that you are disproportionately selected "at random" (presumably) to participate in a search

7. To estimate the number of soft drinks consumed by middle school students each day, Ms. Darnell's class obtains a list of students in the school and writes each name on a card. They put the cards in a box and select the names of 40 students to survey.

8. A television news report said that 80% of adults in the United States support the right of school authorities to open student lockers to search for drugs, alcohol, and weapons. The editors of the school paper want to find out how students in their school feel about this issue. They select 26 students for their survey—one whose name begins with A, one whose name begins with B, one whose name begins with C, and so on.

9. Choose one of the issues in questions 5–8. Write a survey question you could ask about the issue, and explain how you could analyze and report the results you collect.

FIGURE 1.1.2 Critiquing sampling strategies from *Connected Mathematics Project*

of your belongings. It may be that you simply were the unlucky individual selected for this inconvenient—albeit important—scrutiny. Or it may be that particular physical characteristics lead airport security to believe you should warrant closer scrutiny. For example, elderly women do not seem to fit the profile of someone who would carry contraband on a commercial flight as much as, perhaps, say, a younger male. Such occurrences that appear to occur "against the odds" or more frequently than you expect lead some to believe that random sampling is not being used. ◆

Bias in Data Collection

Whenever we set out to collect data, great care must be taken that we do not somehow bias our data. This is particularly true with survey types of data. For example, suppose we wanted to get an idea of which presidential candidate was favored to win the next election? If we decide to poll all people exiting the local Democratic headquarters, our sample is bound to be biased! That was a rather blatant example, but bias can creep in unless we carefully select our sample and use caution in the phrasing of surveys or polls. In some surveys it is possible to determine what response the surveyors wish you to give simply by how the questions are phrased.

A recurrent theme in statistics is the idea that random samples enable us to draw inferences about populations. Sampling techniques can introduce bias and yield inferences that are inappropriate. Additionally, the advent of Internet polls has created a "culture of sound bytes" in which samples and populations are mistakenly interchanged as though they are synonymous. For example, consider the following *Classroom Connection*.

Classroom Connection

An extension to the previous lesson from *Connected Mathematics Project* (page 35) is shown in Figure 1.1.3 and focuses on the use of exit polling. Exit polls are conducted as a means to predict the outcomes of elections.

Extensions

14. Television stations, radio stations, and newspapers often predict the winners of important elections long before the votes are counted. They make these predictions based on polls.

a. What factors might cause a preelection poll to be inaccurate?

b. Political parties often conduct their own preelection polls to find out what voters think about their campaign and their candidates. How might a political party bias such a poll?

c. Find out how a local television station, radio station, or newspaper takes preelection polls. Do you think the method they use is sensible?

FIGURE 1.1.3 Examining bias in polling from *Connected Mathematics Project*

Answer each of the questions (a), (b), and (c). Several factors may introduce bias into the results of the exit polling and lead to conclusions that are inaccurate. ◆

Focus on Understanding

1. Design a sampling strategy for an exit poll that is likely to minimize bias and prevent inaccurate conclusions.
2. Media outlets such as newspapers, television, and radio stations often report results that are not scientific (Figure 1.1.4). Why are results of non-scientific polls reported? Explain.
3. Find examples of scientific or non-scientific polls or surveys in media publications such as your local newspaper or various news magazines. Share your findings with others in your class!

In the next section we will learn that data comes in many types, and the types of data influence how data can be analyzed and displayed.

1.2 TYPES OF DATA

By asking questions and collecting data to answer them, we can learn quite a bit about the world at large and particular topics that interest us. Naturally, different types of questions require different types of answers. For example, consider the questions, "Where were you born?" and "How many siblings do you have?" The first question requires the name of a place for an answer, and the second requires a number. These two types of answers form the overarching types of data: categorical

Tribune erred in printing 'unscientific' election poll

Editor, the Tribune: What is the purpose of an unscientific Tribune poll published on a day of elections before the polls close?

Although the two articles were balanced, the headline was not. "Survey shows taxes favored; Early voters like Props A, B and L" it read. An incredible assertion considering the unscientific nature of the poll.

I revel at the opportunity for pundits, experts and amateurs to express opinions in the Editorial and Opinion pages. Unscientific polls don't belong on the front page on election day.

**Greg Shuck
302 W.Broadway**

FIGURE 1.1.4 Letter to the editor critical of the use of non-scientific polls

and numerical. **Categorical or qualitative data** can be sorted into mutually exclusive groups (i.e., an object can belong to one and only one group). Some examples of variables that are categorical in nature are gender, eye color, and political party affiliation. Categorical data are usually not numbers. **Numerical or quantitative data**, as you may suspect, are numbers. Quantitative data represent a quantity that can be measured, such as how tall a person is or how many siblings a person has. There are situations, however, where a number does not represent a quantity; it does not tell how much or how many of something is present. For example, zip code is a number, but it does not tell "how much" or "how many" of anything. The same is true of numbers on football jerseys. These types of numbers are used as identifiers.

Classroom Connection

A portion of *Data About Us* from the *Connected Mathematics Project* (pages 24 and 25) is shown in Figures 1.2.1a, 1.2.1b, and 1.2.1c. This investigation helps students come to understand the difference between categorical and numerical data. In this case, the data are represented using frequency tables and bar graphs. We will explore ways to display data in detail in Chapter 2, but for right now we would like you to just focus on the types of data represented in the figures. Later in this particular investigation, the authors of *Connected Mathematics* highlight the different analyses that are possible depending upon which type of data is collected. ◆

12 Chapter 1 Elements of Statistics

Focus on Understanding

1. Using Figures 1.2.1a and 1.2.1b, answer questions A through F shown in Figure 1.2.1c.
2. Write two questions that have categorical data as answers.
3. Write two questions that have numerical data as answers.

Data can be further classified by type, beyond what the *Connected Mathematics Project* curriculum requires of middle school students. The type of data collected really determines what kind of analysis is appropriate for that particular bit of data. For example, it would make no sense to talk about the "average" eye color of your class. In order to compute an "average" you need to have some kind of numerical data. The four data types are: nominal, ordinal, interval, and ratio.

Nominal data are a type of categorical data. Data are classified into categories based on some defined characteristic. The data categories are mutually exclusive, and the categories have no logical order. Examples of nominal data include the color of jelly beans, marital status, or gender. Numerical values may be assigned to the nominal data (e.g., 0 for male and 1 for female), but these values are arbitrary. This is an example of categorical data that are composed of numbers. Be careful! Just because numbers have been assigned to categories, it does not mean that it would be appropriate to add all the numbers together and take the average. For example, phone numbers are a type of nominal data. It would make no sense, however, to add up the phone numbers of your ten closest friends and take the average.

The students made tables to show the tallies or frequencies, and then made bar graphs to display the data.

Favorite Kinds of Pets

Pet	Frequency
cat	4
dog	7
fish	2
bird	2
horse	3
goat	1
cow	2
rabbit	3
duck	1
pig	1

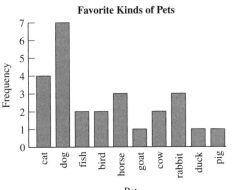

FIGURE 1.2.1a Pet data

Numbers of Pets

Number of pets	Frequency
0	2
1	2
2	5
3	4
4	1
5	2
6	3
7	0
8	1
9	1
10	0
11	0
12	1
13	0
14	1
15	0
16	0
17	1
18	0
19	1
20	0
21	1

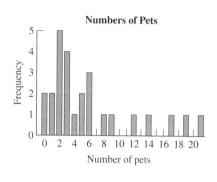

FIGURE 1.2.1b More pet data

Problem 2.2

Decide whether each question below can be answered by using data from the graphs the students created. If a question can be answered, give the answer and explain how you got it. If a question cannot be answered, explain why not and tell what additional information you would need to answer the question.

A. Which graph shows categorical data, and which graph shows numerical data?

B. What is the total number of pets the students have?

C. What is the greatest number of pets that any student in the class has?

D. How many students are in the class?

E. How many students chose cats as their favorite kind of pet?

F. How many cats do students have as pets?

G. What is the mode for the favorite kind of pet?

H. What is the median number of pets students have?

I. What is the range of the numbers of pets students have?

J. Tomas is a student in this class. How many pets does he have?

K. Do the girls have more pets than the boys?

FIGURE 1.2.1c Connected Mathematics Problem 2.2

Ordinal data are classified into categories that have some sort of logical order. The data categories are mutually exclusive, and the categories are scaled according to the amount of the particular characteristic they possess. Ordinal data may be categorical or numerical. One example of categorical ordinal data is birth month. Another example is the reported classroom use of calculators, with the choices being *never, rarely, often*, and *always*. Ordinal data can be numerical when the ranking choices are numbers instead of words. For example, suppose people were asked to read several statements about their self-concept and classify each statement on a rating scale of 1 to 5, with the numbers representing *strongly disagree, disagree, neutral, agree*, and *strongly agree*. A rating of 1 indicates less agreement than a rating of 2, so a logical order is implied. It should be noted that, in ordinal data, the distinction between two adjacent classifications is not always the same. The difference between *never using calculators* and *rarely using calculators* may be quite different than the difference between *rarely using calculators* and *often using calculators*.

Interval data are a type of numerical data. The differences between levels of categories on any part of the scale reflect equal differences in the characteristic measured. With interval data, the data categories are mutually exclusive and have some logical order. These categories are scaled according to the amount of the particular characteristic they possess. Equal differences in the characteristic are represented by equal differences in the numbers assigned to the categories. Additionally, the point 0 is just another point on the scale and does not really indicate the absence of whatever particular characteristic is being measured or described. Interval data values can be added or subtracted, but not meaningfully multiplied or divided. Longitude is an example of interval data. The interval between $20°$ W and $40°$ W is the same as the interval between $40°$ W and $60°$ W, yet the prime meridian ($0°$) does not indicate the absence of longitude; it is simply an agreed-upon reference point. The measurement of time in years is another example of interval data, as the zero point (1 CE) is arbitrary, and time existed before that point. A book written in 2000 CE is not twice as old as a book written in 1000 CE. (If this were true, how many times older would this book be compared to one written in 1000 BCE?) Since multiplication is not meaningful, this is interval data.

Ratio data are another type of numerical data. With ratio data, we are able to make statements not only about the equality of the differences between any two points, but also about the proportional amounts of the characteristics that two objects possess. Ratio data is similar to interval data, except for the fact that the point 0 reflects an absence of the characteristic. The number of pets in Figure 1.2.1 is an example of ratio data. A response of *eight* indicates three more pets than a response of *five*, and four times as many pets as a response of *two*. Additionally, a response of *zero* indicates a lack of pets. Another example of ratio data is the number of minutes needed to complete a 5K run. A finish time of 30 minutes is twice as long as a finish time of 15 minutes.

Focus on Understanding

1. Give two additional examples of nominal, interval, and ratio data. Explain why your examples fit each type.
2. Provide two examples of ordinal data: one that is numerical, and another that is categorical. Provide reasons for classifying these as ordinal data.
3. What type of data are zip codes? Justify your conclusion.
4. Explain why temperature measured in degrees Fahrenheit is an example of interval data. What about degrees Celsius? What about degrees Kelvin?
5. Discuss the difference between elevation and height (both measured in feet). What type of data is elevation? Height?
6. Go back and take a look at the different kinds (or types) of information you generated in response to question 2 of Exploration 1.1. Identify the different data types in your list of information you would like about the workload and responsibilities of middle school teachers in the United States.

Even though it appears as if the distinctions between the four data types are relatively clean, in real life the lines blur a bit depending upon the situation and what actually is being recorded.

A quantitative variable is said to be **discrete** if the possible values are distinct and separate. It is possible to count all possible data values, even though there may be an infinite number of possibilities. The number of telephone calls to 911 per day is an example of a discrete data set. **Continuous** variables may take on any value within a finite or infinite range. It is not possible to enumerate all possible values for the data. Examples of continuous data include the length of time required to run a marathon; the temperature at noon in Columbia, Missouri; and the percent of people per state that have blood type B. The nature of a data set (whether a discrete or continuous variable) determines what type of graphical display would be appropriate. These will be discussed in Sections 2.2 and 2.3.

Focus on Understanding

1. Look back at the examples of nominal, ordinal, interval, and ratio data that you gave earlier. For each example, tell whether the variable is discrete or continuous. Write a convincing argument to support your conclusion.
2. Yaxi and Ashish are collecting data on the finishing times for the "Beat the Clock 5K Run." Yaxi says that this is a continuous variable, since the finishing time can take any value between 0 seconds and 2 hours (when the race is officially over). Ashish believes that the variable is discrete, since they are counting hours, minutes, and seconds. Who do you agree with? Explain your reasoning.

Chapter 1 Summary

Chapter 1 has introduced you to some of the basic vocabulary and sampling techniques used in data collection. Paying attention to the vocabulary associated with samples versus populations will help you in later chapters to understand what type of calculation or statistical method is appropriate for the situation. How a sample is taken has an impact on not only the data collected, but also on what kinds of inference or predictions we can make from the sample. In later chapters you will see that the type of data or level of measurement will determine not only the type of display we can use with the data, but also the type of viable statistical analysis that can be run.

The key terms and ideas from Chapter 1 are listed below:

statistics 2	raw data 4	nominal data 12
data 2	simple random sampling 4	ordinal data 14
descriptive statistics 2	systematic sampling 6	interval data 14
inferential statistics 2	convenience sampling 6	ratio data 14
variable 3	cluster sampling 6	discrete variable 15
data set 3	stratified sampling 7	continuous variable 15
population 3	categorical data 11	
population parameters 3	qualitative data 11	
sample 3	numerical data 11	
sample statistics 3	quantitative data 11	

Assessment is an integral part of every curriculum from the elementary school all the way through college. The question always arises—*What is it that students should be able to do after completing this lesson/unit/chapter?* We have included here our intended learning goals for Chapter 1. Students who have a good grasp of the concepts developed in Chapter 1 should be successful in responding to these items:

- Explain or describe what is meant by each of the terms in the vocabulary list.
- Identify the population, sample, variables, raw data, sample statistics, and population parameters given the description of a survey (like the *Focus on Understanding* on page 4).
- Identify or give examples of each of the sampling techniques and support your reasoning.
- Classify types of data or give examples of the different data types and support your reasoning.

Your course instructor may have additional or different assessable outcomes for your class. As teachers (or future teachers) you should think about the assessment outcomes and learning goals for each chapter as you work through them.

EXERCISES FOR CHAPTER 1

1. Using the *Data Analysis and Probability Standards for Grades 6-8* from the NCTM's *Principles and Standards for School Mathematics* found at www.nctm.org, identify the middle school objectives that are found in Chapter 1.

2. Using the state standards for mathematics for the content area of data analysis and probability for your state, identify the middle school objectives that are found in Chapter 1. The following website may be useful: www.doe.state.in.us. This website will allow you to access the web pages of the state departments of education for the 50 states. From the state web pages you should be able to find the state's mathematical standards.

3. A teacher is interested in the television viewing habits of sixth graders. She chooses a day and has her sixth grade class of 25 students record the amount of time they spend watching television on that particular day. She finds that her class watched an average of 3.15 hours of television that day.

   a. Describe or identify each of the following:

      i. the population
      ii. the sample
      iii. the variable
      iv. the raw data
      v. a sample statistic
      vi. a population parameter

   b. What sampling method did the teacher use? What are the advantages and disadvantages of this method?
   c. What data type is the variable you identified?

4. Classify each of the variables below as categorical or quantitative. If the variable is categorical, decide whether it is nominal or ordinal. If the variable is quantitative, decide whether it is interval or ratio.

   a. Type of television show (comedy, drama, sports, etc. . .)
   b. Length of time watching television in a given day
   c. Olympic medal types (gold, silver, bronze)
   d. Temperature in degrees Fahrenheit
   e. Number of children in a family
   f. Area codes for long-distance calls
   g. Number of minutes spent grading homework papers on a particular night
   h. How much children like broccoli (really like, like, it's okay, don't like, really don't like)

5. Classify each of the quantitative variables below as discrete or continuous.

   a. Number of points scored in a basketball game
   b. Average number of points per game for a basketball player
   c. Number of eggs in a bird's nest
   d. Weight of eggs in a bird's nest
   e. Length of eggs in a bird's nest
   f. Number of hurricanes in a given three-month period
   g. Scores on your most recent statistics quiz
   h. Length of time spent by students riding the bus to and from school

6. The table of contents for a book gives the page number for the start of each chapter.
    a. Suppose that Chapter 3 starts on page 30 and Chapter 4 starts on page 42. The difference in the page numbers is 12. Is this difference meaningful in this situation? If so, what does it mean?
    b. Suppose that Chapter 3 starts on page 30 and Chapter 5 starts on page 60. The quotient of these page numbers is $60/30 = 2$. Is this quotient meaningful in this situation? If so, what does it mean?
    c. Based on your answers to (a) and (b), what type of data (nominal, ordinal, interval, or ratio) would you consider the pages numbers in a table of contents to be?

7. Give an example of each of the following data types, if possible. If it is not possible to give an example of the specified data type, explain why.
    a. discrete
    b. nominal
    c. continuous
    d. ordinal
    e. ratio
    f. ratio, discrete
    g. ratio, continuous
    h. interval, discrete
    i. interval, continuous
    j. nominal, continuous
    k. nominal, discrete
    l. ordinal, discrete
    m. categorical
    n. quantitative

8. Identify the sampling technique used in each of the following situations. Explain why you chose that particular technique for the situation and discuss the advantages/disadvantages of the technique in each situation.
    a. Ms. Kidder assigns a different number to each of her 23 students. She then uses her calculator to randomly select five of those students to put up a new bulletin board display.
    b. Every sixth student in the cafeteria line is given a carrot stick with their lunch instead of a celery stick.
    c. The State Department of Agriculture decides to check on the average number of bushels per acre of corn being harvested in the state of Iowa. There are 99 counties in Iowa. They randomly select 10 of those counties and survey all of the farmers in those counties regarding the corn harvest.
    d. The Internal Review Service decides to audit every 500th income return that they receive in their office.
    e. Mr. Marshall polls his sixth period science class to determine whether more students like football or more like basketball.

f. New Mexico has 33 counties. Suppose those counties are classified by high income, middle income, or low income based on the average annual income of the residents of those counties. From each of those three classifications, two counties are selected at random to be polled on the effects of a proposed tax hike.

g. Parents attending a basketball game at the local high school are asked to complete a survey on the types of snacks they would like to see sold at the concession stand.

CHAPTER 2

Organizing and Displaying Data

2.1 DISPLAYING CATEGORICAL DATA
2.2 DISPLAYING QUANTITATIVE DATA
2.3 MISLEADING GRAPHS

Classroom Exploration 2.1

At some time or another in your youth you were asked to draw various types of graphs.

1. Generate a list of types of data displays (pie graphs and so on). Try to be as exhaustive as you can! You may not remember the exact name of each graph, but that is okay.
2. Sketch a sample of each type of data display.
3. List the kinds of data (e.g., nominal, ordinal) that are appropriate for each type of data display.
4. Generate a list of what you think middle school students should understand about each type of data display.

In this chapter we will explore the various ways to display data and the ins and outs of each type of display.

2.1 DISPLAYING CATEGORICAL DATA

Once data are collected, we need to organize it in some way so that we can begin to get a sense of what it is telling us about our sample or population. The type of data we have collected greatly influences the type of data display we may use. In other words, some types of data displays are more appropriate for certain types of data than others. There are three primary means of displaying categorical data: frequency tables, bar graphs, and circle graphs. In elementary school students may also draw pictographs to display categorical data. Numerical data may be displayed using frequency tables, histograms, stem-and-leaf plots, box-and-whisker plots, line plots (or dot plots), and line graphs. Notice that both categorical and numerical data can use frequency tables as a way of organizing data. There are also additional categories of displays such as double and triple bar graphs, back-to-back stem-and-leaf plots, and more advanced graphs such as frequency polygons and ogives. We will focus on the basic ways that data are displayed in this chapter.

Frequency and Relative Frequency Tables for Categorical Data

One of the first ways to go about organizing a data set is to make a **frequency table**. A frequency table shows the number of data items that fall into each category or class. Figure 2.1.1 shows a frequency table from *MathThematics, Book 2* (page 5).

A frequency table consists of columns that contain the data classes, the tally, and the frequency for each class. In the example above, each class represents the response that students selected from a questionnaire about the amount of time they spent doing homework each day. Since there were six choices from which to choose on the survey, we should have six classes in the frequency table. Sometimes people put in an additional class of "no response" in order to keep track of how many surveys did not have any response to a particular question. The choice is entirely up to the person constructing the frequency table.

The "tally" column is used to keep a running count of each of the responses on the surveys. Once all the responses are recorded in the tally column, the total number of tally marks for each class is recorded as the **frequency** for that class in the frequency column. The finished product would look something like Figure 2.1.2.

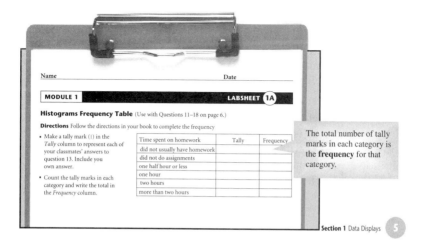

FIGURE 2.1.1 Frequency table for categorical data

Time Spent on Homework	Tally	Frequency
did not usually have homework	||	2
did not do assignments	|||	3
one half hour or less	⊦⊦⊦⊦	5
one hour	⊦⊦⊦⊦ ||	7
two hours	||||	4
more than two hours	|	1

FIGURE 2.1.2 Completed frequency table

Time Spent on Homework	Tally	Frequency	Relative Frequency	Percentage
did not usually have homework	||	2	$\frac{2}{22}$	9.1%
did not do assignments	|||	3	$\frac{3}{22}$	13.6%
one half hour or less	⊦⊦⊦⊦	5	$\frac{5}{22}$	22.7%
one hour	⊦⊦⊦⊦ ||	7	$\frac{7}{22}$	31.8%
two hours	||||	4	$\frac{4}{22}$	18.2%
more than two hours	|	1	$\frac{1}{22}$	4.5%

FIGURE 2.1.3 Frequency table with relative frequency and percentages

The tally column is an aid in finding the frequencies. Once the frequency column is filled in, the tally column is often omitted.

Two additional columns are sometimes added to help give a sense of what proportion of students were in each class. The first column to the right of the frequency column in Figure 2.1.3 represents the **relative frequency**. The relative frequency is simply the ratio of the frequency of the class to the total number of students who responded to the survey. In this case, 22 students responded to the question, so our sample size is $n = 22$. The lowercase letter n is most often used to represent sample size, while f is used to represent the individual frequencies of the classes:

$$\textbf{Relative Frequency} = \frac{f}{n}$$

Percentages are found in the usual way, that is, the relative frequency is written as a decimal and then multiplied by 100. In this case, we rounded our percentages to one decimal point. We could have just as easily rounded to the nearest percent or to two decimal points. Some people even keep all the decimal places. We think this is a bit obsessive. In truth, it depends upon the nature of the data and the degree of precision necessary for a particular statistical project.

$$\textbf{Percentage} = \frac{f}{n} \cdot 100$$

Though the example used categorical data, frequency tables and relative frequency tables are also used for quantitative data. We will look at constructing frequency tables for quantitative data in depth in Section 2.2.

Graphical Displays

Listing raw data is often not the most useful way to report the overall findings of an experiment. Graphical displays help to organize the data and allow for inferences to be made. There are many different types of graphical displays for univariate (one-variable) data. These displays provide different information, so it is important that an appropriate display is selected.

Classroom Connection

Page 566 shown in Figure 2.1.4 from *MathThematics, Book 3*, provides a brief description of several types of data displays for both categorical and numerical data. **Bar graphs** and **circle graphs** (pie charts) are good ways to visually display categorical data. **Stem-and-leaf plots, histograms**, **line plots**, and **line graphs** are used to display quantitative data. **Box-and-whisker plots** are also used to display quantitative data, but involve a bit more statistical information than just the frequencies. Box-and-whisker plots will be discussed in Section 3.4. **Scatter plots** are used for displaying bivariate (two-variable) data and will be discussed in Section 4.1. ◆

Bar Graphs and Circle Graphs

Frequency tables provide some facts, but they may not allow for easy comparison between groups or from a group to the whole. **Bar graphs**, which report the frequency of each category by the height of the bar, are useful for making comparisons between groups. **Circle graphs** show the relative size of each group to the whole.

Several different types of graphs are shown in Figure 2.1.5. These graphs are based on 1990 census data from the Missouri counties of Camden, Laclede, Miller, Morgan, and Pulaski. In this case, the task is to represent the numbers of south central Missouri residents who are married, divorced, separated, widowed, or have never married. This data is available with the Fathom software package. Five hundred individuals from these counties were selected at random from the entire population of these counties. Fathom's display of this data is shown in Figure 2.1.5.

24 Chapter 2 Organizing and Displaying Data

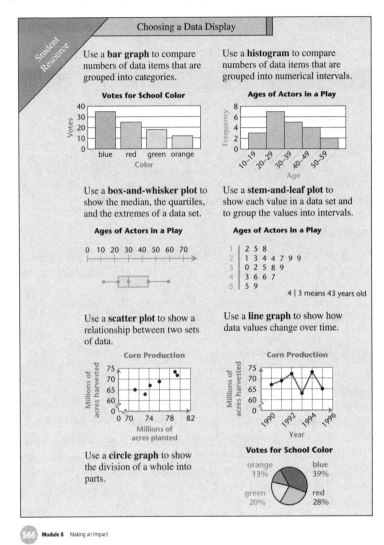

FIGURE 2.1.4 Characteristics of different data displays

Focus on Understanding

The following questions refer to the graphical displays in Figure 2.1.5.

1. Which displays show that about 50% of the residents of this area are married?
2. Which displays show that there are roughly the same numbers of divorced and widowed residents?
3. Describe a situation where a circle graph would not be as appropriate as a bar graph for displaying categorical data.

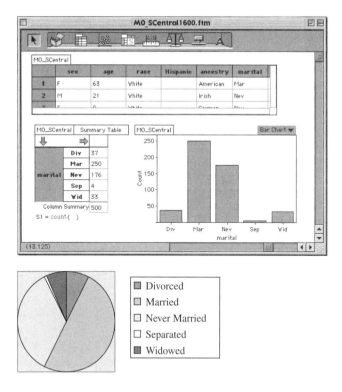

FIGURE 2.1.5 Three representations of the same data set

2.2 DISPLAYING QUANTITATIVE DATA

As we mentioned in the opening paragraphs of this chapter, the type of data you have collected will determine the types of displays you may use. Quantitative data (numerical) uses frequency tables as a method of displaying data much like categorical data; however, the types of graphs that can be used are more varied. In general, the data displays for quantitative data are a bit more involved because we can simply do more with a set of numbers than with descriptive categories like eye color.

Frequency and Relative Frequency Tables for Quantitative Data

Quantitative frequency tables consist of the same organizational columns as categorical frequency tables, but the way in which the classes are determined is a bit more complex. Most of the time in middle school curricula, the classes are predetermined for the students. However, as a teacher you must be prepared to give your students some guidance regarding the construction of the classes for quantitative frequency tables. Figure 2.2.1 shows a homework problem from *MathThematics, Book 3* (page 555). Students are given the following data and then asked to complete a lab sheet that gives two different frequency tables with predetermined classes. The lab sheet is hard to read, but that's okay. We want to focus on how to set up the classes ourselves anyway.

26 Chapter 2 Organizing and Displaying Data

▶ The Garfield computer club creates a school home page and monitors the number of "hits" per day the page receives. (A "hit" occurs each time a person accesses the page.) The numbers of hits per day for one month are shown.

16, 27, 26, 5, 11, 33, 23, 17, 15, 20, 3, 14, 29, 21, 23, 31, 16, 8, 14, 28, 19, 20, 24, 35, 7, 12, 22, 27, 18, 20

Use Labsheet 2B for Questions 10–15.

10 Labsheet 2B shows two frequency tables and two sets of axes for *Histograms*. The first table and set of axes use intervals of 5.

 a. For each interval in the first table, make a tally of the numbers of hits per day that lie within the interval, and record the corresponding frequency. The first two rows have been completed for you.

 b. Use the table to draw a histogram on the first set of axes.

11 The second frequency table and set of axes use intervals of 10. Complete the table and use it to draw a histogram.

12 Compare the shapes of the two histograms you drew. How are they alike? How are they different?

13 Which histogram gives more information? Explain.

14 Can you use the first histogram to make the second? Can you use the second histogram to make the first? Explain.

15 Can you use either histogram to recover the original data values? Why or why not?

FIGURE 2.2.1 Homework from *MathThematics, Book 3*, on frequency tables

We first need to determine the largest and smallest numbers in the data set. In this case the smallest number is 3 and the largest number is 35. Different statistics books will give different instructions on how to set up the classes based on how many classes are considered "optimal," the range or spread of the data, and a host of other things. In general, that level of precision and formality is not needed or expected at the middle school level. Most people looking at the data set given on the previous page would either opt to have the classes go from 0–4, 5–9, 10–14, 15–19, and so on OR go from 0–9, 10–19, 20–29, and so on. Those seem to be the most natural ways of grouping the numbers into classes. You could, if you wanted to, create classes of 3–7, 8–12, 13–17, 18–22, etc. Nothing is wrong with that at all. It is simply a matter of personal taste. There are, however, some rules of thumb:

1. All the classes should be of the same width (or length depending upon the book you are using) **whenever possible**.
2. The classes should **NOT** overlap.
3. There should be no empty classes (classes with a zero frequency) **whenever possible**.
4. There should be enough classes that all the data are not lumped into one or two huge classes.

We will use the data set from Figure 2.2.1 to illustrate what we mean by each of these rules.

Classes Option 1	Frequency	Classes Option 2	Frequency	Classes Option 3	Frequency
0–4		0–9		3–7	
5–9		10–19		8–12	
10–14		20–29		13–17	
15–19		30–39		18–22	
20–24				23–27	
25–29				28–32	
30–34				33–37	
35–39					

FIGURE 2.2.2 Frequency table with optional classes

The **width** can be a little confusing in terms of how it is computed. We set up three different options for classes for the data. Let's put it in tabular form to get a feel for how it would look as a frequency table (see Figure 2.2.2).

The width of the classes under Option 1 is 5. To most students it would seem as if the width should be 4 since $4 - 0 = 4$ and $9 - 5 = 4$. Unfortunately, that is not how we get the width. In this case the data are whole number data and the width represents the numerical values that would fit into each class. In the first class, 0–4, five numerical values would fit into this class: 0, 1, 2, 3, and 4. There are five numerical values that fit in the second class: 5, 6, 7, 8, and 9. Each class under Option 1 has five possible values for that class.

The width of the classes in Option 2 would be 10 as there are ten values possible for each class. Again, we are dealing with whole number data here. The first class contains the values 0, 1, 2, 3, 4, 5, 6, 7, 8, and 9 or ten numerical values.

The width of the classes in Option 3 would be what? Take a second and see if you can find the width of these classes. If you found the width to be 5, then you understand the idea of width.

There are other ways of determining the class width. Take a look at the minimal or smallest value in each class. These are actually called the **lower class limits**. The lower class limits for Option 1 are 0, 5, 10, 15, 20, 25, 30, and 35. Notice that we are counting by five. Take a look at the **upper class limits**: 4, 9, 14, 19, 24, 29, 34, 39. Again, we are counting by fives. In most elementary statistics classes at the college level we talk about the "middle" of each class or the **class mark**. To find the class marks just add the upper and lower limit for each class and divide by 2. The class marks for Option 1 are: 2, 7, 12, 17, 22, 27, 32, 37. Once again, we are counting by fives.

There is still one more way to find the class width that you will NOT see in middle school curricula, but you will see in a college statistics course. This last way involves using the **class boundaries**. We debated about skipping any discussion at all about class boundaries, but a lot of people prefer to use class boundaries rather than class limits because using the boundaries generally makes the corresponding

Classes Option 1	Frequency	Classes Option 2	Frequency	Classes Option 3	Frequency
−0.5–4.5		−0.5–9.5		—	
4.5–9.5		9.5–19.5		—	
9.5–14.5		19.5–29.5		—	
14.5–19.5		29.5–39.5		—	
19.5–24.5				—	
24.5–29.5				—	
29.5–34.5				—	
34.5–39.5					

FIGURE 2.2.3 Frequency table showing class boundaries

graphical representation look a lot less cluttered. Simply put, a class boundary is the halfway value between one class and the next. Look at Option 1 again in the table. The first class goes from 0 to 4. The second class goes from 5 to 9. The third class goes from 10 to 14. What number is halfway between 4 and 5? 4.5! What number is halfway between 9 and 10? 9.5! If we construct our frequency table using class boundaries rather than class limits it would look like Figure 2.2.3.

Do not worry that the first class now goes from NEGATIVE 0.5 to 4.5. That is not a mistake! Remember that we were dealing with whole number data in this example. The boundaries by necessity are numerical values that do NOT actually occur in the data.

Please note that middle school textbooks **do not** make such a big deal about setting up the classes, nor do they typically introduce vocabulary such as class limits, class marks, and class boundaries to the students. Most college-level statistics books do, so we included that information in most of its detail here.

Now that you have totally forgotten what we were doing with that long-winded song and dance about widths, we need to go back to the other three rules of thumb. The second rule of thumb was that the classes should not overlap. Another way to say that is that each data item should go into one and only one class. Suppose we revise Option 1 and Option 2 in Figure 2.2.4.

In Option 1 Revised and Option 2 Revised we now have class limits that overlap. This means that if the data set would happen to contain the value 15, which it does, 15 would get tallied in both the third class and the fourth class under Option 1 Revised. This is what happens when people confuse class limits with class boundaries! Remember that class limits can be values that are present in the data set while class boundaries cannot.

Before we discuss the third and fourth rules of thumb, take a minute and complete the frequency table given in Figure 2.2.5.

Once you get the frequencies for each of the options notice that NONE of the classes are empty (contain no values). Classes 0 to 4 and 35 to 39 under Option 1

Classes Option 1	Frequency	Classes Option 1 Revised	Frequency	Classes Option 2	Frequency	Classes Option 2 Revised	Frequency
0–4		0–5		0–9		0–10	
5–9		5–10		10–19		10–20	
10–14		10–15		20–29		20–30	
15–19		15–20		30–39		30–40	
20–24		20–25					
25–29		25–30					
30–34		35–40					
35–39							

FIGURE 2.2.4 Frequency table with overlapping classes—not good

Classes Option 1	Frequency	Classes Option 2	Frequency	Classes Option 3	Frequency	Classes Option 4	Frequency
0–4		0–9		3–7		0–19	
5–9		10–19		8–12		20–39	
10–14		20–29		13–17			
15–19		30–39		18–22			
20–24				23–27			
25–29				28–32			
30–34				33–37			
35–39							

FIGURE 2.2.5 Frequency table to be completed

each only have one value in them, but that is okay. Sometimes you may have an empty class or two. What you do not want is to have something like four out of eight classes empty. Option 2 packs the data into fewer classes so each class has a higher frequency. Option 3 yields a frequency distribution similar to Option 1, though the exact frequencies are a little different for each class because each of the classes is different. Option 4 only has two classes. The first class has a frequency of 14 and the second class has a frequency of 16. Every data item is lumped into one of these two classes. We really cannot tell much about the original data set from a distribution of this type other than about half the values were below 20. Again, how you set up

30   Chapter 2   Organizing and Displaying Data

the classes is a personal choice. We encourage you to set up the classes in a way that makes sense to you and that follows the four basic rules of thumb.

> ## Focus on Understanding
>
> While middle school texts do not focus on the extensive vocabulary associated with frequency distributions, a lot of college statistics courses do. To strengthen your understanding of the vocabulary and what it means, please do the following:
>
> 1. In Figure 2.2.2, find the class marks for each of the classes under Options 2 and 3.
> 2. In Figure 2.2.2, identify the lower class limits for classes under Options 2 and 3. Identify the upper class limits for Options 2 and 3.
> 3. In Figure 2.2.3, find the class boundaries for each of the classes under Option 3.
> 4. In Figure 2.2.5, find the class widths for Option 4.

Now that you know more than you ever wanted to know about constructing frequency distribution, we will move on to pictorial or graphical ways to represent the quantitative data you so carefully compiled in a frequency distribution!

Stem-and-Leaf Plots

There are several ways to organize and present displays of quantitative or numerical data. **Stem-and-leaf plots** present data in an organized manner. With small data sets, they also provide a simple way to find the center of the data set. Unlike other types of data displays, stem-and-leaf plots also preserve all of the original raw data. For example, when we constructed the frequency table back in Figure 2.2.5, we lost some of what we knew about the original raw data. Only classes or data were represented in our grouped frequency table. We know how many values are in each class, but unless we have access to the original raw data, the individual items in each class are unknown.

Classroom Connection

The stem-and-leaf plot in Figure 2.2.6 is found in *MathThematics, Book 1* (page 219).
    In the stem-and-leaf plot shown, the numbers to the left of the vertical line are called stems and the numbers to the right are called leaves. Each of the recorded lengths in the data set is represented by a single leaf in the display. For example, the 6-leaf next to the stem of 1 represents the number 16 in the data set.   ◆

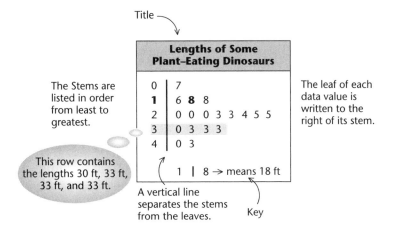

FIGURE 2.2.6 A Stem-and-leaf plot

Constructing a Stem-and-Leaf Plot

1. Find the high and low values of the data.
2. Decide on stems.
3. List the stems in a column from least to greatest.
4. Use each piece of data to create leaves to the right of the stems on the appropriate rows (stems).
5. If the plot is to be ordered, list the leaves in order from least to greatest.
6. Add a legend identifying the values represented by the stems and leaves.
7. Add a title explaining what the graph is about.

You will often find **back-to-back** stem-and-leaf plots in middle school curricula. Back-to-back stem-and-leaf plots are useful for comparing two data sets. The following stem-and-leaf plot is from *MathScape, Looking Behind the Numbers* (page 17) and shows a back-to-back type of plot. Students measured the lengths of their forearms and feet to the nearest one half centimeter. Their data are recorded in Figure 2.2.7.

In this case, the stems are 1, 2, 3, and 4, which represent 10, 20, 30, and 40 centimeters, respectively. The black dots between the stems indicate that the interval from 20 to 29.5 was split into two pieces to make the plot more manageable and easier to read. The leaves would go out a long way on both the left and right of the stems if the intervals were not split!

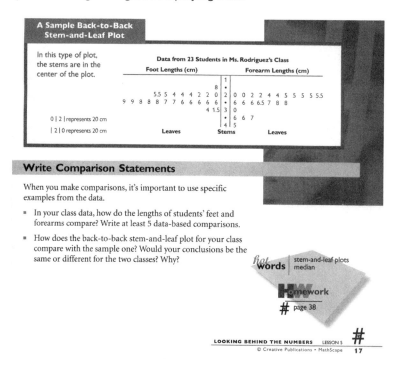

FIGURE 2.2.7 Back-to-back stem-and-leaf plot

Histograms

Histograms look very similar to bar graphs. Histograms should be used for continuous numerical data, while bar graphs are appropriate for categorical data sets. In a histogram, each bar represents the frequency of numerical values in a specified interval; the bars of a bar graph refer to separate categories. The intervals in a histogram all have equal width, and the bars touch each other. Using a frequency table to construct a histogram is relatively simple. Figure 2.2.8 shows a frequency table and histogram from *MathThematics, Book 2* (page 320).

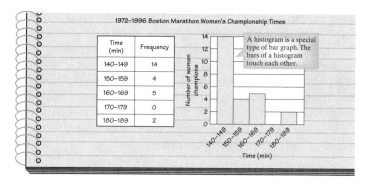

FIGURE 2.2.8 Frequency table and corresponding histogram

Section 2.2   Displaying Quantitative Data   33

On the other hand, constructing a histogram by hand from a raw data set is a very detailed procedure. Careful consideration must be given to the number of intervals and the width of each interval. Additionally, the boundaries for each interval must be selected so that no data value lies on the boundary. (Remember our long, boring discussion of frequency distributions earlier in this chapter?!) Fortunately, computers and calculators can be programmed to carry out this procedure. Using these tools, individuals can focus on the interpretation of the data and not be bogged down with time-consuming tasks.

It should be noted that **circle graphs** can also be used with numerical data sets, though they are not commonly used with this data type. Numerical intervals are treated as categories.

Focus on Understanding

1. Describe a method you could use to find the typical or most representative value from a stem-and-leaf plot.
2. How do the stem-and-leaf plot and histogram relate to each other? (It may help to rotate the stem-and-leaf plot 90° counterclockwise.)
3. Why should the bars touch each other in a histogram? Is the same true for a bar graph?

Classroom Connection

A frequency table and histogram are shown in a lesson from *MathThematics, Book 2* (page 318) in Figure 2.2.9. ◆

Focus on Understanding

The following questions refer to Figure 2.2.9.

1. Why is a stem-and-leaf plot not used to represent this data set?
2. Can either of the displays shown be used to find the slowest finishing time? Explain your answer.
3. If a runner were to finish the race in 35 minutes, would that be considered *fast*, *slow*, or *average*? Explain.

Line Plots

A **line plot**, sometimes referred to as a **dot plot** or **class frequency graph** by some books, is similar to a bar graph in that the frequencies are plotted along either a horizontal (usually) or vertical numerical scale. The scale must include both the smallest (minimum) and largest (maximum) values in the data set. Figure 2.2.10 shows a line plot from *MathThematics, Book 1* (page 188).

34 Chapter 2   Organizing and Displaying Data

GOAL

LEARN HOW TO...
♦ interpret histograms

As You...
♦ examine the results of a 5K race

KEY TERM
♦ histogram

Exploration 3

Histograms

▶ In Exploration 2, you saw running times displayed with a stem-and-leaf plot. Sometimes another type of data display is more convenient.

> The Schoolpower Laguna Beach Classic is run along the Pacific coast and through a scenic canyon. In 1996, runners aged 3–78 helped raise money to aid local schools.

20 **Discussion** Nearly 1000 people ran in the 5 km, or 5K, race of the 1996 Schoolpower Laguna Beach Classic. Their times ranged from the winning time of 15 min 36 s to times slightly over an hour.

  a. Estimate the range of the running times in the 5K race.

  b. Suppose you had a list of all the individual running times for the 5K race. Why would it be difficult to use a stem-and-leaf plot to display the data?

▶ The frequency table below shows how many runners finished the 1996 Schoolpower 5K race in each of the time intervals listed. The *histogram* below displays the same data.

1996 Schoolpower Laguna Beach Classic Finishing Times	
Interval (min:s)	Frequency
15:00–24:59	168
25:00–34:59	462
35:00–44:59	218
45:00–54:59	121
55:00–64:59	21

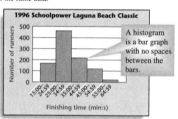

A histogram is a bar graph with no spaces between the bars.

21 a. Why does the first time interval start at 15 rather than at 0?

  b. In which time interval did most of the runners finish the race?

  c. How many runners finished in the 45:00 – 54:59 time interval?

FIGURE 2.2.9 Frequency tables and histograms

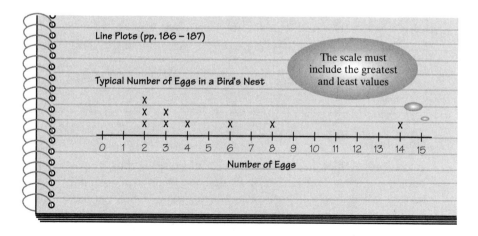

FIGURE 2.2.10 Line plot from *MathThematics*

Classroom Connection

Figure 2.2.11 is from page 6 of *Mathscape: What Does the Data Say?* ◆

How Many Glasses of Soda We Drink

```
                                    X
                                    X
       X                            X
       X                            X
       X                  X         X
       X                  X         X
       X     X            X         X
       X     X            X         X
       X     X            X         X
 X  X        X            X         X
─────────────────────────────────────
 1  2  3  4  5  6  7  8  9  10
        Number of glasses
```

FIGURE 2.2.11 Line plot from *Mathscape*

Focus on Understanding

The following questions refer to Figure 2.2.11.

1. Why is a stem-and-leaf plot not used to represent the data set?
2. Could a histogram be used to represent the data set? Explain.
3. How many glasses of soda does a typical student in this class drink?
4. Could you use a line plot to display categorical data? Explain.

Line Graphs

A **line graph** is used to denote how data values *change over time*. The points on the graph are connected to imply some sort of continuity in the data. The line graph in Figure 2.2.12 is from *Variables and Patterns* from *Connected Mathematics* (page 26).

Focus on Understanding

1. Why was a line graph chosen for the data shown in Figure 2.2.12?
2. Suppose you wanted to create a graphical display of data on the number of babies born in the United States for each day of January, 1978. What type of display would you choose? Why would you choose this display?
3. Would a line graph be appropriate for displaying discrete data? Explain.

36 Chapter 2 Organizing and Displaying Data

Applications

1. Here is a graph of temperature data collected on the students' trip from Atlantic City to Lewes.

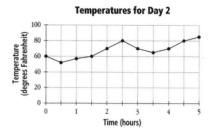

FIGURE 2.2.12 Line Graph from *Connected Mathematics Project*

Classroom Connection

Figures 2.2.13a and b show a lesson from *MathThematics: Book 3* (page 563) with many different data displays all based on the same data set. Different displays provide different information. The questions posed on these pages highlight the strengths and weaknesses of each type of display. ◆

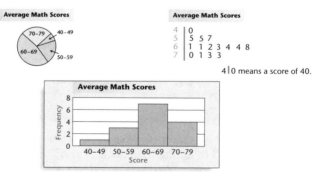

FIGURE 2.2.13a Choosing a data display

3 Which display best shows that almost half the average scores are from 60 to 69?

4 How are the histogram and the stem-and-leaf plot alike? How are they different?

5 Which display gives you the actual scores? List the scores.

6 Given the 15 average scores, can you make a histogram different from the one above? Can you make a stem-and-leaf plot different from the one above? Explain.

7 What information might you want that the displays do not give?

FIGURE 2.2.13b Choosing a data display (*Continued*)

Focus on Understanding

1. Answer questions 3 through 7 from Figures 2.2.13a and b.
2. Suppose you wanted to construct a line plot from the information in the figure. Which display would help you construct a line plot? Explain.

2.3 MISLEADING GRAPHS

As we discussed in the last section, graphical displays of data can be very informative. Different information may be inferred from various displays. A problem arises, however, when a display is modified or misused. A quick glance at such displays could lead the reader to make a false inference about the data. For this reason, these types of displays are called misleading graphs.

There are a few basic ways that a graph can be misleading. Graphs with numerical axes may not begin at zero. Axes with numerical values may not spread out the values proportionately. A display may not contain enough information for it to be properly interpreted. Finally, there may be a different type of display that would better represent the data set. Keep in mind that these graphs are not "bad," they simply require a bit more thought to interpret.

In Figure 2.3.1, it appears that, by 2008, the projected number of Hepatitis C deaths will double the present value. Closer inspection shows that this is really only an increase from 15,000 to 18,000. This misconception occurs because people assume the vertical axis starts at zero; this one begins at 13,000.

A billboard resembling Figure 2.3.2 advertised a certain casino, showing that the expected payoff for that casino was greater than other area casinos. Billboards are usually seen for just a few seconds, so motorists may believe that there is a large difference in payoffs at various casinos in town.

38 Chapter 2 Organizing and Displaying Data

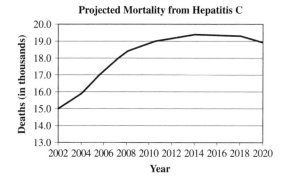

FIGURE 2.3.1 "Projected mortality," adapted from *Newsweek*, April 22, 2002

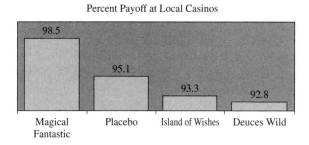

FIGURE 2.3.2 Bar graph used for advertising Magical Fantastic Casino

---

Focus on Understanding

1. What makes Figure 2.3.2 a misleading graph?
2. How could Figure 2.3.2 be changed to make it less misleading?
3. What incorrect assumptions are supported by the graph in Figure 2.3.3?

Section 2.3   Misleading Graphs   39

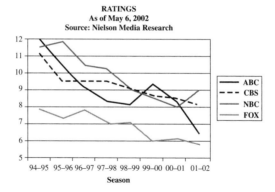

FIGURE 2.3.3 "Stiff competition," adapted from *Newsweek*, May 20, 2002

Classroom Connection

Figures 2.3.4a and 2.3.4b show a few pages from *MathThematics, Book 1* (pages 574 and 575). The authors highlight the importance of selecting an appropriate display to communicate information. In particular, the two line graphs seem to convey different messages with the same data.   ◆

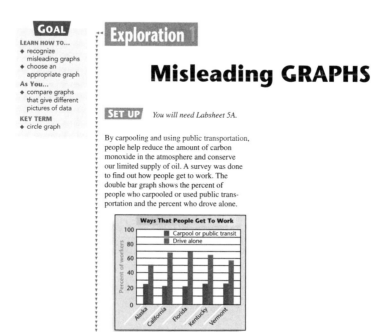

FIGURE 2.3.4a Misleading graphs

4 a. In Kentucky about 26% of workers carpooled or used public transportation. About what percent drove alone?

b. Why do you think the two percents in part (a) have a sum that is less than 100%?

5 The commuter data used to create the graph on page 574 is given on the labsheet. Use the *Commuter Data and Blank Grid* to create a different picture of how people commute.

6 Try This as a Class Use your graphs of the commuter data.

a. In Vermont the percent of commuters who carpooled or used public transportation is about one half the percent of commuters who drove alone. Do both graphs show this relationship? Explain.

b. Which graph seems to show that very few people carpooled or used public transportation? How does it show this?

c. Which graph would you use if you wanted to show that a large percent of workers in these five states carpooled or used public transportation? Why?

d. How are the scales different for each graph?

▶ Line graphs can also show different pictures of data. The two line graphs below show the use of petroleum (oil) in the United States.

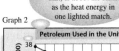

One Btu (British thermal unit) is about the same as the heat energy in one lighted match.

Graph 1

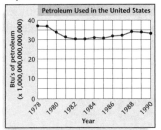

Graph 2

7 a. In which year was the least petroleum used in the United States?

b. Which graph did you use to answer part (a)? Why does this graph make it easier to find the lowest point on the graph?

c. How are the scales for the two graphs different?

d. Which graph seems to show that the amount of petroleum used in the United States has stayed about the same? What does the other graph seem to show?

FIGURE 2.3.4b Misleading graphs (*Continued*)

Focus on Understanding

1. Answer question 4 from Figure 2.3.4b.
2. Answer question 7 from Figure 2.3.4b.

Occasionally, the numerical values on an axis are not evenly distributed. Notice the horizontal axis in Figure 2.3.5. The reader must assume the value for the bar between 1950 and 1960. Furthermore, the space that represents 10 years is different between 1970 and 1980 than it is between 1940 and 1950. Changing the spacing may alter the appearance of a graph drastically.

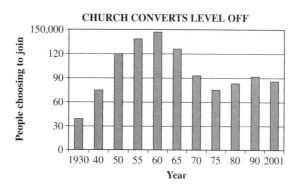

FIGURE 2.3.5 "Church converts level off," adapted from *Newsweek*, March 4, 2002

The graph in Figure 2.3.6 shows the weekly earnings of *A Beautiful Mind*, as reported in www.the-movie-times.com. Weekend 1 corresponds with December 21–23, 2001. Figure 2.3.7 shows how this graph can be manipulated to give a different impression. Note that the same data was used to create both graphs.

FIGURE 2.3.6 Gross earnings of *"A Beautiful Mind"*

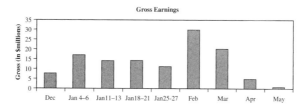

FIGURE 2.3.7 Gross earnings of *"A Beautiful Mind"* with a modified horizontal scale

42 Chapter 2   Organizing and Displaying Data

Focus on Understanding

1. Discuss how to modify the graph in Figure 2.3.5 to be less misleading. What would the modified graph(s) look like?
2. Describe the different messages that the graphs in Figure 2.3.6 and Figure 2.3.7 convey.

Sometimes, graphical displays of data do not show all possible responses for a question. This can be very practical at times. For example, there are a multitude of different responses to the question, "What restaurant makes the best hamburgers?" Many graphs represent the less-frequent responses in a category labeled "other." However, when there are very few options, these should be represented. Graphs should also provide enough information about the data to prevent misinterpretation.

Focus on Understanding

1. How could the graph in Figure 2.3.8 be considered misleading?
2. A line graph has been used in Figure 2.3.9, suggesting a change in time. Is this an appropriate display? What other type of display could be used in this case?

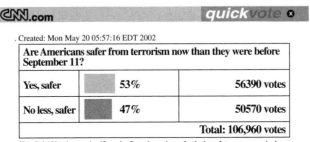

FIGURE 2.3.8 Quickvote poll graph from www.cnn.com

Pictographs sometimes mislead readers by reducing or enlarging a graphic image in more than one dimension. As an example, consider the graph in Figure 2.3.10, which shows the number of breeding pairs of Piping Plovers in various regions of North America. The graph intends to use the height of the bird to indicate the number of breeding pairs, yet the height and width are both changed

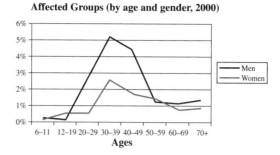

FIGURE 2.3.9 "Affected groups by age and gender, 2000," adapted from *Newsweek*, April 22, 2002

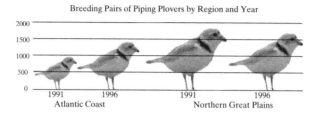

FIGURE 2.3.10 Enlarging an image in two dimension (created by Dustin Jones)

by the same amount. This leads to the impression that the number of breeding pairs of Piping Plovers along the Atlantic Coast in 1991 is about one-fourth of that in the Northern Great Plains in the same year, not one-half. (This is because about four of the smallest bird would "fit" into the largest bird.)

It must be noted that a misleading graph is not necessarily a poor graph. Instead of condemning misleading graphs, we should seek to educate students to be able to read graphical displays of data, determine which information is relevant and which may be misleading, and then draw proper conclusions.

Chapter 2 Summary

In Chapter 2, we focused on several ways to organize and display both categorical and numerical data. As you have learned, certain graphs are more appropriate for certain data types than others such as using bar graphs or circle graphs for categorical data and stem-and-leaf plots for numerical data. Frequency tables are a way of displaying information about the frequencies of the items or classes based on raw data. Additional ways to present data will be introduced once a few more concepts regarding how to describe a data set are developed. Box-and-whisker plots are one such type of display. Box-and-whisker plots use the results of specific numerical computations for their construction and will be discussed in the next chapter. Scatter plots, which are used to display two-variable data, will be examined in Chapter 4.

The last section of the chapter introduced the idea of misleading graphs. Over the last 10 years or so the number of graphs being used in news magazines, newspapers, and other media sources has increased tremendously. Consequently, the study of misleading graphs and how to avoid making graphs that might lead people to the wrong conclusion has also increased.

The key terms and ideas from Chapter 2 are listed below:

frequency table 21	class width 27
frequency 21	lower class limit 27
relative frequency 22	upper class limit 27
percentage 23	class marks 27
bar graph 23	class boundaries 27
circle graph 23	back-to-back stem-and-leaf plot 31
stem-and-leaf plot 23	dot plot 33
histogram 23	class frequency graph 33
line plot 23	misleading graph 36
line graph 23	

Assessment is an integral part of every curriculum from the elementary school all the way through college. The question always arises—*what is it that students should be able to do after completing this lesson/unit/chapter*? We have included here our intended learning goals for Chapter 2. Students who have a good grasp of the concepts developed in Chapter 2 should be successful in responding to these items:

- Explain or describe what is meant by each of the terms in the vocabulary list.
- Construct and interpret frequency and relative frequency tables for both categorical and numerical data.
- Construct, interpret, and compare bar graphs and circle graphs for categorical data.
- Construct, interpret, and compare histograms, stem-and-leaf plots, line plots, and line graphs for numerical data.
- Choose an appropriate graph for a given data set.
- Identify potentially misleading features of given graphs.

Your course instructor may have additional or different assessable outcomes for your class. As teachers (or future teachers) you should think about the assessment outcomes and learning goals for each chapter as you work through them.

EXERCISES FOR CHAPTER 2

1. Using the *Data Analysis and Probability Standards for Grades 6–8* from the NCTM's *Principles and Standards for School Mathematics* found at www.nctm.org, identify the middle school objectives that are found in Chapter 2.

2. Using the state standards for mathematics for the content area of data analysis and probability for your state, identify the middle school objectives that are found in Chapter 2. The following website may be useful: www.doe.state.in.us. This website will allow you to access the web pages of the state departments of education for the 50 states. From the state web pages you should be able to find the state's mathematical standards.

3. The 20 students in Ms. Little's sixth grade class were asked to identify their favorite color. Their responses are shown below:

blue   blue   yellow   green   red   blue   green   blue   red   blue
green   red   red   blue   blue   green   blue   yellow   blue   green

   a. Construct a relative frequency table for the data.
   b. Construct a bar graph for the data.
   c. Construct a circle graph for the data.

4. A regular six-sided number cube is rolled 50 times. The number of dots showing on each roll is recorded below:

   2 2 5 3 4 1 5 3 3 4 2 4 3
   2 6 3 6 5 6 2 2 2 5 5 3 3
   6 6 4 3 1 6 1 1 2 2 3 3 5
   4 3 4 4 2 4 6 1 6 1 3

   a. Construct a relative frequency table for the data.
   b. Construct a bar graph for the data.
   c. Construct a circle graph for the data.

5. The weights in pounds (to the nearest pound) of 25 dogs are shown below:

   15   12   45   60   78   35   62   67   84   91
   6   18   22   36   48   43   44   57   92   72
   8   26   45   61   65

   a. Construct a relative frequency table for the data using the classes 1–10, 11–20, 21–30, 31–40, and so on.
   b. Find the class width.
   c. Find the class marks for each class.
   d. Determine the class boundaries.
   e. Construct a histogram for the data.
   f. Construct a stem-and-leaf plot for the data.
   g. Would a line plot be a good choice to represent this data? Why or why not?

6. The number of cups of coffee consumed per day by a college professor over a 30-day period are shown below:

   5 5 8 3 9 2 3 6 5 4
   4 5 5 6 10 3 6 6 7 8
   6 5 6 4 2 5 6 6 5 4

   a. Construct a frequency distribution for the data.
   b. Construct a line plot for the data.
   c. Would a stem-and-leaf plot be a good choice to represent this data? Why or why not?
   d. Construct a histogram for this data set.

7. Using the frequency table you completed in Figure 2.2.5, answer the following questions:
   a. Construct a histogram for Option 1.
   b. Construct a histogram for Option 2.
   c. Construct a histogram for Option 3.
   d. Compare the shapes of the histograms you constructed for (a), (b), and (c). How are they the same? How are they different?

46 Chapter 2 Organizing and Displaying Data

   e. Which histogram gives more information?
   f. Is it possible to recover the original raw data values from any of these histograms? Why or why not?

8. The data presented in the table below is taken from the website of the Statistical Abstract of the United States. The data show the number of total households (as defined by the Census Bureau) in 1,000 s from 1900 to 2000 (www.census.gov/statab/www).

Year	Total Number of Households (in 1,000 s)
1900	15,964
1910	20,256
1920	24,352
1930	29,905
1940	34,949
1950	42,251
1960	53,024
1970	63,450
1980	80,390
1990	91,947
2000	105,480

Construct a line graph to display the data. Be sure to label your axes! Remember to avoid the pitfalls of misleading graphs that you studied in Section 2.3.

9. When asked to create a graphic which would inform the student body about the number of and type of crimes committed on a college campus, a student journalist constructed the following graphic for use in the student paper. The caption read "Crime on Campus Sky Rockets!"

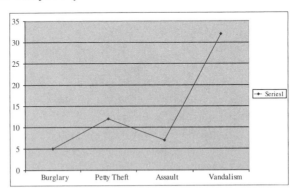

   a. The editor of the student paper is not happy with this graph and feels that it is misleading. Do you agree or disagree? Why?
   b. What type of graph would be a better choice to display this type of data? Why?

10. Suppose the following visual display is used to compare the amount of snow from two consecutive years. Does this graph give an accurate impression of the data? Why or why not?

Exercises for Chapter 2

CHAPTER 3
Describing Data with Numbers

3.1 MEASURES OF CENTER
3.2 MEASURES OF SPREAD
3.3 MEASURES OF POSITION
3.4 BOX-AND-WHISKER PLOTS

Classroom Exploration 3.1

Suppose your boss (principal, supervisor, or whomever) gives you a large set of raw data with the directive, "Tell me what the typical or middle value is for this data!" What would you do? Look at the data set below from *Mathematics In Context: Dealing with Data* (page 40). How would you respond to the directive? What numerical value would you assign to indicate the "middle" of the data set? Why?

Speed of 13 Animals without Hooves in Kilometers per hour

48 113 15 69 40 68 63 48 56 81 56 19 24

3.1 MEASURES OF CENTER

Once we have our data collected and appropriately displayed we are ready to begin analyzing. Typically, we start analysis of a data set by finding the "center" or "middle" of the data set. The center of a data set can be measured several different ways. We will focus on three of those ways in this section: **mode**, **median**, and **mean** (arithmetic mean or average).

The Mode

We will begin with the **mode** as it is the quickest and easiest measure to find in a data set. The mode is simply the item or class that has the highest frequency. It is quite possible for a data set to have more than one mode. In cases like that we say the data is bimodal if it has two modes and multimodal if it has more than two modes. In Exploration 3.1, the values 48 and 56 both occur two times. No other value occurs more than twice so we would say this data set is bimodal.

In a histogram or in a grouped frequency distribution, the class with the highest frequency is called the **modal class**. The mode is symbolized as an x with a dot over it, $\dot{x}$. The mode is always a number or item in the actual data set. This may seem like an odd comment, but we shall see that not every measure of the center can be found in the raw data!

Classroom Connection

We saw the stem-and-leaf plot in Figure 3.1.1 from *MathThematics, Book 1* (page 219) back in Chapter 2. ◆

Focus on Understanding

1. What is/are the mode/s of the stem-and-leaf plot? How can you tell?
2. What is the length of the largest dinosaur length recorded in the plot? What is the length of the smallest?
3. Based on the plot, does it appear that plant-eating dinosaurs are generally more than 20-something feet or less? Explain your reasoning.
4. What is the range of the data in this plot? Do you think this is a large value? A small value? Explain.

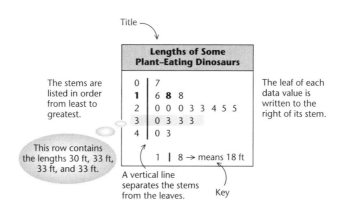

FIGURE 3.1.1 Modes from stem-and-leaf plots

The Median

The **median** is the actual physical middle of a data set. This means that when you arrange the individual items in the data set from largest to smallest the number or item that falls exactly in the middle is the median or $\tilde{x}$. If the number of items in the data set is odd, the median will always be a number from the raw data. If the number of items in the data set is even, the median may or may not be in the set. Whether or not the median is a value from the raw data depends upon the number of items in the set and the distribution of the numbers within the set.

Try This!

List the numbers in the stem-and-leaf plot showing the lengths of 18 plant-eating dinosaurs on lined paper—one number per line starting with the largest value 43 and ending with the smallest value 7. Fold your paper so that the value 43 is on top of the value 7. Crease your paper where the fold occurs. Now open your paper up. There should be a fold line between the numbers 23 and 24. The median number for this data is halfway between 23 and 24 or $\tilde{x} = 23.5$. Notice that 23.5 is not actually one of the numbers in the data set. That's okay. We are looking for ways to describe the center and 23.5 represents the middle of the data itself. What if the fold occurs between two numbers that are not consecutive (one right after the other)? For example, suppose the crease of your paper fell between the numbers 23 and 28. The same idea applies. We are looking for the number that lies halfway between 23 and 28. The easiest way to find this number is simply to add 23 and 28 and then divide by 2. This would give us a median value of 25.5. Again, 25.5 is not in the data set, but it doesn't matter.

Finding the Median of a Data Set

1. Put the data in order from high to low (or from low to high, it doesn't matter).
2. If there is an odd number of items in the data set, the median is the middle item.
3. If there is an even number of items in the data set, the median is the average of the two middle items *whether they are the same numbers, consecutive numbers, or non-consecutive numbers.*

The Mean

The **arithmetic mean** of a data set is simply the sum of all the numbers in the set divided by the total number of numbers in the set. There are several different types of means used in mathematics such as the geometric mean, the harmonic mean, the weighted mean, and the arithmetic mean. In statistics, we generally focus on the arithmetic mean, though the weighted mean is also used. Right now we will focus on the arithmetic mean. The formula for the arithmetic mean is given below:

$$\bar{x} = \frac{\Sigma x}{n} \text{ for samples and } \quad \mu = \frac{\Sigma X}{N} \text{ for populations}$$

The symbol Σ, or sigma, is mathematical shorthand for "add all these things together." Remember that we use Greek letters to stand for population parameters in statistics and lowercase letters to stand for sample statistics. The *x* with the bar over it is read as "x bar" and the funny looking *u* is really the Greek letter mu (pronounced "mew"). Mu is used in statistics to represent the mean of a population. Like the median, the mean may or may not be a number in the raw data set. Again, it depends upon the items in the set and their distribution. Unlike the mode or the median, to compute the mean we must use every single item in our data set.

While the median represented the physical middle of the data set, the mean is the "balancing point" for the data set. What do we mean by "balancing point?" Well, imagine a long "weightless" stick like a meter stick. The long stick is marked off with the values, equally spaced, of course, from our data set. For the data set from Figure 3.1.1 our stick would start with a mark at 7 (the smallest number in our set) and we would mark off one unit increments (8, 9, 10, 11, ...) until we get to 43 (the largest number in our set). If we hang a weight representing each of the values in our data set at the corresponding numbers on the stick (some values like 33 would have three weights at that spot on the stick), the point at which we can balance the meter stick would be the **mean** of the set of values. In physics we would call this the center of mass. In statistics, it is the mean. Take a look at the following example from *MathThematics, Book 1* (page 186). In Figure 3.1.2 we have both the raw data and a line plot.

Imagine the "x's" as weights on a long "weightless" stick marked from 25 to 39. If you used your finger as a fulcrum, the point on the stick at which it balanced would be the mean. This is unfortunately tricky to do in real life, as any type of stick we could use would have some mass (and weight) and throw our mean off a bit. Hence, the need for other ways of finding the mean!

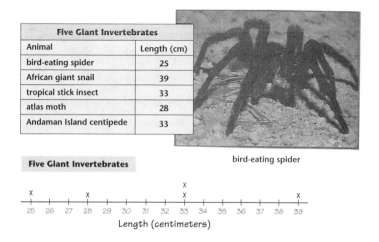

FIGURE 3.1.2 The mean as a balancing point

Focus on Understanding

1. What is the modal length for the giant invertebrates in the data set in Figure 3.1.2?
2. What is the median length?
3. Estimate the mean using the line plot. Do you think the mean is closer to 39 or closer to 25? Explain.
4. Find the mean length of these giant invertebrates using the formula for the mean. How does the mean compare to the mode? The median?
5. Suppose the value of 25 was incorrectly recorded and really should be 20. How does this change affect the mean? The median? The mode?

Line plots like the one shown in Figure 3.1.2 provide another interesting way to explore the mean, as we will see in the next Classroom Connection.

Classroom Connection

Blocks may also be used to give physical meaning to the concept of the mean as shown in Figure 3.1.3, an excerpt from *Connected Mathematics: Data About Us* (page 54). ◆

5.1 Evening Things Out

Six students in a middle-school class determined the number of people in their households using the United States census guidelines. Each student then made a cube tower to show the number of people in his or her household.

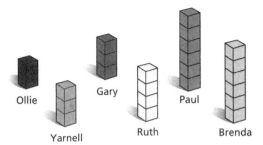

You can easily see from the cube towers that the six households vary in size. The students wondered what the average number of people is in their households. Their teacher asked them what they might do, using their cube towers, to find the answer to their question.

FIGURE 3.1.3 Evening things out

Focus on Understanding

1. What are some ways that you think students might go about finding the average household?
2. Some students might think about rearranging the blocks to even out the towers. Try using this method to find the average. What advantages or disadvantages are there to this method?

Classroom Connection

Figure 3.1.4 is from *MathScape: What Does the Data Say?* (page 37). How would you respond to Frank and Tessie?   ◆

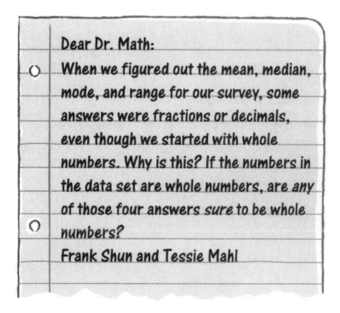

FIGURE 3.1.4 Mean, median, mode, and range

54 Chapter 3 Describing Data with Numbers

> ## Focus on Understanding
>
> 1. Is it possible for the mean, median, and mode to be equal? Construct a data set where this is the case.
> 2. Is it possible for the mean to be the largest of these three measures? The smallest? Construct a data set (if possible) where the mean is larger than the median. Construct a data set where the mean is smaller than the median.
> 3. What affect do outliers (extremely large or extremely small values) have on the mean? The median? The mode?
> 4. In Figure 3.1.4 Frank and Tessie had a specific question about the kind of numbers they could expect for the mean, median, mode, and range. (The range is the highest value in the data set minus the lowest value in the data set.) What guidelines could you give them to help distinguish among these concepts?

The mean turns out to be one of the chief ways to measure the "middle" or "center" of a data set. There are several key properties of the mean that you should keep in mind when you are analyzing a data set.

Key Properties of the Mean

1. **The mean is located between the extreme values.**

   EXAMPLE Suppose your data set is a set of exam scores that go from 45 to 98. The mean of this data set must be somewhere between 45 and 98. ∎

2. **The sum of the deviations from the mean is zero.**

   EXAMPLE The mean of 7, 5, and 3 is 5. A deviation is simply the distance away from the mean a particular number in the data set is. In this case the deviations would be $7 - 5 = 2$, $5 - 5 = 0$, and $3 - 5 = -2$. If we add all these deviations up we get $2 + 0 + (-2) = 0$. ∎

3. **The mean does not necessarily equal one of the values that were summed.**

   EXAMPLE The mean of 1, 2, and 6 is 3. Three was not one of the numbers in the data set. ∎

4. **The mean can be a fraction (or decimal) that has no counterpart in physical reality.**

   EXAMPLE The mean number of children per family is 2.2. It is not physically possible to have two-tenths of a child even though the computed mean value is 2 and *two-tenths*. ∎

5. **When one calculates the mean, the value of zero, if it appears, must be taken into account.**

   EXAMPLE Suppose our data set is: {5,3,0,2,2} Even though it is tempting to "throw out" the zero, we cannot! The mean of our data set is 2.4. While the zero does not affect the sum of our numbers, it does count as a data value for the total number of items being summed. ∎

6. **The mean is affected by extreme scores.**

   EXAMPLE A small business has five employees whose salaries are $24,000, $30,000, $32,000, $36,000, and $80,000. The "average" salary at this company would be $40,400. Notice all of the employees but one (probably the boss) make less than $40,400!

   There are other interesting properties of the mean, but we believe these are the main ones to help you think about data sets. ∎

The Weighted Mean

Earlier in our discussion of the arithmetic mean we mentioned various types of means. The other commonly used type of mean is the **weighted mean**. College students should be very familiar with weighted means as it is precisely the type of measure used by registrars' offices nationwide to compute their grade point average (GPA).

Using a weighted mean simply means that certain values in the data set are given more "weight" than others. Another way to think of it is that they get counted more than once in the overall computation. Let's look at a couple of examples to see what is going on here.

EXAMPLE 3.1.1 Finding a GPA for a given semester

Ron's report card for the fall semester is shown below:

Class	Grade	Credit Hours
History	D	3
Statistics	B	3
Swimming	A	1
English	D	4
Biology with Lab	B	5
		16

Using a typical four-point scale, A's are worth 4 points, B's are worth 3 points, C's are worth 2 points, D's are worth 1 point, and F's are worth 0 points. Classes taken for credit (i.e., pass/fail) are not computed into your overall GPA as a general rule. Some schools even have a credit/no credit system for these types of courses.

These credit hours, assuming you passed the course, are figured into the total hours needed toward graduation, but they are not part of your GPA computation as regular letter grades are. Since History was 3 credit hours, it gets counted three times when we compute the GPA. Biology will count five times since it was worth 5 credit hours. Our computation would be:

$$\text{Grade Point Average} = \frac{(1)(3) + (3)(3) + (4)(1) + (1)(4) + (3)(5)}{3 + 3 + 1 + 4 + 5}$$

$$= \frac{35}{16} = 2.1875$$

The D in English hurt because it had a low value (1) but a high weight (4). The general formula for the weighted mean is shown below:

Formula for the Weighted Mean

$$\overline{x_w} = \frac{(value\ 1)(weight\ 1) + (value\ 2)(weight\ 2) + \cdots + (value\ k)(weight\ k)}{(weight\ 1) + (weight\ 2) + \cdots + (weight\ k)}$$

or, using summation notation $\overline{x_w} = \frac{\sum xw}{\sum w}$ where the x's are the individual data items and the w's are the corresponding weights of the items. BE CAREFUL!! This notation says to multiply each x by its corresponding weight first and then add all those products up.

Let's try using the formula on another example.

EXAMPLE 3.1.2 Professor Smith uses a weighted mean for computing his students' final grades in his statistics course. The first exam counts as 20% of the grade, the second exam counts as 25% of the grade, quizzes count as 10% of the grade, and the final exam counts as 45% of the grade. Amy made 94 on the first exam, 86 on the second exam, 100 on her quizzes, and 75 on the final. What would her final grade be for the class? The percentages serve as the weights in this case and the scores are the values. Amy's final grade would be:

$$\text{Final Grade} = \frac{(94)(0.20) + (86)(0.25) + (100)(0.10) + (75)(0.45)}{0.20 + 0.25 + 0.10 + 0.45}$$

$$= \frac{84.05}{1} = 84.05$$

Notice that the sum of the weights in the denominator adds up to 1. This is because the sum of the percents adds up to 100% of her grade.

The weighted mean shows up in other places in statistics as well. We will use the idea of weighted means again in Chapter 7 when we discuss the means of probability distributions. Weighted means are also used to find the means of the grouped frequency distributions we explored back in Chapter 2.

3.2 MEASURES OF SPREAD

With so many ways to find the center of a data set, you might think we really don't need any other statistics to help us analyze information. We know people, in fact, who believe that the mean is the only number ever needed to describe a data set. The following simple comparison will hopefully help dispel this notion!

Able Andy and Big-Play Bill are two football players. The table below shows the number of yards per carry for a *sample* of three plays for each player.

Able Andy: Number of Yards per Carry	Big-Play Bill: Number of Yards per Carry
8	−2
12	32
10	0

Find the mean number of yards per carry for Andy and Bill. What do you notice? If you are the coach, does it make a difference which of these guys carries the ball? Intuitively, yes. Able Andy seems to be much more consistent than Big-Play Bill, doesn't he? When we talk about measures of spread we are talking about measuring the consistency of the data set.

The Range

The easiest measure of spread is the **range**. In Chapter 2 we needed the range in order to be able to construct a histogram, so you should be at least a little familiar with the concept already. The range is usually defined as the difference between the largest value in the data set and the smallest. Some middle school curricula, *Connected Mathematics* for example (page 74), define the range to be the smallest value to the largest value. In other words, they would say that Andy's range in the problem above would be from 8 to 12. We will use the following definition since it is the one found in college-level statistics books:

Range = Largest Value − Smallest Value

In the example, the range for the number of yards per carry for Andy was 12−8 or 4 yards, while the range for Bill was 32−(−2) or 34. Clearly, the numbers for Bill are more spread out or have more variation than those of Andy. Unfortunately, since we only use the highest and lowest values in the data set, we cannot tell much about how those values are really distributed by just using the range. Because the sample sizes are so small in this case—three values each—distribution of the numbers isn't much of a problem. If the data set contained 100 items though, we would need to know quite a bit more than the range in order to get a handle on how spread out the numbers were.

Variance

It would be nice to have some way of measuring the spread of a data set that actually takes into account every value. Remember, that was one of the advantages of using the mean as a measure of the center—means use every item in the data set. One way we could do this is to look at each item and its relation to the mean or its **deviation** from the mean, as we usually call it. We talked a little bit about deviations when we discussed properties of the mean in the previous section. We are now going to add a bit more to the idea of deviations. We will stick with our very small data sets for Andy and Bill, as we would rather not get bogged down in tedious arithmetic just yet.

To compute the deviations for each number in a data set, take the raw score and subtract the mean. Raw scores are typically represented by x and the mean by $\bar{x}$, so a formula for the deviation of a particular value is:

$$\text{Deviation} = x - \bar{x}$$

The deviations for the number of yards per carry are shown in the table below based on the means of 10 that you computed earlier for each set of numbers:

Deviations for Able Andy Data	Deviations for Big-Play Bill Data
8 − 10 = −2	−2 − 10 = −12
12 − 10 = 2	32 − 10 = 22
10 − 10 = 0	0 − 10 = −10

Each of these deviations would be measured in yards per carry just like the original values. A deviation of negative 12 means that the value of −2 in Bill's data set was 12 below the mean of 10 or that it is 12 away from the mean. A curious thing happens with deviations. If all of the deviations for Andy's data are added (summed) the result is −2 + 2 + 0 = 0. The same is true for the sum of Bill's deviations. As it turns out, this is always true. The sum of the deviations for a data set should always be zero. For people who do computations by hand rather than with a calculator or computer, this provides a nice way to check for computational errors. If the sum isn't zero, then something is wrong somewhere.

What we would really like is some way of getting at the average amount each number in the data set differs or varies from the mean. Clearly adding the deviations together and dividing by the total number of items in the set isn't going to help much, as we are going to get zero every time. We have two options. We can either take the absolute value to force all the deviations to be positive—which makes sense as we are dealing with distances from the mean—or we can square the deviations to make them positive. Both of these methods are valid. The first method, taking the absolute value of the deviations, yields what is called the **mean deviation**, although mean absolute deviation might be a better name.

$$\textbf{Mean Deviation} = \frac{\Sigma |x - \bar{x}|}{n}$$

The mean deviation for the deviations of Andy's data would be $\frac{|-2|+|2|+|0|}{3}$ or $\frac{4}{3}$ or $1\frac{1}{3}$ yards per carry. In other words, the average distance of the data items from the mean is $1\frac{1}{3}$ yards per carry. The idea of the mean deviation is very straightforward and is easily accessible for middle school students as the arithmetic involved is not too intense. The mean deviation is also nice for developing the idea of finding some way of measuring spread that involves every item in the data set and helps lay the groundwork for what is called the **standard deviation** of a data set.

While certainly valid, the first method is not the "standard" way of finding the average amount each number differs or varies from the mean. The second method, squaring the deviations, is more prevalent as it includes an adjustment for possible statistical bias. (Statistical bias will be discussed in more detail in later chapters.) This method yields what is called the **variance** of the data set. Variance is generally denoted by s^2 for samples or σ^2 (sigma squared) for populations. The formula for variance is given by:

$$s^2 = \frac{\Sigma(x - \bar{x})^2}{n - 1} \text{ for samples or } \sigma^2 = \frac{\Sigma(X - \mu)^2}{N} \text{ for populations}$$

Notice the change in notation once again. We use Greek and capital letters to designate population measures and lower-case letters to indicate sample values. One other thing might strike you as odd. For samples, our divisor is $n - 1$, while for populations the divisor is N. What is the deal here? Remember that ultimately our goal is to use samples to make inferences or predictions about a population. As it turns out (for reasons we will not inflict upon you at this point), dividing by $n - 1$ gives a much better estimate of the variance of the population than simply dividing by n.

Focus on Understanding

1. Which football player do you think has the largest variance (remember you are working with a sample here)? Why?
2. Find the variance for both Andy's data and Bill's data.
3. What are the units of variance? Are units a problem? Explain.

Standard Deviation

As you discovered, the problem with variance is that the units you end up with to describe it are NOT the same as the units for the mean. The variance for Andy's data set was 4 yards-per-carry *squared* and the variance for Bill's data set was 364 yards-per-carry *squared*. Since the mean in both cases was 10 yards-per-carry it would be nice to measure the average amount each item differs from the mean in yards-per-carry as well. This is relatively easy to accomplish. All we need to do is take the square root of the variance and our units will be the same as the units for the mean. The square root of the variance is called the **standard deviation**. In the case of Andy and Bill, Andy's standard deviation would be $\sqrt{4} = 2$ yards per carry while Bill's standard deviation would be $\sqrt{364} \approx 19$ yards per carry. The larger the value of the standard deviation, the more spread out the data is within the data set.

The common way to describe the average amount each item in a data set varies from the mean is with the standard deviation:

$$s = \sqrt{\frac{\Sigma(x - \bar{x})^2}{n - 1}} \text{ sample standard deviation}$$

$$\sigma = \sqrt{\frac{\Sigma(X - \mu)^2}{N}} \text{ population standard deviation}$$

There are other equivalent formulas for computing the standard deviation. The other formula which is commonly used for standard deviation is called the computing formula or the raw score formula:

$$s = \sqrt{\frac{n\Sigma x^2 - (\Sigma x)^2}{n(n - 1)}} \text{ raw score formula for sample standard deviation}$$

$$\sigma = \sqrt{\frac{N\Sigma X^2 - (\Sigma X)^2}{N^2}} \text{ raw score formula for population standard deviation}$$

It has the advantage that the mean does not have to be computed first and sometimes the numbers are not as messy to work with—particularly if you have whole number data. In reality, most people now use either calculators or computers to do these kinds of computations, so which formula you prefer doesn't matter.

Using Calculator Technology

Let's go back to our original exploration at the beginning of this chapter and use the TI-83 plus to compute the basic measures of center and spread that we've discussed so far. We are going to give the directions for using the TI-83 plus, but any graphing calculator will be able to give you the same basic statistical information. The TI-83 plus was designed to incorporate more statistical analyses so that is what we will use throughout this text.

The data set shown below is the same one from Exploration 3.1. It is taken from *Mathematics In Context: Dealing with Data* (page 40).

Speed of 13 Animals without Hooves in Kilometers per hour

48 113 15 69 40 68 63 48 56 81 56 19 24

We will take you through the problem using the STAT menu step-by-step.

Calculator Directions

Turn your calculator on using the ON key located in the lower, left-hand corner. Then press the CLEAR key, which is on the right-hand side of your calculator under the arrow keys.

1. Locate the STAT key on the TI-83 plus. It is in the row with the green button labeled ALPHA. It should be the third key toward the right in that row. This key is the one you will use to perform statistical analyses. Press the STAT key. You should now see a menu that looks like this:

   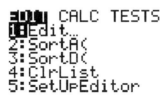

   There will be a black rectangle over EDIT on the top line and a smaller black rectangle over "1:". The larger black rectangle tells you that you are in the EDIT menu. The smaller one indicates the current menu selection. To move to the CALC menu, press the RIGHT ARROW key located at the upper right-hand portion of your calculator. When you press the RIGHT ARROW key, the large black rectangle should now be over CALC and you should see the following:

   ```
   EDIT CALC TESTS
   1:1-Var Stats
   2:2-Var Stats
   3:Med-Med
   4:LinReg(ax+b)
   5:QuadReg
   6:CubicReg
   7↓QuartReg
   ```

   The smaller black rectangle should be over "1:" again. The arrow after the "7" means that there are more choices on that menu. To see the rest of the choices you would press the DOWN ARROW. The DOWN ARROW will let you scroll through the rest of the choices. Right now we do not need to worry about all of the choices for the CALC menu nor do we need anything from the TESTS menu, so use the LEFT ARROW key to put the black rectangle back over EDIT. You should see the original menu again.

2. We are not going to assume that your calculator has been used for statistics before now, so from the EDIT menu select SetUpEditor. There are two ways to do this. Either use the DOWN ARROW key to move the black rectangle from 1: Edit to 5: SetUpEditor or just press the "5" key. Either way will work. There are several ways to accomplish a lot of the same things on this calculator. In general, we will show you one way to do something and let you explore others on your own. SetUpEditor turns on internally some programs that we will need.

   SetUpEditor should now appear at the top of the home screen with a small, flashing black rectangle following it. Press the ENTER key at the bottom right-hand corner of your calculator to have it execute the command. The word DONE should now appear on the line after SetUpEditor. This means that it has done what it needed to do internally to get everything turned on for you. This step does not have to be completed every time you turn on your calculator. Once should be enough unless you inadvertently turn the Editor off or something tragic happens to your calculator.

3. Now we finally get to the heart and soul of using your calculator for statistics. We need to enter in the data from our example. The first thing we need to do is wipe out any old data. Press the STAT key to get back to the EDIT menu. Again, there are a bunch of ways to do what we are going to do next. We are going to discuss one of those ways. The small black rectangle should be over the "1:" which is Edit. We want to Edit our data, so press ENTER. You should now see:

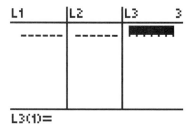

There may or may not be values in the columns labeled L1, L2, and L3. It does not matter. To clear out a column so that you can enter your new data use the UP ARROW key to move the black rectangle until it is over L1 at the top of your screen. Once the black rectangle is over L1, press the CLEAR key and then press ENTER. This should clear out any old data. The black rectangle is now on the first line of column L1. It is waiting for you to enter in the first data value. The first number in our example was "48" so type 48 using the number keys and press ENTER. You must press ENTER after each value in order to move to the next line in the column. You should enter only one value per line. Continue entering the values in from the example. You should have 13 values in column L1.

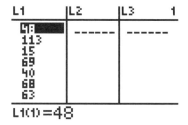

Section 3.2 Measures of Spread 63

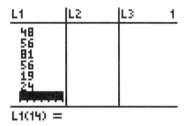

There is black rectangle on the line after the last entry. Ignore it. The hardest part of using either a calculator or a computer to do statistical analysis anymore is getting the data entered. Take a moment now to double check your entries! If you entered a value incorrectly it is easily fixed. Incorrect data will give you incorrect results so be very careful in checking your numbers. If you find that you have messed up an entry, use the UP or DOWN ARROW keys to move the black rectangle to that entry. Type the correct value (it will appear after the L1() = prompt at the bottom of the screen) and press ENTER. The correct value should now appear in your column of numbers.

4. Now that the values are entered into L1, we are ready to do the fun part! Press the STAT key again. Do not worry about losing the data. Unless you clear it out of the memory it will be there! Use the RIGHT ARROW key to move the black rectangle to the CALC menu. The small black rectangle should now be over "1:" of 1-Var Statistics. This is SELECTION to get all of the basic descriptive statistics that we have discussed so far (plus a lot more). Press ENTER. 1-Var Stats should now appear on your home screen with the black rectangle flashing after it. By the way, the flashing black rectangle is called the "cursor." You probably already knew that. It is an excellent idea to get in the habit of telling the calculator *EXACTLY* what column of data you wish it to process. If you look above the "1" key you will see a tiny "L1" written in yellow. Our data for this example is stored in L1, so press the yellow key marked 2nd and then press 1. You should now see L1 following 1-Var Stats on your screen with the cursor flashing after L1. Press ENTER. The calculator has now computed a whole host of basic statistics for you to read and interpret. Your display should look like this:

```
1-Var Stats
 x̄=53.84615385
 Σx=700
 Σx²=46326
 Sx=26.8230192
 σx=25.77072326
↓n=13
```

The first value is the mean, $\bar{x}$, of our sample. The second line shows us the sum of all of the data values. The third gives us the sum of each data value squared. The fourth line shows the sample standard deviation. This is typically the standard deviation value that you want. Generally we work with samples and this is the sample standard deviation. The fifth line shows the adjusted population standard deviation. The last line with the arrow gives the sample size. Do not forget the units here! The data was in kilometers per hour so the mean and standard deviation are also measured in kilometers per hour.

The down arrow on the last line means scroll down for more information. If you use the DOWN ARROW key to scroll to the end of the list, you will see the following additional information:

```
1-Var Stats
↑n=13
  minX=15
  Q₁=32
  Med=56
  Q₃=68.5
  maxX=113
```

"minX = 15" means that the minimum or smallest value in the data set was 15. "Med" tells us the median value is 56. "maxX" gives us the largest or maximum value in the data set or 113 in this case.

Your calculator is not smart enough to determine the mode by itself. You will have to find the mode on your own. In this case our data set is bimodal as we discussed at the beginning of the chapter.

Q1 and Q3 are new pieces of information that we have not yet discussed. Q1 is shorthand notation for the "first quartile" and Q3 is shorthand for the "third quartile." The easiest way to think about quartiles is to think about quarters in a dollar. Four quarters are equivalent to a dollar bill or 100 pennies. Each quartile corresponds to 25% of the data in a set. Essentially Q1 = 32 means that 25% of the data is below 32. Q3 = 68.5 means that 75% of the data is below 68.5. The median value is the same as Q2, or the second quartile. Remember that 50% of the data is below the median and 50% is above the median. Different textbooks and calculators compute quartiles differently. We will go into more detail about quartiles in Sections 3.3 and 3.4, but we are not going to make a big deal out of all the different ways that people find quartiles. Again, most people do not make such computations by hand any more so we are going to use what the calculator gives us.

Notice that the calculator does NOT give the variance for the data set. This is not a huge problem. Remember that the standard deviation is the square-root of the variance. We can get the variance then by squaring the value for the standard deviation, $(s_x)^2$. For this data set the variance would be $(26.8230192)^2$ or 719.474359 kilometers squared per hour squared.

The steps for finding the basic descriptive statistics using a TI-83 plus calculator are summarized in the box below.

Using the TI-83 Plus to Compute Basic Descriptive Statistics

1. Enter the data into the list of your choice by going to the STAT menu and selecting Edit from the EDIT menu under STAT. Generally for a single data set use L1.
2. Go to the STAT menu and select CALC. From the CALC menu, select 1-Var Stats.
3. When 1-Var Stats shows up on the home screen, enter the list where the data is located and press ENTER.

To use a statistical computer application such as Minitab, Fathom, or Excel you would use a similar process. First you must enter in your data and then select the appropriate statistical analysis from one of the menus. We will say more about this in another chapter.

Dealing with Data

8. a. Approximately how many 12-ounce cans of soda do you drink per day?
   b. Do you think this number is typical? Explain.

Students at Fontana Middle School surveyed their classmates to find out approximately how many cans of soda students drink per day. In a sixth-grade class, they found the following results.

Number of Cans of Soda Sixth-Graders Drink per Day
0, 1, 5, 1, 0, 0, 5, 4, 5, 0, 4, 2, 1, 3, 3, 0, 1, 0, 4, 5, 5, 2, 4, 5

 9. a. Find the mean number of cans of soda the class drinks per day. Explain how you calculated the mean.
   b. Do you think that the mean is a good way to describe the amount of soda a sixth-grader drinks per day? Why or why not?

Below are the results of the survey for an eighth-grade class.

Number of Cans of Soda Eighth-Graders Drink per Day
1, 0, 0, 0, 1, 2, 3, 4, 2, 0, 1, 1, 0, 0, 13, 3, 2, 2, 0, 0, 1, 11, 1, 2, 3, 3, 13

FIGURE 3.2.1 Comparing soda consumption

66 Chapter 3 Describing Data with Numbers

Classroom Connection

The information in Figure 3.2.1 is from *Mathematics in Context: Dealing with Data* (page 26). While standard deviation is not a typical middle school concept, as teachers or future teachers it is an important concept for you to understand. ◆

Focus on Understanding

1. Using a calculator or a computer, find the mean, median, and mode for both data sets in Figure 3.2.1. Please put the data for the sixth grade in L1 and the data for the eighth grade in L2. How does the soda consumption for the sixth grade compare to the soda consumption for the eighth grade?
2. Find the range of each data set. How do the ranges compare?
3. Compute the sample standard deviation for each data set. Which class is the most variable in terms of soda consumption? Why do you think that might be?

Using Technology to Generate a Histogram

Now that you know how to enter data into the TI-83 plus, let's go back and look at generating a histogram using the calculator. Since you just entered the data for the sixth grade class into L1, we will generate the histogram for that data set using the calculator. You should have 24 values entered into L1. The next thing we need to do is go to the STAT PLOT menu and turn on one of the plots. To do this, press the yellow 2nd key (located in the upper left-hand portion of your calculator) and then the Y = key, which is directly above the yellow 2nd key.

You should now see a menu that looks like this:

You probably also see a line of weird stuff under each plot line like a little graph of some kind followed by L1, L3, or whatever. That is normal. The dark rectangle should be over "1:". This is the plot we want to turn on, so press ENTER.

You should now see the following with a dark rectangle over Plot1 and Off with the cursor flashing on On:

Section 3.2 Measures of Spread 67

Since the cursor is flashing over On, all we need to do is press ENTER to move the cursor to On. This turns Plot1 on for us. Now use the down arrow key to move the cursor to the Type: line. You should see two rows of graphs following Type:. The first graph after Type: is a scatter plot. We will work with scatter plots in Chapter 4. The second graph in the row is a kind of line graph. The third graph on the first row is a histogram. That's the one we want! Down below you see two graphs that look almost the same. Both are box-and-whisker plots. We'll use those at the end of this chapter. To select a particular graph, use the left or right arrow keys to move the cursor to the graph you want. In this case, we need to move the cursor to the histogram on the far right of the top row. Press ENTER to select the histogram.

When you selected the histogram by pressing the ENTER key, the screen should have changed to look something like this:

This tells you that the data it is going to use to make the histogram is in L1. The cursor should be flashing over L1. This means you can tell it a different data location if your data is not stored in L1. For example, to graph a histogram of the eighth grade data that is stored in L2, you would press the yellow 2nd key and then the "2" key to change the data location to L2. Since the sixth grade data is in L1, we do not need to change anything. We've turned the plot on, told it what type of graph we wanted and where the data is stored, so we are ready to actually generate the histogram. The quickest way to do this for most people is to use the ZOOM menu. The ZOOM key is located on the top row right in the middle. If you press ZOOM, the following menu should appear:

68 Chapter 3 Describing Data with Numbers

Remember, the down arrow after the 7 means that there are more choices, so use the DOWN arrow key to move the cursor down the menu. We want ZoomStat, which is choice number 9 from the menu. ZoomStat automatically previews your data and sets a reasonable (usually) window for your graph. You could set the window by hand. Let's just let the calculator do it for us. Once you get the cursor moved so that the dark box is over the "9:", press ENTER. You should get a very lovely histogram that looks something like this:

At first glance this histogram seems pretty good. Look a little closer! Notice that we have a gap between the fifth bar of our histogram and the sixth one. What is going on? Look along the top row of keys again and you should see a button marked TRACE. Press the TRACE key to find out what happened. Your screen now looks like this:

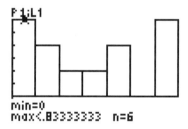

If you use the right arrow keys to move the cursor to the top of each bar, you see that the information at the bottom gets stranger and stranger. Essentially the calculator is telling you what choices it made for the boundaries for the classes and, as it turns out, it did not make particularly smart choices! To get a better looking histogram, we need to actually tell it what boundaries (remember this from Chapter 2) and what class width we would like it to use. To set the window by hand, press the WINDOW key on the top row. Your screen now looks like:

```
WINDOW
  Xmin=0
  Xmax=5.8333333...
  Xscl=.8333333...
  Ymin=-1.80414
  Ymax=7.02
  Yscl=1
  Xres=1
```

We want to change these values to get a nicer graph. The cursor should be flashing over the 0 after Xmin. Change this value to negative 0.5 by typing (−) 0.5. Please make sure to use the negative key (to the left of ENTER) and NOT the subtraction key. After you change each value, press the ENTER key to move the cursor to the next number. Use the following values for your histogram window:

```
WINDOW
 Xmin=-.5
 Xmax=5.5
 Xscl=1
 Ymin=0
 Ymax=10
 Yscl=1
 Xres=1
```

Once you get the values changed, press the **GRAPH** key in the upper right-hand portion of your calculator. Now, if you press the **TRACE** key you will see the lower and upper boundaries for each class appear at the bottom of your screen along with the frequency for that class.

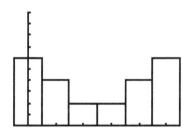

The steps for constructing a histogram using your calculator are summarized below.

Generating a Histogram with a TI-83 Plus Calculator

1. Enter the data into the list of your choice. L1 is generally a good choice unless you are working with more than one data set at a time.
2. Go to the STAT PLOT menu, turn the plot on, select histogram as the type, and tell the calculator what list contains the data.
3. Either use the ZOOM menu and ZoomStat or set your own window by using WINDOW and then GRAPH to generate the histogram.

Calculators and computers are marvelous tools for doing tedious chores like long computations for means and standard deviations of large data sets and for generating quick graphs of data.

3.3 MEASURES OF POSITION

So far we've discussed measures of the center and measures of spread. There is one other type of measurement we use to evaluate the items in a data set—position. While measures of center and spread work with the entire data set, measures of position are used to describe the position of an individual data item within the set. Percentile, decile, and quartile rank are three ways used to describe the position of an item in a set. Essentially, a **percentile** (or quartile or decile) rank is the position in a data set where some percentage of the data is above that value and some percentage is below. For example, if a student scores at the 80th percentile on a standardized exam it means that he or she did better than 80% of the people who took the exam. In reality, it makes very little sense to compute percentile or decile ranks for small data sets. Quartiles are used to construct box-and-whisker plots so we will explore quartiles in more detail in the next section. Be careful. It is very easy to confuse percentages with percentile rank! Percentages tell you nothing about the position in a data set! A person could score 75% on a test, but be at the 70th percentile. Take a look at the following example, even though it really doesn't have enough data items to warrant finding percentile ranks. Hopefully, it will give you an idea of what we are talking about!

EXAMPLE 3.3.3 Ms. Smith gave a short test in her geography class. The total number of points possible was 60. Here are the scores arranged from lowest to highest:

$$13 \quad 18 \quad 24 \quad 32 \quad 36 \quad 40 \quad 40 \quad 45 \quad 48 \quad 51 \quad 56$$

We will focus on the score of 45. First we will find what percentage corresponds to a score of 45. We worked with percentages back in Chapter 2 when we constructed frequency tables and the idea is the same here. The score of 45 is really 45 out of 60 possible points. So,

$$\frac{45}{60} = 0.75 = 75\%$$

If we look at the position of 45 within the ranked data set, however, we see that there are seven items below 45 and three items above it. Since there are only 11 items in the set, we can see that approximately 70% of the items are below 45 and 30% are above, so a score of 45 would **approximately** be at the 70th percentile for this data set. Computations for finding percentile ranks are quite messy and generally only approximations. As we mentioned earlier, measures of percentile and decile really only make sense for large data sets. We simply do not compute these types of measures for large data sets by hand anymore, so we are not going to inflict the formula(s) for finding percentile and decile ranks upon you! We will get to quartiles in the next section. ∎

While percentile, decile, and quartile ranks are three ways to measure the position of an individual data item within a set, probably the most important way to measure the position of an item is by what is called a **z-score**. Did it seem odd to you that we discussed measures of center and spread but then totally ignored both of those ideas for position? Well, z-scores are a measure of position that take into account both the mean and standard deviation of a data set. They provide a useful tool for comparing data sets with different means and standard deviations. Z-scores are also the primary computation that we use for making conjectures about data sets as you will see in Chapters 7, 8, 9, and 10!

So what is a z-score? A z-score is a measure of position that tells you how many standard deviations a particular data item is relative to the mean of the data set. You have probably heard people talk about professors who determine the grades in their classes using what is called a bell curve or a normal curve. A normal curve is shown below:

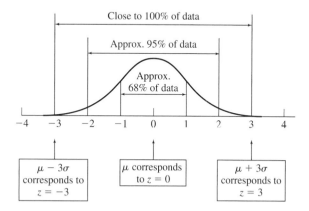

Students whose total points for the class fall within one standard deviation of the mean would get a C for the course. Students whose points are between one and two standard deviations above the mean would get a B. Students between one and two standard deviations below the mean would get a D and so on. As you can see, normal curves are symmetrical with 50% of the data above the mean and 50% below the mean. Since the mean is zero standard deviations away from itself, it has a z-score of zero. Why do people use one standard deviation above and below the mean as a cutoff? Basically this decision has to do with something called **The Empirical Rule**.

The Empirical Rule says that for a *normal distribution* (more about this in Chapter 7) roughly 68% of the data values will lie within one standard deviation of the mean, 95% within two standard deviations, and close to 100% within three standard deviations. Hence, some people use these values of ±one standard deviation, ±two standard deviations, and ±three standard deviations to determine cutoffs. The catch to this is that you have to assume that the distribution of the data is indeed bell-shaped or normal.

Let's take a look at an example to see exactly how to find the z-score for a particular number in a data set. We'll use our previous data from Ms. Smith's class.

EXAMPLE 3.3.4 Here were the scores in Ms. Smith's class:

13   18   24   32   36   40   40   45   48   51   56

Since we worked with finding the position of 45 before, we will use it again in this example to find its z-score. The first thing we need to do to compute the z-score, which corresponds to 45, is to find the mean and standard deviation of the data set. You've done this before, so hopefully this should not cause major problems for you. You should get a mean of 36.64 (rounded to the nearest hundredth) and a standard deviation of 13.11 (rounded to the nearest hundredth). Again, a z-score simply tells you how many standard deviations above or below the mean your data value is. How far away from the mean is 45? Well, this is the mean deviation we discussed in the previous sections:

$$\text{Deviation from the mean} = x - \bar{x} = 45 - 36.64 = 8.36$$

Notice that we did not put absolute value signs around our subtraction this time. This is important! We are trying to find how far *above or below* the mean a value is. Since we want to know where the value is, just looking at the distance (absolute value) will not help. Values below (less than) the mean will have negative z-scores while values above (greater than) the mean will have positive z-scores.

In statistics, we measure distances by standard deviations. How many standard deviations of 13.11 are there in a deviation of 8.36?

$$\text{z-score for 45} = \frac{8.36}{13.11} = 0.6376811594 \approx 0.64$$

So, a score of 45 is less than one standard deviation above the mean. If Ms. Smith uses a normal curve to assign grades then this student would get a C. ∎

Generally z-scores are rounded to two decimal places, though because of the ever-increasing use of calculators and computers, some people prefer to carry the values out more places. We also rounded our mean and standard deviation values to two decimal places. A general rule of thumb is to round your mean and standard deviation values to no less than two decimal places beyond your data values. What this means is that if your data set is in whole numbers, round to the hundredths place. If your data items were to the tenths place (like 15.6), you would round your mean and standard deviation to the thousandths place. Some people, however, insist on keeping eight or nine decimal places worth of digits for every computation. For what we are doing, we really do not need that level of precision.

Z-score Formula

To compute the z-score for an individual item, x, in a data set that has a mean, $\bar{x}$, and a standard deviation, s:

$$\text{z-score} = \frac{x - \bar{x}}{s}$$

A negative z-score indicates that x was below or less than the mean value. A positive z-score indicates that x was above or greater than the mean value.

We mentioned earlier that z-scores may be used to compare different data sets. Let's look at an example to illustrate what we mean by this.

EXAMPLE 3.3.5 Suppose you had the misfortune to have a statistics test and a history test on the same day. When you got your tests back, here is the information given to you by your professors regarding your performance and the performance of the class on these exams.

	Statistics	History
Your Score	82	93
Mean	71.06	85.43
Standard Deviation	10.32	18.91

At first glance it would appear that you did much better on the history exam because your score on the history exam was higher, but to really compare these two tests we need to look at the z-scores for both exams!

$$\text{z-score for Statistics score of } 82 = \frac{82 - 71.06}{10.32} = 1.06$$

$$\text{z-score for History score of } 93 = \frac{93 - 85.43}{18.91} = 0.40$$

Relatively speaking, you did much better on the statistics test because your score was more than one (barely) standard deviation above the mean. The higher score of 93 only had a corresponding z-score of 0.40, which does not make it out of the "C" range for those folks who use normal curve grading. Weird, huh? ∎

Focus on Understanding

Suppose the mean and standard deviation on a biology test were given as 72 and 12, respectively. Use this information to answer the following questions.

1. Susan's z-score on a test in biology was 2.34. Was Susan's test score above or below the mean? How do you know?
2. David's z-score on the same test in biology was -1.25. Was David's test score above or below the mean? How do you know?
3. Dakota had a z-score of 0.08 on the biology test. What does this z-score tell you about Dakota's test score relative to the mean?
4. You know the z-scores for Susan, David, and Dakota. Find their actual test scores based on the information you have about the mean and standard deviation.
5. Rebecca made an 80 on the biology test. Find her z-score.
6. In Example 3.3.3, the mean of the history scores was 85.43 while the standard deviation was 18.91. What, if anything, does this tell you about the spread of the data? Does this seem likely to happen?

We will do much, much more with z-scores in the chapters ahead, but always keep in mind that simply speaking a z-score just tells you where a value falls relative to the mean based on the standard deviation of the data set. They are nice measures to work with in that they use both the mean and the standard deviation in their computation.

3.4 BOX-AND-WHISKER PLOTS

Box-and-whisker plots (also called **box plots**) show how the values in a data set are distributed. The construction of a box plot requires the **median, lower quartile, upper quartile, maximum value**, and **minimum value** of the data set. Box plots are commonly found in middle school curricula and are fairly easy to construct. Notice that box plots do NOT use the standard deviation in their construction.

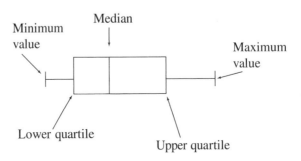

FIGURE 3.4.1 Five-number summaries and their relation to box plots

We promised in the previous section we would say more about quartiles. Quartiles are two of the five measures we need for the construction of a box plot.

Definition 3.4.1 The **lower quartile** (or Q_1) of a data set is the median of the data values below the median. The **upper quartile** (or Q_3) of a data set is the median of the data values above the median.

Some box plots indicate only the four quartiles of data distribution, with the first and fourth quartiles ending at the minimum and maximum data values, respectively. Other box plots prescribe a maximum length for the first and fourth quartiles. **Outliers** are marked separately in this type of box plot. The median, lower quartile, upper quartile, minimum value, and maximum value are oftentimes referred to as a **five-number summary**. The box plot in Figure 3.4.1 shows the relationship of each of the numbers in the box plot. Box plots also include a scaled, horizontal axis (or vertical axis if the box plot is vertical). Where each of these values falls in relation to the horizontal axis depends upon the data set. We did not include a horizontal axis in Figure 3.4.1, just the five-number summary.

While five-number summaries are really all we need to construct a box plot, occasionally we use another measure to help us further analyze what our data look like. The interquartile range is a measure used to identify extreme values within our data set.

Definition 3.4.2 The **interquartile range** (or **IQR**) is the difference between the upper quartile and the lower quartile.

Classroom Connection

The following *Classroom Connection* is from *Connected Mathematics, Samples and Populations* (page 33). ◆

11. Samples of adults and eighth grade students were asked how much time they spend on the telephone each evening. The results are displayed in the box plots below.

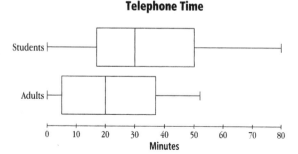

a. What are the median, lower quartile, upper quartile, and range of the telephone times for the students?

b. What are the median, lower quartile, upper quartile, and range of the telephone times for the adults?

c. Describe the similarities and differences between the distribution of telephone times for adults and the distribution of telephone times for students.

FIGURE 3.4.2 Determining the five-number summary from a plot

Focus on Understanding

1. For the box plot for student telephone time, about what percentage of the data set is above the median? About what percentage is below? Is this also true for the box plot representing the adult telephone time? Why or why not?
2. Answer the questions in parts a, b, and c in the classroom connection shown in Figure 3.4.2.
3. About what percentage of the data set should be in each quartile (minimum to first quartile, first quartile to median, median to third quartile, and third quartile to maximum)? Explain why!
4. What percentage of the data set should be contained within the box part of a box plot? Why?
5. Find the interquartile range (IQR) for this data set. What part of a box plot is represented by the IQR?

Definition 3.4.3 Outliers are data values that are much greater than or less than most of the other values in the data set. A data value is considered an outlier if it is less than the lower quartile minus 1.5 times the IQR, or if it is greater than the upper quartile plus 1.5 times the IQR. These values, the lower quartile minus $1.5 \times$ IQR and the upper quartile plus $1.5 \times$ IQR, are respectively referred to as the lower fence and upper fence.

Constructing a Box Plot

1. Find the Upper Extreme (highest value) and Lower Extreme (lowest value) of the data set.
2. Determine the values of Q_1, Median (Q_2), and Q_3.
3. Use Q_1, Median (Q_2), and Q_3 to construct the "Box" containing the middle half of the data set.
4. Determine the IQR and use it to find the Upper Fence and Lower Fence:

$$\text{Upper Fence} = Q_3 + 1.5 \times \text{IQR}$$
$$\text{Lower Fence} = Q_1 - 1.5 \times \text{IQR}$$

5. Construct "Whiskers" containing the upper- and lower-quartiles. Use largest values and smallest values that lie between the Upper and Lower Fences. (Note: the Upper and Lower Extreme values may not lie between the Upper and Lower Fences).
6. Identify all Outliers with asterisks.

Classroom Connection

The *Classroom Connection* shown in Figure 3.4.3 is from *MathThematics, Book 3* (page 154). Notice that in the tables the data are already ordered from least to greatest values. Notice that *MathThematics* refers to these types of displays as box-and-whisker plots as opposed to *Connected Mathematics*' use of the term box plot. As we mentioned earlier, both terms apply to the same type of graph. It is simply a matter of preference, so don't let it fool you! ◆

Focus on Understanding

Using the data set from the 1995 T-shirt prices shown in Figure 3.4.3:

1. Find the five-number summary and construct the box-and-whisker plot.
2. Compute the IQR for the data set.
3. Determine any outliers for the data set. If there are outliers, redraw your plot to show the outliers.

9. The price of a T-shirt at 25 amusement parks in the United States in 1995 and 1996 is given below.

1995 T-shirt prices (dollars)				
5	6	6	6	7
7	7	7	8	8
8	8	8.50	9	9
10	10	10	10	10
10	11	12	12	13

1996 T-shirt prices (dollars)				
5	6	6	7	8
8	8	8	8	8
9	9	9	9	9
10	10	10	10	10
10	11	11	13	13

a. Make a box-and-whisker plot for the 1995 prices and one for the 1996 prices below the same number line. Label each box-and-whisker plot.

b. Does either set of data contain outliers? If so redraw the box-and-whisker plot using asterisks to show the outliers.

c. Do the box-and-whisker plots show that the price of T-shirts at amusement parks changed significantly from 1995 to 1996? Explain.

FIGURE 3.4.3 Constructing box-and-whisker plots

As you might have guessed by now, box plots are frequently used to compare multiple data distributions. For example, both of the classroom connections you have explored so far involved two data sets. Box plots are a handy, visual way to compare two or more data sets. The census data used in Figure 3.4.4 has been analyzed to compare the age distribution of residents for each marital status category. This is the same data set that was represented in Figure 2.1.5 back in Chapter 2. The categories on the vertical scale are "Divorced," "Married," "Never Married," "Separated," and "Widowed."

Focus on Understanding

1. About what percentage of the values in a data set are in the box part of a box-and-whisker plot? In the whisker? (This question should look familiar!)
2. Compare the two types of box plots, those that show outliers and those that do not. What information can be inferred from each type?

The following questions refer to Figure 3.4.4.

3. Why do you think the box for "Widowed" is so far to the right? What does this mean?
4. Why do you think there are so many outliers in the "Never Married" category?
5. Do these box plots reflect the number of people in each category? (See Figure 2.1.5 for the original data set.)

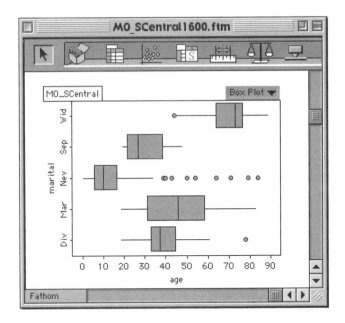

FIGURE 3.4.4 Comparing distributions using box plots

Back in Chapter 2 you constructed many different types of data displays, but not box plots as they make use of descriptive statistics unlike other types of displays. As part of Chapter 2, you also discussed the advantages and disadvantages of one type of graph over another. It might seem that box plots are superior simply because they enable the reader to see the five-number summary. However, depending upon what information you would like to read from your graph, this may not be the case.

Classroom Connection

Take a look at the excerpt from *MathThematics, Book 3* (page 154). Here middle school students are asked to compare to data displays. ◆

10. Choosing a Data Display Dana keeps track of the catalogs her family receives in the mail. Each week for 26 weeks, she records the number of catalogs. She uses the data to make a line plot and a box-and-whisker plot.

  a. For 3 of the 26 weeks, Dana counted 6 catalogs. How is that shown in the line plot? Does the box-and-whisker plot give this information as well? Explain.

  b. What does the box-and-whisker plot show that the line plot does not?

  c. Open-ended Which data display do you think shows more information? Explain your choice.

FIGURE 3.4.5 Comparing box plots and line plots

Section 3.4   Box-and-Whisker Plots   79

> ### Focus on Understanding
>
> 1. Using the two plots in Figure 3.4.5, answer questions a, b, and c.
> 2. For the data shown in the line plot, what other graphs might be appropriate? For example, would it make sense to make a stem-and-leaf plot? A histogram? Why or why not?

Using Technology to Generate Box Plots

Remember the really, really long section on how to use a TI-83 plus to generate a histogram? Well, if you can remember back that far, one of the choices for graph type under the statistical plot (STAT PLOT) menu was a box-and-whisker plot. In fact, two of the choices were box-and-whisker plots. The first and second choices on the second row for TYPE are both box plots. The first choice with the little dots trailing behind is a box plot showing outliers. The second is just a regular box plot without showing outliers.

We will give you a chance to play with the box plot graph on your calculator in the homework!

Chapter 3 Summary

Chapter 3 turned out to be a really long chapter filled with lots of information and a lot of formulas. The focus of the chapter was on the different ways used in statistics to describe a data set. Ways of measuring the center of a data set, the spread of a data set, and measures of position for individual items were explored. These measurements are often referred to as descriptive statistics and will be used over and over again throughout the book. Calculators also played a big role in this chapter. As we mentioned in the preface, the GAISE College Report funded by the American Statistical Association has recommended the use of calculators in college statistics classes and the National Council of Teachers of Mathematics has been a proponent of the appropriate use of technology for years. In the real world, technology is used to work with data.

As promised at the end of Chapter 2, a new type of graph—a box-and-whisker plot—was introduced to visually display descriptive information from five-number summaries. We also introduced the idea of a z-score which, along with the normal curve, plays an important role in upcoming chapters. Descriptive statistics, along with some basic ideas from probability, provide the foundation for all of inferential statistics.

The key terms and ideas from Chapter 3 are listed below:

mode 49	sample standard deviation 61	lower quartile 75
median 51	population standard deviation 61	upper quartile 75
mean 51	percentile 71	five number summary 75
weighted mean 56	decile 71	outliers 75
range 58	z-score 71	interquartile range 75
deviation 59	box-and-whisker plot/box plot 75	upper fence 77
mean deviation 59		lower fence 77
variance 60		
standard deviation 60		

Assessment is an integral part of every curriculum from the elementary school all the way through college. The question always arises—*What is it that students should be able to do after completing this lesson/unit/chapter?* We have included here our intended learning goals for Chapter 3. Students who have a good grasp of the concepts developed in Chapter 3 should be successful in responding to these items:

- Explain or describe what is meant by each of the terms in the vocabulary list.
- Given a data set, compute the mean, median, mode, range, sample standard deviation, and sample variance.
- Create a data set that has a given mean, median, mode, range, and sample standard deviation.
- Predict what will happen to the mean and median when values are added to or deleted from a data set.
- Apply and compute a weighted mean.
- Find the z-score for an item or items in a data set.
- Given a z-score and the corresponding mean and standard deviation for a data set, determine the value of the original data item.
- Find the five-number summary for a given data set and use that information to construct a box-and-whisker plot.
- Determine the outliers for a data set using the interquartile range.
- Given a box-and-whisker plot, determine the median, lower quartile, upper quartile, maximum value, and minimum value.
- Compare multiple data distributions using box-and-whisker plots.

Your course instructor may have additional or different assessable outcomes for your class. As teachers (or future teachers) you should think about the assessment outcomes and learning goals for each chapter as you work through them.

EXERCISES FOR CHAPTER 3

1. Using the *Data Analysis and Probability Standards for Grades 6–8* from the NCTM's *Principles and Standards for School Mathematics* found at www.nctm.org, identify the middle school objectives that are found in Chapter 3.
2. Using the state standards for mathematics for the content area of data analysis and probability for your state, identify the middle school objectives that are

found in Chapter 3. The following website may be useful: www.doe.state.in.us. This website will allow you to access the web pages of the state departments of education for the 50 states. From the state web pages you should be able to find the state's mathematical standards.

3. For each of the following data sets, find the mean, median, mode, range, sample standard deviation, and sample variance.
   a. 94 68 68 89 73 79 40 42
      74 64 56 74 50 63 74 94
   b. 2   12 12 20 6   7   8   8   12 14
      18 18 14 12 10 10 14 13 17
   c. 50 47 46 44 41 37 36 34
      32 30 28 26 25 23 22 20
   d. 12 23 14 14 14 10 10 10 10

4. For each of the following problems, create a data set with the given characteristics. Write at least one sentence describing how you solved the problem.
   a. 10 quiz scores, mean = 6
   b. 12 quiz scores, mean = 12.5
   c. 10 test scores, median = 80
   d. 10 test scores, median = 82.5
   e. 13 test scores, median = 76
   f. 5 test scores, mean = 75, median = 80
   g. 10 data items in the set, median = 22, but 22 is not an element of the set
   h. 8 data items, median = 12, mean = 12, mode = 12
   i. 6 bowling scores, median = 110, mean is 110, but the mode is not 110
   j. 9 data items, median = 14, mean = 11, mode = 8
   k. 10 data items, median = 8, mean = 8, mode = 8, standard deviation = 0

5. Ms. Hanson has 16 students in her class. On the day of her last test in statistics, one student was gone. Here are the scores for the 15 students who took the test:
   28 60 68 71 72 73 73 74
   76 77 79 80 82 82 85

   a. Find the mean, median, mode, and sample standard deviation for these test scores.
   b. James, the student who was absent, returned and took the test the following day. Ms. Hanson noticed that when she added James' score to the data set, the mean of the data set increased slightly, but the median stayed the same. How is this possible?
   c. Ms. Hanson really wanted the median of the test scores to be 75. What would James have needed to score in order for the median of the data set to be 75? Why?
   d. Suppose that instead of having a median of 75, Ms. Hanson really wanted the mean to be 75. What would James have needed to score in order for the mean of the data set to be 75? Why?

6. Students often come up with their own algorithms for computations. As teachers, we have to be able to decide if a particular algorithm is mathematically correct or not. Take a look at the following student algorithm:
   Rachael's class was working on finding the mean of a small set of numbers. The class was given four test scores—70, 80, 90, and 80—and asked to find the mean of these four numbers. Rachael's class found that these four numbers had a mean of 80. Rachael's teacher then added a fifth score, 50, to these scores and asked

the class to find the new mean. The class determined that the new mean of these five numbers was 74.

Rachael has an idea, however, about how to determine the new mean after the fifth score is added. "I noticed that 50 was 30 less than 80. I divided 30 by five and got 6. I then subtracted 6 from 80 and got 74 and that's the new average!"

  a. Does Rachael's strategy for finding the "new average" (of five scores) work for any fifth score? Use alternatives for the fifth score to support or reject her conjecture.
  b. Formulate a mathematical argument that uses algebra to prove "Rachael's Theorem."
  c. Does this strategy work for larger data sets (with more than five scores)? How would you go about determining the answer to this question?

7. Sharla pretty much knows what grades she is going to get at the end of the semester. She expects her report card to look like this:

Class	Grade	Hours
History	A	3
College Algebra	C	3
Biology Lecture	B	3
Biology Lab	A	1
English Composition	B	4
Bowling	D	1

  a. Find Sharla's semester GPA based on this information.
  b. As it turned out, Sharla surprised herself and earned a B in College Algebra, but she only earned a B in Biology Lab. Will her GPA be higher or lower than the GPA you computed in part a? Why?

8. One activity to help students understand the concept of the mean and what happens to the mean when numbers are added that are above or below the mean is the "Sticky Note Activity." In this activity, students construct a line plot using sticky notes. A target mean is selected and students have to place sticky notes on the line plot representing various numbers in the data set. The effect of these numbers on the mean is then analyzed by the class.

  a. Draw a line plot to represent the following values from a sticky note activity. The scale starts at 0 on the left and goes to 22 on the right. Label the line plot. Start with sticky notes on 10, 14, 18, 18, 12, and 12. The target mean is 14. (You should verify that the mean of these six numbers is actually 14 before moving on!)

  b. A single new sticky note is placed on 9. Where would you place a single sticky note in order to make the mean 14 again?
  c. Bert next puts one sticky note on 2. Did the mean go up or down? Why?
  d. Mary needs to rebalance the mean at 14. What would you tell her to do? Does she need more than one sticky note in order to get the mean to 14 again? Why or why not?
  e. Mary was successful in getting the mean back to 14 in part c. Patrick is given only one sticky note and is told that he may put the sticky note anywhere he wants as long as the mean stays at 14. Where should Patrick put his sticky note and why?

Exercises for Chapter 3   83

9. The following data represent the number of miles driven per day by a government employee over a 20-day period.

   56   78   120   145   120   96   34   254   120   137
   120   78   78   206   83   43   122   187   158   62

   a. Find the five-number summary for this data set.
   b. Construct a box plot using your five-number summary in part a.
   c. Determine the interquartile range for this data set.
   d. Determine any outliers for this data set. If there are outliers, redraw your box plot to show the outliers.

10. In Mr. Harris' class, test scores make up 40% of the final grade, the final exam makes up 45% of the final grade, and homework makes up 15% of the final grade. Use this information to find the final percentage grade for each of the following students:

    a. Bob: Tests—84%, Final—88%, Homework—64%
    b. Christy: Tests—92%, Final—86%, Homework—100%
    c. Reggie: Tests—73%, Final—80%, Homework—90%

11. Given the box plot below:

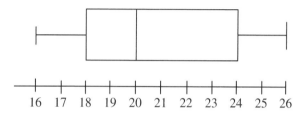

Create a data set (using as many items as you would like) that would generate the same box plot.

12. Use your calculator to generate the box plot for the 1996 T-shirt data from Figure 3.4.3. Use TRACE to see the five-number summary values from your display and then sketch a copy of the graph in your homework.

13. Mike and Debbie were comparing utility bills for two different states over a 12-month period. The box plots for their data shown below are the Fathom-generated box plots.

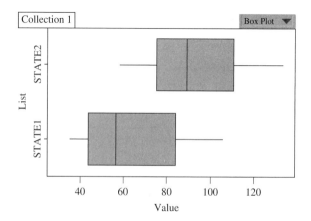

	Jan	Feb	Mar	Apr	May	June	July	Aug	Sept	Oct	Nov	Dec
State 1	$98	$106	$83	$56	$48	$38	$42	$45	$35	$57	$66	$85
State 2	$120	$134	$120	$94	$67	$58	$86	$92	$84	$68	$82	$102

Using the box plots from Fathom, compare the utility costs from the two states. Which state would you rather live in and why?

14. A prospective new car buyer decided to compare the reported miles per gallon on 12 mid-sized sedans. His data are given below:

    28   32   36   24   24   25   22   38   22   24   18   43

    a. Find the mean and standard deviation for this data set. Round your sample standard deviation to two decimal places.

    b. Using the mean and standard deviation from part a, compute the z-scores for the following data values:

        i. 28   **ii.** 22   **iii.** 43   **iv.** 18

    c. Based on the z-scores from part b, which of the values listed in part b is the farthest away from the mean?

15. On the first statistics test in Mrs. Upshaw's class, the mean was 73.5 and the standard deviation was 12.

    a. If Ben had a z-score of 1.56 on this test, what was his original raw test score?

    b. If Susan had a z-score of -2.08 on this test, what was her original raw test score?

    c. If Juan had a z-score of 2.35 on this test, what was his original raw test score?

    d. If Bernice had a z-score of -0.25 on this test, what was her original raw test score?

Data with Two Variables

CHAPTER 4

4.1 SCATTER PLOTS AND CORRELATION
4.2 PEARSON'S CORRELATION COEFFICIENT
4.3 SLOPES AND EQUATIONS OF FITTED LINES
4.4 THE LEAST SQUARES LINE
4.5 THE MEDIAN-MEDIAN LINE

In Chapter 2, we discussed methods for displaying data in tables and charts. Chapter 3 dealt with methods for describing data with numbers. So far, however, our focus has been on **univariate** data sets, those with only one variable. This chapter will focus on methods for displaying and describing **bivariate** data sets, specifically those with two quantitative variables.

Why deal with two variables at once? Why not just look at the data on each variable separately, using the methods we already know? The reason is that variables are sometimes related to each other: as one variable increases, the other tends to change in a particular direction, either increasing or decreasing. This type of relationship between variables is called **correlation**. It is almost impossible to see correlation between variables by looking at each variable separately.

The kind of correlation we have just described is associated with straight lines. Variables can also have more complex relationships, involving parabolas or other curves, but these higher-order relationships are beyond our scope here. This book, like the middle school curricula, will focus only on *linear* relationships between variables.

4.1 SCATTER PLOTS AND CORRELATION

Scatter Plots

The most common type of graph for data on two quantitative variables is the **scatter plot**. Figure 4.1.1 shows an example of a scatter plot of data from *Math in Context: Statistics and the Environment, Murre Island Bats* (page 5).

86 Chapter 4 Data with Two Variables

Temperature in Celsius	Minutes Outside
20	16
22	21
19	15
22	24
26	30
23	23
19	14

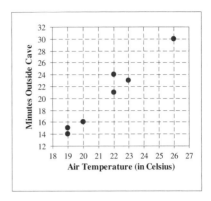

FIGURE 4.1.1 Bat data with scatter plot

Each pair of values in the data set represents a single observation. For example, the first line in the table (20, 16) represents an observation where the temperature was 20°C and the bat (or bats—the source is not clear which) spent 16 minutes outside the cave. The scatter plot consists of a vertical axis and a horizontal axis (one for each variable), forming a coordinate plane in which observations are plotted as data points in the same way that solution points are plotted to graph equations in an algebra class.

There are only two important differences between plotting data points in a scatter plot and plotting solution points. One difference is that, in an algebra class, the axes almost always correspond to the lines $x = 0$ and $y = 0$, meeting at the point (0, 0). In scatter plots, the axes are often moved closer to the data points. In the scatter plot shown, the axes are at $x = 18$ and $y = 12$, meeting at the point (18, 12).

The other difference between plotting points in statistics and in algebra is that, in statistics, the data points should not be expected to lie exactly on a nice straight line or a simple curve such as a parabola. In the scatter plot shown, the points do not exactly lie on a straight line, but they do appear to be close to a line. This brings us to the subject of **correlation**.

Correlation

To understand the idea behind correlation, we will start by revisiting a topic from algebra: positive and negative slope, as shown in Figure 4.1.2.

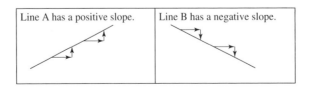

FIGURE 4.1.2 Lines with positive and negative slope

Classroom Exploration 4.1

Refer to the lines in Figure 4.1.2.

1. As you move from left to right along either Line A or Line B, are the **x-values** increasing or decreasing?
2. Line A has a **positive** slope. As you move from left to right along this line, are the **y-values** increasing or decreasing?
3. Line B has a **negative** slope. As you move from left to right along this line, are the **y-values** increasing or decreasing?
4. As *x* increases along a line with positive slope, *y*_____.
5. As *x* increases along a line with negative slope, *y*_____.

Now look at the scatter plots in Figure 4.1.3. They are taken from *MathScape: Looking Behind the Numbers* (page 21).

Notice that in the first two scatter plots, those with perfect correlation, the data points lie exactly in a straight line. If the line has positive slope, the correlation is positive. If the line has negative slope, the correlation is negative. When the correlation is less than perfect, the points do not lie exactly on a line, but might be thought of as being near a line. The more scattered the points are, the farther they are from this imaginary line, the weaker the correlation.

A different way to look at correlation relates to Exploration 4.1. For a line with positive slope, the points get higher as we move from left to right. In other words, *y* increases as *x* increases. When looking at a scatter plot, if the points tend to rise as you look from left to right, the correlation is positive. The stronger this tendency is, the stronger the correlation.

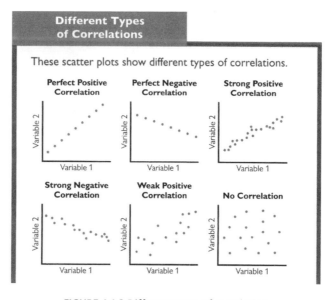

FIGURE 4.1.3 Different types of correlation

88 Chapter 4 Data with Two Variables

Likewise, for a line with negative slope, the points get lower as we move from left to right; y decreases as x increases. If a scatter plot has this same tendency, the correlation is negative. The stronger this tendency is, the stronger the correlation.

As *MathThematics, Book 3* (page 538) defines it:

> Two variables that are related in some way are said to be *correlated*. There is a **positive correlation** if one variable tends to increase as the other increases. There is a **negative correlation** if one variable tends to decrease as the other increases.

Figure 4.1.4 reproduces page 42 from *Mathematics in Context: Insights Into Data*.

Focus on Understanding

1. Answer the questions in Figure 4.1.4.
2. Do you think that, in general, it will always be clear from a scatter plot whether the correlation between two variables is strong or weak? If we added a third category, moderate, would that eliminate disagreement about the strength of correlations?

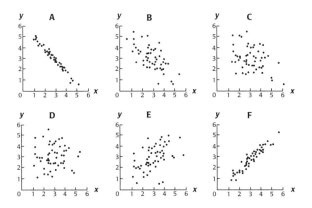

FIGURE 4.1.4 Describing correlation

Correlation and Cause-and-Effect

It is important to understand that correlation does not imply a cause-and-effect relationship between the variables. In other words, just because one variable tends to increase (or decrease) as the other increases, that does not necessarily mean that a change in one variable **causes** the other to change. Consider the exercise in Figure 4.1.5. It is taken from *MathThematics, Book 3* (page 539).

13 The scatter plot below shows the number of Florida manatees killed by boats and the price of stocks as measured by the Dow Jones Industrial Average for the years 1975–1990.

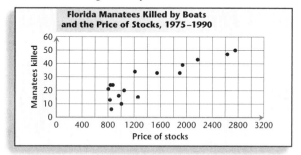

a. Is there a *positive correlation,* a *negative correlation,* or *no correlation* between the price of stocks and the number of manatees killed by boats?

b. Would it be correct to say that a rise in stock prices tends to *cause* an increase in the number of manatees killed by boats?

c. Discussion If a correlation exists between two variables, does that necessarily mean there is a cause-and-effect relationship between the variables? Explain.

FIGURE 4.1.5 Stock prices and manatees killed

Notice that the price of stocks and the number of manatees killed appear to have a positive correlation, but there is no cause-and-effect relationship between them. An increase in the number of manatees killed does not *cause* stock prices to rise, nor do rising stock prices cause more manatees to be killed. Correlation does not imply a cause-and-effect relationship.

So far, everything discussed in this section is well within the capabilities of middle school students. In the newer middle school curricula, students are expected to:

- Understand the idea of positive and negative correlation and be able to identify examples of these from a scatter plot.
- Understand that correlations between variables can have different strengths. Some variables are strongly correlated; others are weakly correlated. They should be able to identify examples of strong and weak correlation from scatter plots.
- Understand that correlation does not necessarily imply a cause-and-effect relationship. Two variables can be correlated without a change in one causing a change in the other.

College students and middle school teachers should have a deeper understanding of correlation as something that can be measured. The next section deals with Pearson's correlation coefficient, a formula statisticians use to measure correlation between two variables. The value of this measurement indicates not only whether one variable tends to increase or decrease as the other increases, but also how strong this tendency is.

4.2 PEARSON'S CORRELATION COEFFICIENT

This section will explore the ideas behind **Pearson's correlation coefficient**, a formula used to measure the strength and direction of linear relationships between variables. Exploration 4.2 will prepare you to understand how this formula works.

Classroom Exploration 4.2

Two lines have been added to the scatter plot at the right, one vertical and the other horizontal, both passing through the center of the data points. These two lines divide the plane into four sections. In this case, the scatter plot shows no correlation between the variables. Notice that the number of data points is approximately the same for each of the four sections in the scatter plot.

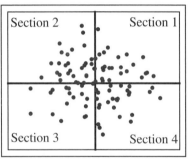

1. Suppose that the data had showed a positive correlation. Which section or sections would you expect to have the most data points? Which would you expect to have the fewest? Explain why you think so.
2. Suppose that data had showed a negative correlation. Which section or sections would you expect to have the most data points? Which would you expect to have the fewest? Explain.

As we develop the formula for Pearson's correlation coefficient, you will see how the relationship you discovered in Exploration 4.2 is used to determine whether the correlation is positive or negative.

To illustrate Pearson's formula without getting too bogged down in computations, we'll use the small, highly rigged data set at the right, where most of the computations work out nicely. By the end of the section, you should be able to calculate the value of Pearson's coefficient for more complex data sets, interpret what this value indicates about the data, and have some understanding of why it works.

We'll start with what we'll call the **three S's**:

x	y
1	1
2	2
3	4
5	2
6	4
7	5

The Three S's

$$S_{xx} = \sum (x - \bar{x})^2 \qquad S_{yy} = \sum (y - \bar{y})^2 \qquad S_{xy} = \sum (x - \bar{x})(y - \bar{y})$$

You've seen S_{xx} before. In Chapter 3 that variance is given by the formula:

$$s^2 = \frac{\Sigma(x - \bar{x})^2}{n - 1}$$

S_{xx} is the numerator of this variance formula. S_{yy} is similar, but for the y-values instead of the x-values. We need both of these because we are now working with two variables instead of one (as in Chapter 3).

The third of the S's, S_{xy}, is the key to measuring correlation. To understand S_{xy}, we start with a scatter plot of the data. Just as in Exploration 4.2, we draw in two additional lines, called mean lines: a horizontal line $y = \bar{y}$, where the y-value is the average of the y-values from the data points, and a vertical line $x = \bar{x}$, where the x-value is the average of the x-values from the data points. In this case, $\bar{y} = 3$ and $\bar{x} = 4$, so the two mean lines are $y = 3$ and $x = 4$. The two mean lines divide the xy-plane (and the data) into four sections, as shown in Figure 4.2.1.

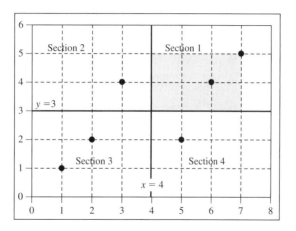

FIGURE 4.2.1 Dividing the plane into four sections

Consider the data point (7, 5) in Section 1. For this point, $x - \bar{x} = 7 - 4 = 3$, a positive number. Also, for this data point, $y - \bar{y} = 5 - 3 = 2$, also a positive number. The product, $(x - \bar{x})(y - \bar{y}) = 3 \cdot 2 = 6$, represents the area of the rectangle between the data point and the two mean lines.

For any data point (x,y) in Section 1, $x > \bar{x}$, so $(x - \bar{x})$ will be positive. Likewise, $y > \bar{y}$, so $(y - \bar{y})$ will be positive. Therefore, $(x - \bar{x})(y - \bar{y})$ will be positive. In the same way, you can determine the sign of $(x - \bar{x})(y - \bar{y})$ in each of the other sections. Fill in the blanks for Sections 2, 3, and 4 in Figure 4.2.2, just as we have already done for Section 1.

You should have concluded that $(x - \bar{x})(y - \bar{y})$ is positive for data points in Sections 1 and 3, and $(x - \bar{x})(y - \bar{y})$ is negative in Sections 2 and 4. For each data point, $(x - \bar{x})(y - \bar{y})$ represents the area of the rectangle between the data point and the two mean lines, except that in Sections 2 and 4, this area is counted as negative, as shown in Figure 4.2.3.

In this case, there are two rectangles in Section 1 with areas 6 and 2, for a total area of 8. Likewise, Section 3 has a total area of 8, while Sections 2 and 4 have a rectangle area of 1 each. The value of S_{xy} is:

$S_{xy} = \Sigma(x - \bar{x})(y - \bar{y}) = 8 + 8 + (-1) + (-1) = 14$, a positive number

Section 2	Section 1
$(x - \bar{x})$ _____	$(x - \bar{x})$ positive
$(y - \bar{y})$ _____	$(y - \bar{y})$ positive
$(x - \bar{x})(y - \bar{y})$ _____	$(x - \bar{x})(y - \bar{y})$ positive
Section 3	Section 4
$(x - \bar{x})$ _____	$(x - \bar{x})$ _____
$(y - \bar{y})$ _____	$(y - \bar{y})$ _____
$(x - \bar{x})(y - \bar{y})$ _____	$(x - \bar{x})(y - \bar{y})$ _____

FIGURE 4.2.2 The sign of $(x - \bar{x})(y - \bar{y})$

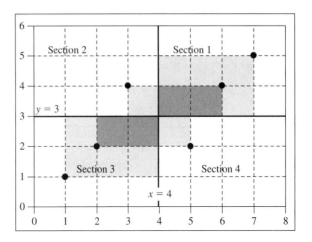

FIGURE 4.2.3 Rectangle areas represented by $(x - \bar{x})(y - \bar{y})$

As you found in Exploration 4.2, when the correlation is positive, most of the points will tend to be in Sections 1 and 3, so most of the rectangle area will be counted as positive. Whenever there is positive correlation, S_{xy} will work out to be positive. Conversely, if the correlation is negative, most of the points will tend to be in Sections 2 and 4, so most of the rectangle area will be counted as negative. Hence, S_{xy} will work out to be negative whenever there is negative correlation.

S_{xy} might be sufficient to determine whether two variables have a positive or negative correlation, but S_{xy} alone is not a good measure of the *strength* of the linear relationship between the variables. There are two reasons for this:

1. The magnitude of S_{xy} is affected by how much spread (variance) x and y have. For example, if x was a distance, we would get different values for S_{xy}, depending on whether we measured x in feet or in inches. Switching from feet to inches multiplies the value of S_{xy} by twelve. We would hesitate to say that converting x from feet to inches strengthens the linear relationship between x and y by a factor of twelve.
2. The magnitude of S_{xy} tends to increase as the number of data points increases. We wouldn't want to say that the linear relationship between two variables should be twice as strong if we collect twice as much data.

For these reasons, S_{xy} is divided by something so that the resulting value is not affected by the variances of x and y or by the number of data points. The result is known as Pearson's correlation coefficient.

Pearson's Correlation Coefficient: $\quad r = \dfrac{S_{xy}}{\sqrt{S_{xx} \cdot S_{yy}}}$

Pearson's correlation coefficient, r, has the following properties:

1. The value of r is always between -1 and 1. If the data points lie exactly on a line with positive slope, then $r = 1$. If the data points lie exactly on a line with negative slope, then $r = -1$. In general, the closer the value of r is to 1 or -1, the stronger the linear relationship between the variables. The table in Figure 4.2.4 might be helpful in classifying the strength of the relationship between two variables (Peck, 2001, p. 157). However, sample size should also be taken into account. A correlation coefficient of $r = 0.6$ does not mean the same thing for a sample of size 3 as it does for a sample of size 300.
2. The value of r is not affected by changing the units that the variables are measured in. For example, changing measurements from feet to inches has no effect on the value of r.
3. The value of r is not affected by which variable is called x and which variable is called y.

The formulas for the three S's given previously are most useful in interpreting what they represent, but for computing the values of the three S's from data, there are some other formulas which are usually easier.

Computing Formulas for the Three S's

$$S_{xx} = \sum x^2 - \frac{(\sum x)^2}{n}$$

$$S_{yy} = \sum y^2 - \frac{(\sum y)^2}{n}$$

$$S_{xy} = \sum xy - \frac{(\sum x)(\sum y)}{n}$$

$-1 < r < -0.8$	$-0.8 < r < -0.5$	$-0.5 < r < 0.5$	$0.5 < r < 0.8$	$0.8 < r < 1$
strong negative correlation	moderate negative correlation	weak or no correlation	moderate positive correlation	strong positive correlation

FIGURE 4.2.4 Values of r and strength of correlation

x	y	$(x - \bar{x})^2$	$(y - \bar{y})^2$	$(x - \bar{x})(y - \bar{y})$	x^2	y^2	xy
1	1	9	4	6	1	1	1
2	2	4	1	2	4	4	4
3	4	1	1	−1	9	16	12
5	2	1	1	−1	25	4	10
6	4	4	1	2	36	16	24
7	5	9	4	6	49	25	35
24	18	28	12	14	124	66	86

FIGURE 4.2.5 Computations for Pearson's correlation coefficient

Let's compute the value of Pearson's correlation coefficient both ways and compare the results. The table in Figure 4.2.5 shows the raw data and some of the computations. The bottom row gives the total for each column. Remember from earlier that $\bar{x} = 4$ and $\bar{y} = 3$.

Without going through all of the details, let's just highlight where a few of the numbers come from. The first number in the $(x - \bar{x})^2$ column uses $x = 1$ and $\bar{x} = 4$:

$$(x - \bar{x})^2 = (1 - 4)^2 = (-3)^2 = 9$$

The next number in that column uses the x-value for that row ($x = 2$):

$$(x - \bar{x})^2 = (2 - 4)^2 = (-2)^2 = 4$$

The total for that column, 28, is $\sum (x - \bar{x})^2$, which is S_{xx}. In the same way, 12, the total for the column headed $(y - \bar{y})^2$, is S_{yy}, and 14, the total for the $(x - \bar{x})(y - \bar{y})$ column, is S_{xy}.

Let's check that we get the same values by using the computing formulas for the three S's:

$$S_{xx} = \sum x^2 - \frac{(\sum x)^2}{n} = 124 - \frac{24^2}{6} = 28$$

$$S_{yy} = \sum y^2 - \frac{(\sum y)^2}{n} = 66 - \frac{18^2}{6} = 12$$

$$S_{xy} = \sum xy - \frac{(\sum x)(\sum y)}{n} = 86 - \frac{(24)(18)}{6} = 14$$

As you can see, the results are the same. For those of you thinking that the original formulas are easier than the computing formulas, remember two things. First, we've done most of the computations for you; it's different when you do them all yourself. Second, this data is highly rigged so that the numbers work out nicely. Ordinarily, $\bar{x}$ and $\bar{y}$ are not nice whole numbers, but long, messy decimals. The computing formulas avoid the problem of plugging in these long, messy decimals for $\bar{x}$ and $\bar{y}$.

Finally, let's compute Pearson's coefficient:

$$r = \frac{S_{xy}}{\sqrt{S_{xx} \cdot S_{yy}}} = \frac{14}{\sqrt{28 \cdot 12}} = .764$$

This value indicates moderate (almost strong) positive correlation.

A Shortcut Using the TI-83 Plus Calculator

You can use a calculator like the TI-83 plus to get the values for $\sum x, \sum y$, etc. Hit STAT and choose 1: EDIT. Enter the *x*-values as L1 and the *y*-values as L2. Then hit STAT and then the right arrow key (to move to CALC). Notice that "2: 2-Var Stats" is on the list of options. Choose that option, and "2-Var Stats" appears on your screen. Hit ENTER and scroll down to see values for $\sum x, \sum x^2, \sum y, \sum y^2$, and $\sum xy$.

Note 1: The value S_x given by the calculator is not one of the three S's in this section, but rather the standard deviation of the *x*-values. We will discuss σ_x in Chapter 7.

Note 2: You'll see an even shorter shortcut in the next section.

The table in Figure 4.2.6 shows some computations based on data from *Math in Context: Statistics and the Environment, Murre Island Bats* (page 5). The first column (*x*) gives the air temperature in Celsius. The second column (*y*) shows the number of minutes that bats spend outside their caves. Some of the computations have been done for you, but there are some blank spaces left for you to fill in.

x	y	$(x - \bar{x})^2$	$(y - \bar{y})^2$	$(x - \bar{x})(y - \bar{y})$	x^2	y^2	xy
20	16	2.469	19.612	6.959	400	256	320
22	21	0.184	0.327	0.245	484	441	462
19	15	6.612	29.469	13.959	361	225	285
22	24	0.184	12.755	1.531	484	576	528
26	30	19.612	91.612	42.388	676	900	780
23	23	2.041	6.612	3.673	529	529	529
19	14	(a)	(b)	(c)	361	196	266
151	143	37.714	201.714	85.286	3295	3123	3170

FIGURE 4.2.6 Data and computations for Murre Island bats

Focus on Understanding

1. Using the information in Figure 4.2.6, compute the values of $\bar{x}$ and $\bar{y}$.
2. Compute the values that go in the three shaded spaces in the table.
3. Find the three S's using the original formulas.
4. Compute the three S's using the computing formulas. Check to see whether you got the same values as in #3.
5. Use the three S's to compute Pearson's correlation coefficient, r. What does this value indicate about the relationship between temperature and the time that bats spend outside their caves? Does this value seem reasonable, based on the scatter plot in Figure 4.1.1?
6. What would happen to the value of r if the temperatures were given in Fahrenheit, rather than in Celsius?
7. In this case, do you think that there is a cause-and-effect relationship between x and y? Explain.

4.3 SLOPES AND EQUATIONS OF FITTED LINES

When two variables are correlated, a **fitted line** is often used to both describe the data and to predict values of one variable from the other. In middle school, the most commonly used method for fitting a line to data is similar to the way that most people hang pictures on a wall. It's not a matter of calculation, but appearance. (Does this look straight to you?) Even so, there are reasons to prefer one fitted line over another.

The exercise in Figure 4.3.1 is taken from *MathThematics, Book 2* (page 338).

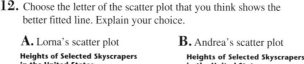

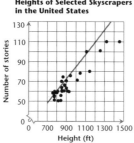 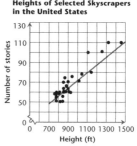

FIGURE 4.3.1 Comparing fitted lines

Section 4.3 Slopes and Equations of Fitted Lines

Classroom Exploration 4.3

1. Answer the question in Figure 4.3.1.
2. If we are going to compare fitted lines as we did in #1, it would be helpful to have some criteria for deciding whether a line fits the data well. List some general guidelines for judging by appearance (no lengthy calculations) whether a line fits the data well.
3. Does your choice for #1 fit your criteria? Explain.

Slope

Figure 4.3.2 will be helpful in understanding the concept of **slope**. If (x_1, y_1) and (x_2, y_2) are two points on a line, the change in x when moving from the first point to the second is $x_2 - x_1$. This horizontal change is also referred to as the *run*. The change in y, or *rise*, is $y_2 - y_1$. The slope m of a line is the ratio of change in y to change in x. Slope can be thought of in a number of different ways.

The Slope m of a Line

$$m = \frac{rise}{run} \qquad m = \frac{y_2 - y_1}{x_2 - x_1}$$

Slope is the change in y that corresponds to a 1-unit increase in x.
(To understand the last one, look at the similar triangles in Figure 4.3.2.)

For example, the points (800, 60) and (1200, 110) appear to be on Lorna's line in Figure 4.3.1. The slope of Lorna's line would be:

$$m = \frac{y_2 - y_1}{x_2 - x_1} = \frac{110 - 60}{1200 - 800} = \frac{50}{400} = \frac{1}{8} = 0.125$$

On Lorna's line, when x increases by 1 unit, y increases by $\frac{1}{8}$ or 0.125. Since $x =$ *the height of a skyscraper in feet* and $y =$ *the number of stories*, if we were using Lorna's

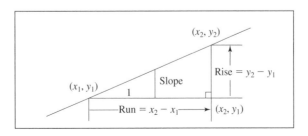

FIGURE 4.3.2 Slope of a line

line to make predictions about the number of stories in a skyscraper, we would expect an increase of 1 foot in height to go with a change of $\frac{1}{8}$ in the number of stories. In this case, it doesn't make much sense to talk about fractions of a story, so a better interpretation might be that an increase of 8 feet in height would add 1 additional story.

Ideally, we'd like to have a formula for predicting one variable from another, such as predicting the number of stories in a skyscraper from its height. To do this, we need to understand some ideas about equations of lines.

Equations of Lines

There are two forms for equations of lines that will be useful to us. They are:

$$\text{The Point-Slope Equation: } y - y_1 = m(x - x_1)$$

$$\text{The Slope-Intercept Equation: } y = mx + b$$

The **point-slope equation** is used for writing the equation of a line. In order to use this equation, we first need to know two things about the line: a point on the line and the slope of the line (hence the name of the equation). The point on the line tells us the values for x_1 and y_1 in the equation; the slope tells us the value of m.

The **slope-intercept equation** is the most useful form to have the equation in. It is set up just the way we would want to predict the value of y from the value of x. Also, by looking at the equation, we can see the slope m and the y-intercept b (hence the name slope-intercept). The **y-intercept** b is the value of y that goes with an x-value of zero. In an algebra class, it tells us where the line will cross the y-axis (the line $x = 0$). In statistics, we have to be a little more careful. An x-value of zero may not be reasonable in the situation. Even if it is, the line $x = 0$ may not be the same as the y-axis. If the y-axis in a scatter plot is not aligned at $x = 0$, the y-intercept will not indicate where the line meets the y-axis, but rather where it would meet the line $x = 0$.

Let's find the equation of Lorna's line. We'd like to use the point-slope equation, since it is used for exactly this purpose, writing equations for lines. Earlier, we noted that the point (800, 60) is on Lorna's line, and we found its slope to be .125. So, we have exactly the information we need to use the point-slope equation. The point (800, 60) tells us $x_1 = 800$ and $y_1 = 60$; the slope gives us $m = 0.125$. Substituting in the equation, we get:

$$y - 60 = 0.125(x - 800)$$

This is an equation for Lorna's line, but we usually would not leave it in this form. Typically, we'd be expected to put the equation in slope-intercept form. Working toward this goal, we get:

$$y - 60 = 0.125x - 100$$

$$y = 0.125x - 40$$

This is the equation of Lorna's line. Would we have obtained the same result if we had used the other point (1200, 110)? We should! You can check for yourself and see.

We could use the equation for Lorna's line to predict, for example, the number of stories in a skyscraper that was 1,000 feet tall. Since x represents the height of the skyscraper in feet, and y is the number of stories, we would substitute 1000 for x and find the value of y.

$$y = 0.125(1000) - 40 = 125 - 40 = 85$$

So, we'd predict that a 1,000-foot skyscraper would have 85 stories. Looking back on Figure 4.3.1, this seems a reasonable estimate, but it is only an estimate. Notice that, of course, the points are not exactly on Lorna's line, but are scattered both above and below the line (mostly below, in Lorna's). So we wouldn't be surprised if a 1,000-foot building had 80 or even 75 stories, but we'd probably be shocked if it had only 40. There are methods for determining how much variation from the predicted value to expect, but we will not deal with them now.

Extrapolation versus Interpolation

Suppose we used the equation for Lorna's line to predict the number of stories in a building that was 200 feet high. Working the same way as above, we get:

$$y = 0.125(200) - 40 = 25 - 40 = -15$$

Negative 15 stories! That's some weird building! What went wrong? Well, some people might blame Lorna for not drawing a better line, but the truth is that we might get a very poor prediction even from the best possible line. The reason is that all of the skyscrapers in the data set were between 700 and 1,500 feet tall. When we used the equation to make a prediction about a 1,000-foot building, we were **interpolating** within the range of heights in the data. When we try to get predictions about a 200-foot building, we are **extrapolating** well outside the range of heights in the data. In general, **interpolation** is making a prediction within the range of the data, while **extrapolation** is making a prediction outside that range. There is often no reason to expect that the same linear relationship between variables holds true outside the range of the data set. In general, extrapolation often leads to very poor predictions; interpolation is a much better bet.

Focus on Understanding

Look back at Figure 4.3.1. Andrea's line appears to go through (1000, 70) and (1500, 110).

1. Find the slope of Andrea's line.
2. Based on the slope, how many feet of additional height would be predicted for 1 additional story. Hint: Slope = (change in number of stories) ÷ (change in height). Plug in the slope and the change in number of stories (1) and solve for change in height.
3. Find an equation for Andrea's line. Put your answer in slope-intercept form.

100 Chapter 4 Data with Two Variables

> 4. What is the *y*-intercept of Andrea's line? Is it meaningful in this situation? Does it indicate where the line would meet the vertical axis in the scatter plot? Why or why not?
>
> 5. Use the equation to predict the number of stories in a skyscraper that is 1,200 feet tall. Do you think that a 1,200-foot building must have exactly this number of stories? Explain.
>
> 6. Use the equation to predict the number of stories in a 100-foot building. What do you think about the accuracy of this prediction and why?
>
> 7. Use the equation to predict the height of a building that has 65 stories.

Middle school students also use automatic features of calculators like the TI-83 plus to find equations for fitted lines (see *Math in Context, Insights into Data*, pages 49 and 50, for example). There are two types of lines commonly computed from data: the least squares line and the median-median line (or median fit line). We'll start with the least squares line.

4.4 THE LEAST SQUARES LINE

The fitted line most commonly used by statisticians is the **least squares line**. In some ways, it is considered to be the best possible fitted line, sometimes referred to as the line of best fit. So, what is the least squares line and what does it have to do with squares? To answer that question, we'll use some data from *MathThematics, Book 2* (page 339) as an illustration. The data are given in Figure 4.4.1.

The Idea Behind the Least Squares Line

Figure 4.4.2 consists of a scatter plot of the tent data from Figure 4.4.1 with $x = $ *floorsize in square feet* and $y = $ *the number of sleepers*. It also includes a fitted line that is *not* the best. The line is a bit low, especially at the right end. This scatter plot was produced using Fathom™ statistical software. Fathom has some features that are extremely helpful in explaining the idea of least squares.

Imagine this: We want to measure the vertical distance from each data point to the line. We would like to square these distances and add them up. Actually computing this sum of squares would be quite a chore, but all we really want to do right now is picture the result. Fathom does this task automatically. The results are shown in Figure 4.4.3.

Each of the squares in Figure 4.4.3 represents the squared distance between a data point and the line. The total area or sum of squares is given below the plot. The idea of "least squares" is to make this total area as small as possible.

Notice the large square at the right end. Moving the right end of the line up would make those squares smaller. Fathom allows us to drag the line around and watch what happens to the squares in the plot and to the sum of the squares (the total area of the squares) given below the plot.

14. The first two columns of the table below show the floor sizes of different tents and the number of sleepers each tent can hold.

   a. Use the data in the first two columns of the table to make a scatter plot.

   b. Draw a fitted line for your scatter plot.

   c. Using your graph, predict the floor size of a tent that can sleep 12 people.

Tent Sizes and Prices		
Floor size (ft^2)	Number of sleepers	Price
135	4	$162.99
140	6	$289.99
160	8	$309.99
200	10	$299.99
108	5	$159.99
110	6	$184.99
64	4	$99.99
81	5	$139.99
100	6	$199.99
109	6	$229.99
49	3	$89.99
60	3	$109.99

15. The last column of the table above shows the price of each tent.

   a. Use the data in the last two columns to make a scatter plot.

   b. Draw a fitted line for your scatter plot.

   c. Using your graph, predict the price of a tent that can sleep 12 people.

FIGURE 4.4.1 Some data on tents

Figure 4.4.4 shows the line with the smallest sum of squares, the **least squares line**. Notice that the total area of the squares is less than the total area in the previous plot (10.13 as opposed to 16.23).

Finding the Least Squares Line

So, now that we know what the least squares line is, how would we find it without using Fathom? In Section 4.2, we introduced the three S's and their computing formulas. They are reproduced in the box on page 103 for your reference.

102 Chapter 4 Data with Two Variables

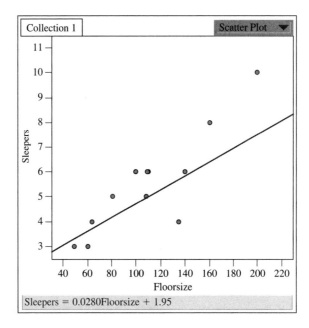

FIGURE 4.4.2 Scatter plot of the tent data

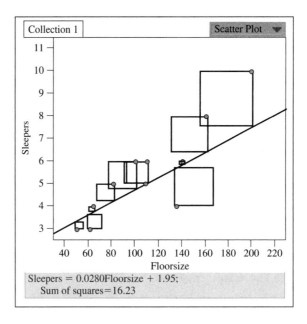

FIGURE 4.4.3 Illustrating the sum of squares

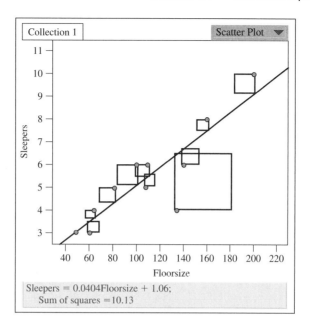

FIGURE 4.4.4 The least squares line

The Three S's and Their Computing Formulas

$$S_{xx} = \sum (x - \bar{x})^2 = \sum x^2 - \frac{(\sum x)^2}{n}$$

$$S_{yy} = \sum (y - \bar{y})^2 = \sum y^2 - \frac{(\sum y)^2}{n}$$

$$S_{xy} = \sum (x - \bar{x})(y - \bar{y}) = \sum xy - \frac{(\sum x)(\sum y)}{n}$$

Two of the three S's are used in computing the slope and y-intercept of the least squares line.

The Slope and y-intercept of the Least Squares Line: $y = mx + b$

The **slope** m is given by the formula: $\quad m = \dfrac{S_{xy}}{S_{xx}}$

The **y-intercept** b is given by: $\quad b = \bar{y} - m\bar{x}$

Caution

The notation for the least squares line varies from book to book and instructor to instructor. We have chosen to use the equation $y = mx + b$ to be consistent with the notation typically used in algebra classes. Some people use the equation $y = ax + b$, so their slope a is our m.

Still others use $y = a + bx$. In this case a is the y-intercept and b is the slope! Their formula for b is our formula for m, and our formula for b is their formula for a.

The TI-83 plus has options for both $y = ax + b$ and $y = a + bx$. The makers of this calculator, like us, realize that there is unfortunately not a universally accepted notation for the least squares line. Just don't get confused if another textbook's notation is different than ours. We're really talking about exactly the same things; we're just using different letters to stand for them.

We will now use the formulas given previously to find the least squares line for the tent data and compare it to the result from Fathom. Figure 4.4.5 contains the data and some of the computations with $x = $ *floorsize* and $y = $ *number of sleepers*. You should know to compute each of the values in the other columns. As before, the bottom row contains the total for each column.

To find the slope of the least squares line, we need S_{xy} and S_{xx}. From Figure 4.4.5:

$$S_{xy} = \sum(x - \bar{x})(y - \bar{y}) = 864.00$$

$$S_{xx} = \sum(x - \bar{x})^2 = 21,406.67$$

Therefore: $m = \dfrac{S_{xy}}{S_{xx}} = \dfrac{864.00}{21,406.67} = 0.04036$

x	y	$(x - \bar{x})^2$	$(y - \bar{y})^2$	$(x - \bar{x})(y - \bar{y})$	x^2	y^2	xy
135	4	641.78	2.25	−38.00	18225	16	540
140	6	920.11	0.25	15.17	19600	36	840
160	8	2533.44	6.25	125.83	25600	64	1280
200	10	8160.11	20.25	406.50	40000	100	2000
108	5	2.78	0.25	0.83	11664	25	540
110	6	0.11	0.25	0.17	12100	36	660
64	4	2085.44	2.25	68.50	4096	16	256
81	5	821.78	0.25	14.33	6561	25	405
100	6	93.44	0.25	−4.83	10000	36	600
109	6	0.44	0.25	−0.33	11881	36	654
49	3	3680.44	6.25	151.67	2401	9	147
60	3	2466.78	6.25	124.17	3600	9	180
1316	66	21406.67	45.00	864.00	165728	408	8102

FIGURE 4.4.5 Data and computations for the least squares line

To compute the y-intercept, we need to know the means, $\bar{x}$ and $\bar{y}$.

$$\bar{x} = \frac{\Sigma x}{n} = \frac{1316}{12} = 109.67$$

$$\bar{y} = \frac{\Sigma y}{n} = \frac{66}{12} = 5.5$$

Therefore, the y-intercept is:

$$b = \bar{y} - m\bar{x} = 5.5 - (0.04036)(109.67) = 1.074$$

Putting things together, the least squares line for the tent data is:

$$y = 0.04036x + 1.074$$

There is a slight difference between our computations and the results from Fathom, but this is just due to round-off error.

Calculators like the TI-83 plus can compute the slope and y-intercept of this line automatically.

1. Enter the x-values as L1 and the y-values as L2.
2. Hit STAT and the right arrow key (to move to CALC). The type of calculation we want is called *linear regression*. On the TI-83 plus, this is 4: LinReg($ax + b$). Choose that option and hit ENTER. The calculator display should look something like this:

```
LinReg
y=ax+b
a=.0403612582
b=1.073715353
r²=.774936157
r=.8803045819
```

In this case, a is the slope, and b is the y-intercept. The results are consistent with what we computed earlier.

Note: If you do not see values for r^2 and r, follow the steps listed here:

1. Hit 2nd and CATALOG, the second function for the 0 key. The catalog is an alphabetical listing of calculator functions.
2. Hit the key with a green letter D above it (the x^{-1} key). This will take you to D in the alphabetical listing.
3. Scroll down until the pointer (▷) is next to the command DiagnosticOn.
4. Hit ENTER twice.

The calculator's screen should show:

```
DiagnosticOn
              Done
```

Now hit STAT, move to CALC, and select 4: LinReg(*ax* + *b*) as you did before. This time, with "diagnostics on," you should see values for r^2 and r. The diagnostics should stay on, so you should not have to turn them on again next time.

The value of r^2 has an important interpretation for statisticians, but we won't get into that in this book. The value of r should agree (except for a little round-off error) with the value of Pearson's correlation coefficient that you found in Section 4.2. This is the shorter shortcut for finding the correlation coefficient that we promised you earlier.

Focus on Understanding

1. Suppose that the table in Figure 4.4.5 did not show the columns labeled $(x - \bar{x})^2, (y - \bar{y})^2$, and $(x - \bar{x})(y - \bar{y})$. Compute S_{xy} and S_{xx} from totals in the other columns and compare your results to the values we got.

2. Use the equation of the least squares line to predict the number of sleepers for a tent with 135 square feet of floor space. Compare your answer to the first tent in the table. Do you think that this tent might comfortably fit more than four sleepers?

3. *MathThematics* asks students to predict the floorsize of a tent that sleeps 12 people. Use the equation of your least squares line to make this prediction.

4. At the end of Section 4.2, you computed $\bar{x}, \bar{y}$, and the three S's for the *Murre Island Bat* data (*Math in Context: Statistics and the Environment,* page 5). You should have found that $\bar{x} = 21.571, \bar{y} = 20.429, S_{xx} = 37.714, S_{yy} = 201.714$, and $S_{xy} = 85.286$. Recall that, in the bat data, x = air temperature in Celsius and y = the number of minutes that bats spend outside their caves.

   a. Find the equation of the least squares line for the *Murre Island Bat* data.
   b. How long would you expect bats to spend outside their caves if the air temperature was 25°C?
   c. Should you use the least squares line to predict how long bats stay outside if the temperature is 100°C? Explain.
   d. Enter the *Murre Island Bat* data into your calculator, and use it to find the least squares line. Compare your answer with your result from (a) above.

4.5 THE MEDIAN-MEDIAN LINE

Although the least squares line is the most common fitted line used by statisticians, it is not the only one. Another fitted line that is often presented to young students is the **median-median line**. The median-median line has several advantages:

1. It tends to satisfy all of the criteria most people would require for the *appearance* of a fitted line (provided the variables are correlated in the first place), while not involving as much computation as the least squares line.
2. As long as students know how to find medians and plot points, it is fairly easy to graph this line. This is well within the capability of middle school students. With a little more knowledge about finding slopes and equations of lines, they can find its equation.
3. Since medians are not affected by outliers, neither is the median-median line. So it may be the preferred fitted line for a data set with outliers.

Graphing the Median-Median Line

The table in Figure 4.5.1 shows the tent data we used in the last section. This time, we've sorted the data set in ascending order by x-values, and divided it into three groups: Left, Middle, and Right.

Figure 4.5.2 will clarify why we named the three groups the way we did. It shows the scatter plot with the three groups marked. In this case, there are four points in each group. It is not always possible to give every group the same number of points. In general, we try to make the groups as equal as possible. If we must have unequal groups, it's best to have more points in the low or high group than in the middle group.

In each group, we want to find a median point, a point whose x-coordinate is the median x-value and whose y-coordinate is the median y-value. For example, in the left group, the x-values arranged in order are: 49, 60, 64, 81

Group	Floorsize	Sleepers
Left	49	3
	60	3
	64	4
	81	5
Middle	100	6
	108	5
	109	6
	110	6
Right	135	4
	140	6
	160	8
	200	10

FIGURE 4.5.1 The tent data, sorted and grouped

108 Chapter 4 Data with Two Variables

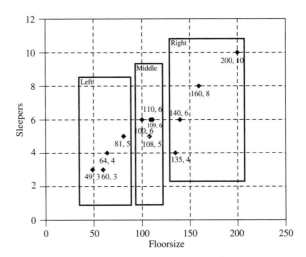

FIGURE 4.5.2 Scatter plot showing groups

Since there is not a middle value, the median is the average of the two middle ones:

$$\tilde{x} = \frac{60 + 64}{2} = 62$$

Likewise, the *y*-values arranged in order are:

$$3 \quad 3 \quad 4 \quad 5$$

The median *y*-value is:

$$\tilde{y} = \frac{3 + 4}{2} = 3.5$$

Therefore, the median point for the left group is (62, 3.5).

In the same way, we find the median points for the middle and right groups; they are (108.5, 6) and (150, 7).

Caution

The *y*-values in the second group are: 6 5 6 6

These must be arranged in order before finding the median: 5 6 6 6

So the median is $\tilde{y} = \frac{6+6}{2} = 6$, **not** $y = \frac{5+6}{2} = 5.5$.

Figure 4.5.3 shows these three median points, together with three lines. Line 1, the lowest of the three, is the line through the two outer median points. Line 2 is parallel to Line 1, but passes through the middle median point. Line 3 is the median-median line. It is the same as Line 1, but moved one-third of the distance in the direction of Line 2.

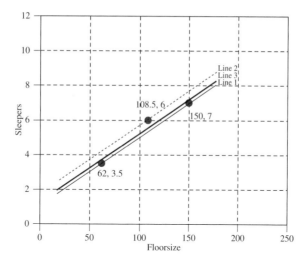

FIGURE 4.5.3 The median points and three lines

The following steps summarize the procedure for graphing the median-median line:

1. Sort the data from the smallest x-value to the largest. Use the x-values to divide the data into three groups, left, middle, and right. The groups should have as close to the same number of points as possible. If equal groups are not possible, it is better to have more points in the left group or the right group than the middle group.
2. Find the median point for each group. Remember to put the y-values in order within each group before finding the median.
3. Find the line through the two outer median points. Move this line one-third of the distance toward the middle median point. This is the median-median line.

Finding the Equation of the Median-Median Line

Continuing the previous example, we will find the equation of the median-median line. We start by finding the slope of Line 1, the line through the two outer median points. Since we know two points on this line, we can use them to find the slope:

$$m = \frac{y_2 - y_1}{x_2 - x_1} = \frac{7 - 3.5}{150 - 62} = .0398$$

Now, we can use the point-slope equation to write an equation for this line and put it into slope-intercept form. We'll use (150, 7) for the point on the line.

$$y - y_1 = m(x - x_1)$$
$$y - 7 = .0398(x - 150)$$

$$y - 7 = .0398x - 5.97$$
$$y = .0398x + 1.03$$

This is the line through the left and right median points.

We next want to find Line 2, the parallel line through the middle median point. Remember that parallel lines have the same slope. We'll use the point-slope formula again, using the same slope, but the middle median point (108.5, 6).

$$y - y_1 = m(x - x_1)$$
$$y - 6 = .0398(x - 108.5)$$
$$y - 6 = .0398x - 4.318$$
$$y = .0398x + 1.682$$

Finally, we know that the median-median line is parallel to the other two, but moved one-third of the distance from Line 1 toward Line 2. We can find the y-intercept for the median-median line by first finding the distance d between the y-intercepts of Lines 1 and 2:

$$d = 1.682 - 1.03 = .652$$

We want to move Line 1 up (toward Line 2) by one-third of this distance, so we add one-third of this distance to the y-intercept of Line 1.

$$y = .0398x + 1.03 + \frac{1}{3}(.652)$$
$$y = .0398x + 1.247$$

This is the median-median line.

Another way to get the y-intercept for the median-median line is to do a weighted average of the y-intercepts for Line 1 and Line 2, counting Line 1 twice and Line 2 once.

$$b = \frac{1.03 + 1.03 + 1.682}{3} = 1.247 \quad \text{(the same as before)}$$

The TI-83 plus calculator can also compute the median-median line. Follow the same directions as for the least squares line, except that instead of selecting 4: LinReg, choose 3: Med-Med. You should get something like the display at the right. Just as with the least squares line, the results are slightly different because the calculator did not round off as much in its calculations as we did.

```
Med-Med
y=ax+b
a=.0397727273
b=1.25094697
```

Focus on Understanding

The table at the right shows tent data from *MathThematics*, Book 2 with $x = $ *price* (to the nearest dollar) and $y = $ *number of sleepers*.

Price	Sleepers
163	4
290	6
310	8
300	10
160	5
185	6
100	4
140	5
200	6
230	6
90	3
110	3

1. Draw a scatter plot of this data.
2. Divide the data into three groups and find the median point for each group. Mark the median points in your scatter plot.
3. Graph the median-median line in your scatter plot.
4. Use the median-median line in your graph to predict the number of people that should be able to sleep in a $110 tent. How does the $110 tent in the data set stack up against this prediction?
5. Based on your graph, which tent appears to be the best bargain? Explain.
6. Find the equation of the median-median line.
7. Use your equation to predict the price of a tent that sleeps ten people. Compare your answer to #5.

Chapter 4 Summary

Unlike the previous chapters, Chapter 4 explores bivariate (two variable) data sets. Scatter plots are typically used to display such data. From the distribution of the data points on the scatter plot graph, it is possible to make an intuitive judgment about the strength or weakness of the relationship (correlation) between the two variables. The strength of the relationship between the two variables can be numerically determined by using Pearson's (r) Correlation Coefficient. Quite a bit of time was spent developing the concept of where the correlation coefficient comes from and why it works the way it does. We believe that it is important for students to see where formulas come from and the ideas behind their development in order to add understanding and meaning to the complex computations. Fortunately, nowadays we let calculators and computers handle most of the routine, but long, calculations!

Ultimately, one of the outcomes of identifying a strong negative or positive relationship between two variables is to use that information to make predictions about one of the variables. Two types of predictor equations were developed in this chapter: the least squares line and the median-median line. The least squares line is the traditional statistical tool for making predictions from linear kinds of relationships. The median-median line is not as commonly used as the least-squares line in college-level elementary statistics courses, but it is introduced in some middle school mathematics courses. Median-median lines are oftentimes more intuitive for students at this level than least squares lines. The computations are also not quite as daunting. We felt it was important to include the median-median line in this chapter on linear relationships for these reasons.

The key terms and ideas from Chapter 4 are listed below:

univariate data 85	Pearson's correlation	slope 97
bivariate data 85	coefficient 90	y-intercept 98
scatter plot 85	S_{xx} 91	interpolation 99
correlation 86	S_{yy} 91	extrapolation 99
positive correlation 88	S_{xy} 91	least squares line 100
negative correlation 88	fitted line 96	median-median line 106

Assessment is an integral part of every curriculum from the elementary school all the way through college. The question always arises—*what is it that students should be able to do after completing this lesson/unit/chapter?* We have included here our intended learning goals for Chapter 4. Students who have a good grasp of the concepts developed in Chapter 4 should be successful in responding to these items:

- Explain, describe, or give an example of what is meant by each of the terms in the vocabulary list.
- Construct a scatter plot from a given data set.
- Determine whether a scatter plot shows positive, negative, or no correlation.
- Calculate the Pearson's correlation coefficient for a data set.
- Use Pearson's correlation coefficient to determine the strength and direction of the linear relationship between two variables.
- Find the equation of a line using two points or a point and a slope.
- Determine the least squares line for a data set.
- Use the least squares line to make predictions about values of the variables.
- Determine the median-median line for a data set.
- Use the median-median line to make predictions about values of the variables.

Your course instructor may have additional or different assessable outcomes for your class. As teachers (or future teachers) you should think about the assessment outcomes and learning goals for each chapter as you work through them.

EXERCISES FOR CHAPTER 4

1. Using the *Data Analysis and Probability Standards for Grades 6–8* from the NCTM's *Principles and Standards for School Mathematics* found at www.nctm.org, identify the middle school objectives that are found in Chapter 4.
2. Using the state standards for mathematics for the content area of data analysis and probability for your state, identify the middle school objectives that are found in Chapter 4. The following website may be useful: www.doe.state.in.us. This website will allow you to access the web pages of the state departments of education for the 50 states. From the state web pages you should be able to find the state's mathematical standards.
3. Julie has noticed that as the temperature goes down, the number of cups of coffee sold at the snack bar goes up. She told Keith there is a positive correlation between

the temperature and the number of cups of coffee since the number of cups of coffee is increasing. Keith says that it must be a negative correlation because the temperature is decreasing. What type of correlation might this be and why?

4. Nine scatter plots are shown below. For each scatter plot identify graphs with positive correlations, negative correlations, or weak (or no) correlations. Explain why you made the choice you did for each graph.

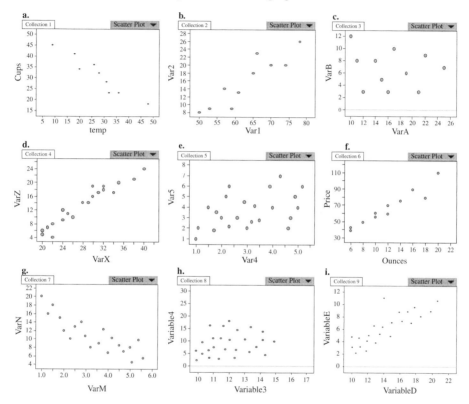

5. Julie and Keith (from problem 3) decided to keep track of the temperature and the number of cups of coffee sold over a 9-day period. Their data are shown here:

Temperature (°F)	9	18	20	26	28	31	32	36	48
Cups of Coffee	45	41	34	36	32	28	23	23	18

   a. Construct a scatter plot of the data. Based on your scatter plot, does there appear to be a positive correlation, a negative correlation, or no correlation between the temperature and the number of cups of coffee sold? (Sketch a graph of your scatter plot for later use.)

   b. Based on your scatter plot, does this data set appear to have any outliers? If so, which points appear to be outliers?

   c. Compute the Pearson's correlation coefficient for the data set. What does this value tell you about the strength and direction of the relationship between the two variables?

   d. Find the least-squares regression line for the data.

e. Use your regression line from part (d) to predict the number of cups of coffee sold if the temperature was 22°F.
f. Keith would like to use the regression line to predict the number of cups of coffee sold if the temperature was −10°F. Is this possible? Is this an appropriate use of your regression line?
g. Using the process described in Section 4.4, split the data set into thirds and find the median point for each of the three groups. Label these points on your scatter plot.
h. Find the equation of the line between your left median point and your right median point.
i. Find the equation of the median-median line.
j. Use your equation from part (i) to predict the number of cups of coffee sold if the temperature was 22°F. How does this number compare with your answer to part (e)?
k. Graph the median-median line and the regression line on your scatter plot.
l. How does your median-median line of part (g) compare to your regression line from part (d)? Are the slopes close? The y-intercepts?

6. Principal King was pretty sure that the number of school days missed by students was having an impact on the end-of-year grade-level assessment for her sixth graders. She collected data from one of her school's sixth grade classes. Her data are shown here.

Days Absent	0	0	0	1	1	2	2	2	2
Performance on EOY assessment	98	93	86	94	89	73	82	80	79

Days Absent	2	4	5	5	6	9	11	15	20
Performance on EOY assessment	84	81	75	72	78	70	65	56	47

a. Construct a scatter plot of the data. Based on your scatter plot, does there appear to be a positive correlation, a negative correlation, or no correlation between the days absent and the assessment performance? (Sketch a graph of your scatter plot for later use.)
b. Based on your scatter plot, does this data set appear to have any outliers? If so, which points appear to be outliers?
c. Compute the Pearson's correlation coefficient for the data set. What does this value tell you about the strength and direction of the relationship between the two variables?
d. Find the least-squares regression line for the data.
e. Use your regression line from part (d) to predict the performance on the end-of-year assessment for a student who missed 7 days of class.
f. Use your regression line from part (d) to predict the performance on the end-of-year assessment for a student who missed 21 days of class. Is this possible? Is this an appropriate use of your regression line? Why or why not?
g. Using the process described in Section 4.4, it is not possible to split the data set into three equal groups. Why not?
h. Use 5 points in the left group, 6 in the middle, and 7 in the right. Find the median point for each of the three groups. Label these points on your scatter plot.

i. Find the equation of the line between your left median point and your right median point.
j. Find the equation of the median-median line.
k. Use your equation from part (j) to predict the performance on the end-of-year assessment for a student who missed 7 days of class. How does this number compare with your answer to part (e)?
l. Graph the median-median line and the least squares line on your scatter plot.
m. How does your median-median line of part (j) compare to your regression line from part (d)? Are the slopes close? The y-intercepts?
n. Do you think that in reality there is a cause-and-effect relationship between the number of days a student is absent and their performance on such an assessment? Why or why not?

7. Ann wants to buy a house. She has been working with the local realtors to find a home that has the floor space she wants in her price range. She has been keeping track of the price of various houses on the market based on the square footage of the house. Her data are shown here.

Square Footage of House	Cost of House
800	$200,000
1,750	$80,000
2,500	$120,000
2,200	$135,000
2,650	$175,000
1,500	$70,000
1,250	$65,000
1,875	$160,000
2,600	$185,000
3,000	$250,000
2,150	$155,000
2,400	$172,000
2,810	$125,000
1,000	$54,000

a. Construct a scatter plot of the data. Based on your scatter plot, does there appear to be a positive correlation, a negative correlation, or no correlation between the variables? (Sketch a graph of your scatter plot for later use.)
b. Plotting numbers on a graph where the vertical scale is in the hundred thousands is pretty ugly. Ann decided to use 200, 80, 120, and so on for the vertical scale rather than 200,000, etc. Is making such a change okay or not? Why or why not? How will such a change affect the correlation coefficient? The regression equation?
c. Based on your scatter plot, does this data set appear to have any outliers? If so, which of the points appear to be outliers?

d. Using y-values in thousands as Ann did in (b), compute Pearson's correlation coefficient for this data set. What does this value tell you about the strength and direction of the relationship between the two variables?
e. Find the least-squares regression line for the data.
f. Use your regression line from part (e) to predict the cost of a house of 2,350 square feet.
g. Use your regression line from part (e) to predict the cost of a house of 4,000 square feet. Is this possible? Is this an appropriate use of your regression line? Why or why not?
h. Using the process described in Section 4.4, split the data set into thirds and find the median point for each of the three groups. Label these points on your scatter plot.
i. Find the equation of the line between your left median point and your right median point.
j. Find the equation of the median-median line.
k. Use your equation from part (j) to predict the cost of a house of 2,350 square feet. How does this number compare with your answer to part (f)?
l. Graph the median-median line and the regression line on your scatter plot.
m. How does your median-median line of part (j) compare to your regression line from part (d)? Are the slopes close? The y-intercepts? Which line do you think is the best fit in this case? Why?
n. Would it make sense to switch the two variables in this case and use the cost of the house to predict the square footage? Why or why not?

8. Professor Henry really likes using spreadsheets. His spreadsheet for two variables, x and y, is shown here.

x	y	$(x - \bar{x})$	$(y - \bar{y})$	$(x - \bar{x})^2$	$(y - \bar{y})^2$	$(x - \bar{x})(y - \bar{y})$	x^2	y^2	xy
15	42	−6.25	10.25	39.0625	105.0625	−64.0625	225	1764	630
16	38	−5.25	6.25	27.5625	39.0625	−32.8125	256	1444	608
18	40	−3.25	8.25	10.5625	68.0625	−26.8125	324	1600	720
19	36	−2.25	4.25	5.0625	18.0625	−9.5625	361	1296	684
23	30	1.75	−1.75	3.0625	3.0625	−3.0625	529	900	690
24	23	2.75	−8.75	7.5625	76.5625	−24.0625	576	529	552
26	25	4.75	−6.75	22.5625	45.5625	−32.0625	676	625	650
29	20	7.75	−11.75	60.0625	138.0625	−91.0625	841	400	580
170	254	0	0	175.5	493.5	−283.5	3788	8558	5114

a. Using Professor Henry's data from the spreadsheet, compute Pearson's correlation coefficient.
b. What does the correlation coefficient in part (a) tell you about the nature and strength of the relationship between x and y (whatever they might be)?

9. Mr. Numoto has been teaching statistics for a long, long time. He believes that student performance on the second exam in statistics is a much better predictor of a student's final exam performance than the first exam. His data are shown here.

First Exam Score	Second Exam Score	Final Exam Score
56	75	82
100	96	98
94	86	90
93	82	86
90	74	76
96	84	85
91	70	73
88	72	74
86	68	72
85	70	73
83	90	88
80	65	72
78	52	55
75	54	60
72	80	78
70	84	82
68	64	60
65	50	46
63	30	42
60	54	52
58	36	35
55	40	42
81	75	78
84	92	93
87	75	75

a. Find the correlation coefficient between the first and second exams. What is the nature and strength of the relationship between these two variables?

b. Find the correlation coefficient between the first and final exams. What is the nature and strength of the relationship between these two variables?

c. Find the correlation coefficient between the second and final exams. What is the nature and strength of the relationship between these two variables?

d. Is Mr. Numoto correct? Is the second exam a better predictor of final exam performance than the first exam?

e. What would happen to the correlation coefficient between the second exam and the final exam if the two variables were switched? Would it make sense to do this in real life? Why or why not?

CHAPTER 5
Probability

5.1 WHAT IS PROBABILITY?
5.2 OUTCOMES AND EVENTS
5.3 BASIC PROBABILITY RULES
5.4 CONDITIONAL PROBABILITY AND INDEPENDENCE
5.5 MULTIPLICATION RULES
5.6 GEOMETRIC PROBABILITY

Classroom Exploration 5.1

Figure 5.0.1 is from *Mathscape: Chance Encounters* (page 6).

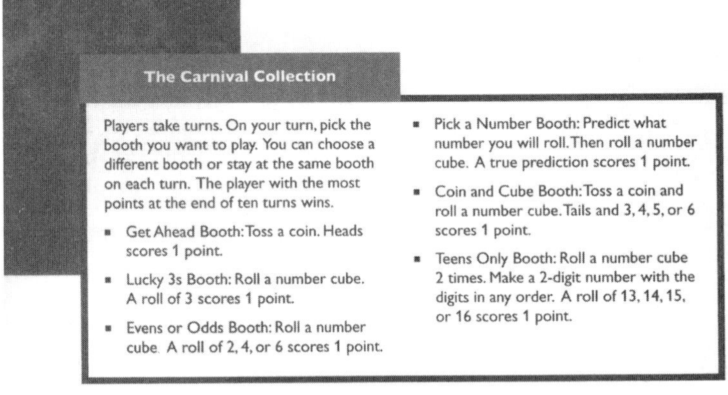

The Carnival Collection

Players take turns. On your turn, pick the booth you want to play. You can choose a different booth or stay at the same booth on each turn. The player with the most points at the end of ten turns wins.

- Get Ahead Booth: Toss a coin. Heads scores 1 point.
- Lucky 3s Booth: Roll a number cube. A roll of 3 scores 1 point.
- Evens or Odds Booth: Roll a number cube. A roll of 2, 4, or 6 scores 1 point.
- Pick a Number Booth: Predict what number you will roll. Then roll a number cube. A true prediction scores 1 point.
- Coin and Cube Booth: Toss a coin and roll a number cube. Tails and 3, 4, 5, or 6 scores 1 point.
- Teens Only Booth: Roll a number cube 2 times. Make a 2-digit number with the digits in any order. A roll of 13, 14, 15, or 16 scores 1 point.

FIGURE 5.0.1 The Carnival Collection

1. Choose three of the booths described in the Carnival Collection. Play the game described at each of the three booths you selected 10 times. Be sure to keep track of what happens on each turn!

2. Of the three booths you selected, which booth appears to be the most difficult in terms of being able to score a point? Which one is the easiest? Why do you think this seems to be the case?

3. Looking at all six game booths, which one do you think would be the most difficult to score points? Which one do you think would be the easiest? Explain your thinking!

5.1 WHAT IS PROBABILITY?

The activity in Exploration 5.1 is called a simulation. Students take turns playing the games of chance at each booth and keep track of the number of points they win. The more a particular game is played, the more information they gather about how likely or unlikely it is that they can score a point at that particular booth. In other words, they are trying to determine the **probability** that they can win points by playing a particular game.

Probability is a term that strikes fear into the hearts of many people, but there is really nothing to fear! A probability is essentially a number between 0 and 1 that characterizes the likelihood of some particular event happening. For example, your insurance company is interested in how many accidents young men between the ages of 16 and 25 have in a year. The companies gather data on the event (having an accident if you are a young man between 16 and 25). They use the data collected and adjust the insurance rates for this age group of men based on the probability of an accident occurring. As you might guess, sometimes it is very easy to find the probability of an event, such as the event of getting a head on one toss of a fair coin, and other times it is very difficult. Finding the probability of winning the lottery or being hit by lightning would certainly be more difficult to determine than getting a head on one toss of a fair coin. In this chapter we will explore the foundations of simple probability. We will explore how to determine more complex probabilities at the end of Chapter 6.

Why do we need to study probabilities? Good question. Probabilities are a type of numerical way to make decisions. While we didn't actually find the probabilities of winning points at the game booths described in Exploration 5.1, we used the data that we gathered to make a decision about which game would be the easiest at which to score points. Insurance companies, automobile manufacturers, pharmaceutical companies, teachers, and school-aged children make decisions based on such data. For example, everyone who was ever a student in the Midwest during the winter months watched the evening news on Sunday nights to hear the weather forecast. Most of us kept our fingers crossed that the weatherperson would predict inches and inches of snow. Why? Because if there were lots of snow then there would be a pretty good chance that school would be cancelled the next day. Both the forecast and the possibility that school would be cancelled are examples of using

120 Chapter 5 Probability

probabilities to make decisions. In the case of the weather forecast, data is collected every day for years and years and years. Based on the data, meteorologists know when particular conditions are met, in terms of the amount of moisture in the air, the air flow, the temperature, and so on, then a certain percentage of the time we get so many inches of snow. Are they always right? No, but that's okay because with probabilities there is no guarantee. We can only say that under these conditions this particular outcome is likely (or unlikely) to happen.

School principals and superintendents use the same ideas to make decisions about whether or not to cancel school. They use their knowledge of how much snow or ice has to fall in order to make the bus routes unsafe for traveling. Based on their prior knowledge and experiences they make a decision about how safe or unsafe the roads are likely to be the next day.

The ideas of likely, unlikely, probable, possible, not possible, sure to happen, sure to *not* happen are introduced in the elementary grades and are introductory ways of developing ideas about probability. The excerpt in Figure 5.1.1 is from *Mathematics in Context: Take a Chance* (page 11). In this activity students are simply asked to predict whether or not a particular event will take place.

Some of the events in the table are relatively easy to predict—you will have (or give) a test in mathematics sometime this year. You are probably pretty sure that will happen. Others, like "New Year's Day will come on the third Monday in January," you are positive will not happen. There are some events, however, that are less certain one way or the other. We simply do not have enough information to make an informed decision, so you might mark those events as "Not sure."

5.2 OUTCOMES AND EVENTS

Before we go much further, we need to introduce some terminology that is used with probability. Each of the situations described in Figure 5.1.1 depicts what is called an **experiment**. In order to have a non-boring experiment we must have some element of chance associated with the outcomes of the experiment. In the situation marked G, the experiment is that of tossing a coin. There are two possible **outcomes** associated with this experiment. You can get heads or you can get tails. If the coin is two-headed, then the only possible outcome would be heads. While this is still an experiment, it is a rather boring one as there is only one possible outcome.

In situation D, the experiment would be rolling a number cube (this is an experiment that happens frequently in the study of statistics, though we normally refer to it as rolling a die rather than a number cube). The possible outcomes would be 1, 2, 3, 4, 5, and 6. The number 7 is not one of the possible outcomes on a regular number cube so we can be really sure that the event (or outcome) of rolling a 7 is sure not to happen.

The set of all possible outcomes is called the **sample space**. The sample space for rolling one number cube is $\{1, 2, 3, 4, 5, 6\}$. We used set notation here to indicate that these outcomes form a collection or set. A **subset** of the sample space is referred to as an **event**. Rolling an even number would be an event in the experiment of rolling a single number cube. The event of rolling an even number consists of the

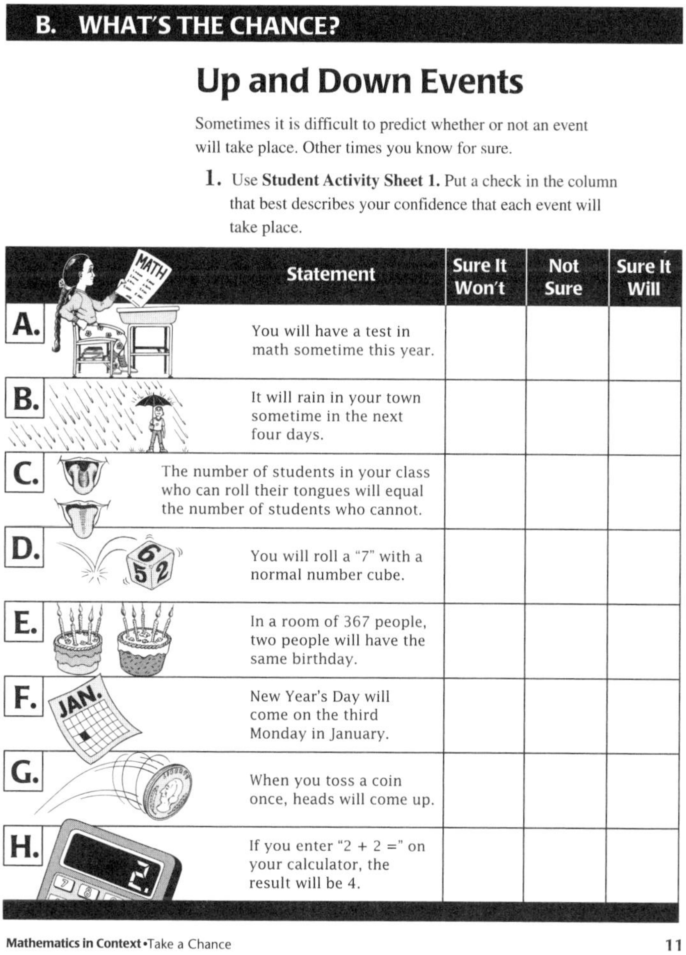

Section 5.2 Outcomes and Events 121

B. WHAT'S THE CHANCE?

Up and Down Events

Sometimes it is difficult to predict whether or not an event will take place. Other times you know for sure.

1. Use **Student Activity Sheet 1**. Put a check in the column that best describes your confidence that each event will take place.

	Statement	Sure It Won't	Not Sure	Sure It Will
A.	You will have a test in math sometime this year.			
B.	It will rain in your town sometime in the next four days.			
C.	The number of students in your class who can roll their tongues will equal the number of students who cannot.			
D.	You will roll a "7" with a normal number cube.			
E.	In a room of 367 people, two people will have the same birthday.			
F.	New Year's Day will come on the third Monday in January.			
G.	When you toss a coin once, heads will come up.			
H.	If you enter "2 + 2 =" on your calculator, the result will be 4.			

Mathematics in Context • Take a Chance

FIGURE 5.1.1 Beginning concepts about probability

outcomes of rolling a 2, 4, or 6. The event of rolling a 7 would be the empty set as it is not one of the possible outcomes in the sample space.

Focus on Understanding

1. Describe the sample spaces for each of the game booths described in Exploration 5.1.
2. Are some of the sample spaces harder to describe than others? Why?

Events or outcomes that are **sure to happen** are assigned a **probability of 1 or 100%**. Outcomes that are **sure not to happen** are assigned a **probability of 0 or 0%**. All other outcomes have a probability between 0 and 1. It is very important that you keep this in mind as we continue on with this chapter! You have probably heard people say that there is a 50–50 chance of something happening. Well, what they are saying is that the probability of that particular outcome occurring from the set of all possible outcomes (the sample space) is 50% or 0.5. Probabilities can be written as fractions, decimals, or percents. It does not matter. But, keep in mind that they must all be between 0 and 1!!

The probability of a particular event or outcome is simply the number of favorable outcomes or events that you are interested in divided by the possible number of outcomes in the sample space.

$$\textbf{Probability} = \frac{\textit{number of favorable outcomes}}{\textit{number of possible outcomes}}$$

This probability is typically referred to as **theoretical probability**. Theoretical probabilities can be computed without ever doing the experiment because you actually know or can count the number of outcomes and construct the sample space directly. We can directly compute the theoretical probability of getting a head on one toss of a fair coin (as in the first game booth in Exploration 5.1) because we know there are two possible outcomes (a head or a tail) and we are interested in the favorable outcome of getting a head. Getting a head is one of two possible outcomes, so the theoretical probability of getting a head on one toss of a fair coin would be $\frac{1}{2}$.

Experimental probabilities are found when an experiment is conducted over and over and the results are recorded and summarized. In Exploration 5.1 you were asked to keep track of your win/loss record for the three booths you picked. We can take the data you collected and get an experimental probability for each booth. For example, if you won 7 times out of 10 at the first booth, then the experimental probability associated with that booth (based on your simulation) would be

$$\frac{\textit{number of wins}}{\textit{total number of tries}} = \frac{7}{10} = 0.70$$

Is the experimental probability the same as the theoretical probability? In this case we know it is not the same since we previously computed the theoretical probability to be $\frac{1}{2}$ for this particular booth. Conditions like the number of times you tried the booth affect the experimental probability. Suppose you only tried the first booth one time and you won. Based on your data of one try, the experimental probability would be:

$$\frac{\textit{number of wins}}{\textit{total number of tries}} = \frac{1}{1} = 1.00$$

Does this seem reasonable to you? Do you really think that if you play the first booth that you will win *every* time? Clearly the number of times you try something (run a simulation), the more data you have to base your experimental probability on and the more likely it is that it won't radically differ from the theoretical probability.

Why don't we always just compute the theoretical probability of an event and be done with it then? Well, there are times when there is no handy or even possible way to determine a theoretical probability. At the beginning of this chapter we discussed the insurance rates of young men between the ages of 16 and 25. The probability of this age group of men having an accident would be an experimental probability. It is based on data collected on the number of accidents for this particular age group of men as compared to the number of accidents occurring in the population as a whole. There is simply no other way to determine the probability of such an event without collecting data over a large number of years for each age group and gender and then making a prediction based on that data. The sample space is just too messy to handle! The process associated with collecting the data and determining the probability is much more complicated than we have made it out to be, but hopefully you get the general idea.

EXAMPLE 5.2.1 Let's go back to our experiment of rolling a number cube (or a regular six-sided die). Without ever rolling the number cube, we know that the total number of possible outcomes is 6. We can roll a 1, 2, 3, 4, 5, or 6. We would like to find the probability of rolling an even number, a 2 or a 4 or a 6. From our sample space, we see that three of the outcomes involve rolling an even number (2, 4, or 6) so the number of favorable outcomes is 3. Consequently, then, the **theoretical probability** of rolling an even number must be:

$$\frac{number\ of\ favorable\ outcomes}{number\ of\ possible\ outcomes} = \frac{3}{6}$$

Could you write that as 0.5 or $\frac{1}{2}$ or 50%? Sure! The idea we are trying to get across here though is that it is the number of favorable outcomes or events divided by the total number of outcomes. If we write $\frac{3}{6}$ in some other way, while mathematically equivalent, it is harder to make that connection. You probably know people who insist that all fractions should be written in lowest terms or converted to decimals or percents. We are not part of that group. When working with probability, it is easier for children (and adults) to get the hang of what is going on by leaving it in fraction form—particularly for familiar fractions. When the fractions get truly obnoxious, like $\frac{351}{837}$, then it is a good time to switch to decimals or percents so that folks can get some idea of how big or how small the probability really is. You do not have to agree with us! ∎

Caution:

It continues to be common practice to talk about "reducing" fractions to lowest forms even though the fraction is really not reduced (smaller). In probability, however, having something occur 4 times out of 5 is NOT the same as having something occur 20 times out of 25. We know this seems counterintuitive since you learned in elementary school that four-fifths is equivalent to twenty twenty-fifths. Think about our experiment of rolling a number cube. Getting a six 20 out of 25 rolls is much less likely than getting a six 4 out of 5 times. We will explore this idea more in Chapter 7 so that you can understand exactly what we are talking about here!

While we are noting strange things, a number cube is simply another way of saying a six-sided die (singular of dice). Some states specifically prohibit referring to dice as "dice" in their curricula or on state assessments since dice are generally connected with gambling. Your authors happen to live in such a state, so we have gotten used to referring to dice as number cubes. Other states do not have such restrictions. Other people make the distinction that number cubes have "numerals" painted on their faces while dice have spots painted on theirs. Who cares? Just be aware that the two terms exist and that they both refer to the same thing. If we interchange the two terms do not panic!

Back to the problem at hand. We were attempting to determine the probability of rolling a 2 or a 4 or a 6 on one toss of a fair number cube. So far we have figured out that the theoretical probability is 0.5. What about the experimental probability? To determine an experimental probability for rolling an even number we could sit down and roll a number cube 10, 20, 50, 100, or 1,000 times or we could simulate those rolls using a graphing calculator. Back in Chapter 2 we discussed the construction of frequency tables as a way of organizing data. Frequency tables are used to organize the outcomes of experiments as well. In the frequency table shown in Figure 5.2.1 we have shown the theoretical probabilities and then the results of using the random number generator on the TI-83 plus calculator. To simulate rolling a number cube, we went to the MATH menu, then moved to the PRB menu. The fifth selection from this menu is randInt. This is the random integer generator. We need to generate random integers from 1 to 6, all of which are equally likely. After selecting randInt we then enter 1, then a comma, and then 6. So randInt(1,6) will generate as many random numbers as we desire just by pushing ENTER over and over. Here are our results. If you try this on your calculator you should NOT expect to get the same results! It is random after all!

From the frequency table, we can get the experimental probability of rolling an even number for the first experiment by adding the relative frequencies for those outcomes. We have 0.2 + 0.1 + 0.2 = 0.5. In this case, our theoretical probability and our experimental probability agree exactly. This should NOT be expected. Notice what happens in the next two experiments. The experimental probability of rolling an even number in Experiment #2 is 0.20 + 0.10 + 0.25 = 0.55. The experimental probability of rolling an even number in Experiment #3 is 0.22 + 0.08 + 0.26 = 0.56. It appears as if our experimental probability is going up the more times we roll! As it turns out, rolling the number cube 50 times is not really a very large sample size. Odd things can happen for a small number of trials. If we conducted this experiment 10,000 times, we are confident that the experimental probability would be very, very close to 0.50! Statisticians rely upon something called the **Law of Large Numbers**. Simply stated, the Law of Large Numbers (as it applies to this example) says that if we continue to conduct the experiment over and over and over again, the experimental probability will get closer and closer to the

Outcome	Theoretical Probability	Experimental Probability #1 (10 rolls)	Experimental Probability #2 (20 rolls)	Experimental Probability #3 (50 rolls)
Rolling a 1	$\frac{1}{6} \approx 0.167$	$\frac{1}{10} = 0.1$	$\frac{4}{20} = 0.20$	$\frac{12}{50} = 0.24$
Rolling a 2	$\frac{1}{6} \approx 0.167$	$\frac{2}{10} = 0.20$	$\frac{4}{20} = 0.20$	$\frac{11}{50} = 0.22$
Rolling a 3	$\frac{1}{6} \approx 0.167$	$\frac{2}{10} = 0.20$	$\frac{3}{20} = 0.15$	$\frac{5}{50} = 0.10$
Rolling a 4	$\frac{1}{6} \approx 0.167$	$\frac{1}{10} = 0.1$	$\frac{2}{20} = 0.10$	$\frac{4}{50} = 0.08$
Rolling a 5	$\frac{1}{6} \approx 0.167$	$\frac{2}{10} = 0.20$	$\frac{2}{20} = 0.10$	$\frac{5}{50} = 0.10$
Rolling a 6	$\frac{1}{6} \approx 0.167$	$\frac{2}{10} = 0.20$	$\frac{5}{20} = 0.25$	$\frac{13}{50} = 0.26$

FIGURE 5.2.1 Frequency table for theoretical and experimental probabilities

theoretical probability. For those of you who have had some calculus, this is exactly the idea of a limit!

Focus on Understanding

1. With a partner, conduct the experiment either by rolling a number cube or by simulating the rolls on your calculator 50 times. Organize your results in a frequency table and find the experimental probabilities for each outcome.
2. Find the experimental probability of rolling an even number.
3. Compare your results with another pair of students. Are there any big discrepancies between your two tables as compared to the theoretical probabilities? If so, where?
4. Combine the information in your frequency table with that of another pair of students. How do your results compare to the theoretical probabilities now? Any big discrepancies?
5. What do you think would happen if you pooled your results with those of the entire class? Try it and see! Make a line graph of the number of trials versus the experimental probabilities of rolling an even number.

Classroom Connection

The *Classroom Connection* shown in Figure 5.2.2 is from *Mathscape: Chance Encounters* (page 37). ◆

9. Answer the letter to Dr. Math.

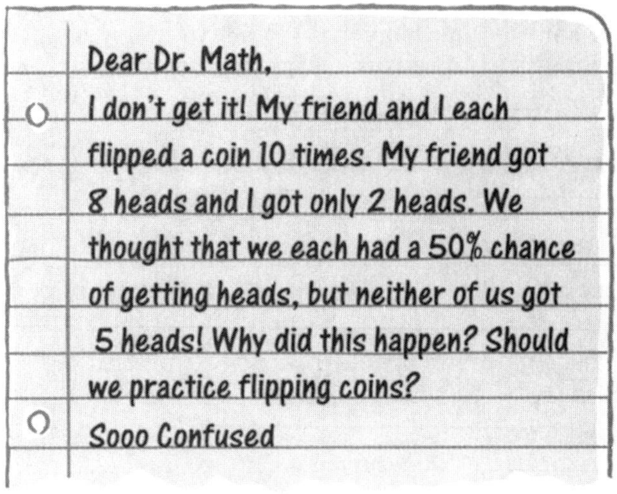

> Dear Dr. Math,
> I don't get it! My friend and I each flipped a coin 10 times. My friend got 8 heads and I got only 2 heads. We thought that we each had a 50% chance of getting heads, but neither of us got 5 heads! Why did this happen? Should we practice flipping coins?
> Sooo Confused

FIGURE 5.2.2 Thinking about the Law of Large Numbers

Focus on Understanding

1. How would you respond to Sooo Confused?
2. What if Sooo Confused and the friend had flipped a coin 100 times and the friend had gotten 80 heads and Sooo Confused had gotten 20? Would you change your response to them? Why or why not?
3. What if they flipped the coin 1,000 times and got 800 and 200 heads, respectively? Would these outcomes change your response? Why or why not?

A Brief Word About Notation

As with every other mathematical concept, it seems we have a shorthand system for denoting the probability of a particular outcome or a particular event. If we want to write the probability of rolling a 3, we could simply say ***P*(rolling a 3)** or ***P*(3)**. We use capital P to stand for "the probability of" and use parentheses to indicate the outcome or event in which we are interested. In our preceding experiment we wanted to find the probability of rolling an even number. We could write this several ways.

$$\text{Probability of rolling an even number} = P(\text{even number})$$
$$= P(2 \text{ or } 4 \text{ or } 6) = P(2 \cup 4 \cup 6)$$

The symbol "∪" in mathematics means union (hence the ∪ symbol), but it is also interpreted as the word "or" as in this outcome **or** that outcome. While we are at it, we may as well introduce the other symbol used in more complex probability situations, the intersection symbol, ∩. Intersection corresponds to the word "and," as in both this outcome **and** this outcome both happening at the same time. We will discuss this much further in Section 5.3 when we get to the rules governing probability. Most of the probability experiments explored in middle school curricula involve what are called **mutually exclusive events**. Rolling a 2 and rolling a 4 are mutually exclusive events because both outcomes cannot happen on the same roll. In one roll of a number cube, it is impossible to roll both a 2 and a 4. All of the outcomes in our number cube experiment were mutually exclusive as it would be impossible to get more than one outcome on one toss of a die. Standard middle school experiments of drawing colored cubes or counters from a bag are also examples of experiments where the outcomes are mutually exclusive. You can draw a green cube from a bag or a red cube, but not one that is both red and green at the same time unless you have some weird looking cubes!

Classroom Connection

Figure 5.2.3 is from *Mathematics in Context: Take a Chance* (page 13). It is a follow-up activity to the one portrayed in Figure 5.1.1. Notice that the ladder goes from

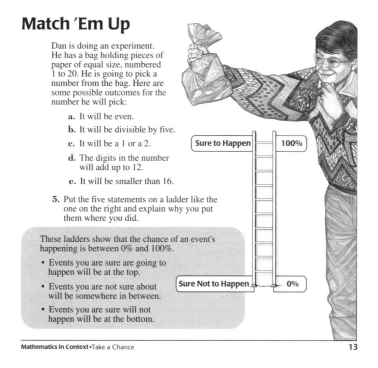

FIGURE 5.2.3 Finding probabilities from an experiment

128 Chapter 5 Probability

0% to 100% so in this activity students need to think about the probabilities of the particular outcomes as percents. ◆

Focus on Understanding

1. Identify the sample space for the experiment. Are the outcomes all mutually exclusive?
2. Why do you think that in the experiment all of the numbers are written on pieces of paper the same size? Is this important? Why or why not?
3. Find the probabilities associated with each of the events described in parts (a) through (e) in Figure 5.2.3. Write your answers as fractions.
4. What is the probability that a number drawn from the bag will be greater than or equal to 16? That it will not be a 1 or a 2? That the digits will not add up to 12? That it will not be divisible by 5? That it will be an odd number? How could you use shorthand notation to write these probabilities?
5. Describe in words any patterns you see from comparing the probabilities that an event will happen and the probability that an event will not happen.
6. Are any of the events described in the activity mutually exclusive? Identify any mutually exclusive events. For the events that are not mutually exclusive, identify the overlaps or intersections between the two events.

Ultimately, we use experimental probabilities from samples to make predictions about populations. In the Classroom Connection in Figure 5.2.3, we knew exactly what the sample space looked like. We knew that Dan had put 20 pieces of paper into a paper bag. We knew the theoretical probability of each possible outcome. What if we know nothing about the sample space other than just general information? Take a look at the next Classroom Connection in Figure 5.2.4 from *Connected Mathematics: How Likely Is It?* (pages 29 and 30).

 Predicting to Win

In the last 5 minutes of the Gee Whiz Everyone Wins! television game show, all the members of the studio audience are called to the stage to select a block randomly from a bucket containing an unknown number of red, yellow, and blue blocks. Before drawing, each contestant is asked to predict the color of the block he or she will draw. If the guess is correct, the contestant wins a prize. After each draw, the block is put back into the bucket.

FIGURE 5.2.4 Predicting to win

Section 5.2   Outcomes and Events   129

Think about this!

Suppose you are a member of the audience. Is there an advantage to being called to the stage first? Is there an advantage to being called last? Why?

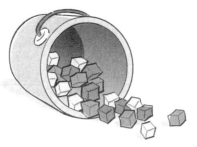

Problem 4.1

Play the block-guessing game with your class. Your teacher will act as the host of the game show, and you and your classmates will be the contestants. Keep a record of the number of times each color is drawn. Play the game until you think you can predict with certainty the chances of each color being drawn.

A. In your class experiment, how many blue blocks were drawn? Red blocks? Yellow blocks? What was the total number of blocks drawn?

B. The probability of drawing a red block can be written as P(red). Find all three probabilities based on the data you collected in your experiment.

P(red) =   P(yellow) =   P(blue) =

Now, your teacher will dump out the blocks so you can see them.

C. How many of the blocks are red? Yellow? Blue? How many blocks are there altogether?

D. Find the fraction of the total blocks that are red, the fraction that are yellow, and the fraction that are blue.

FIGURE 5.2.4 Predicting to win *(Continued)*

Classroom Connection

Focus on Understanding
1. Conduct the experiment in Figure 5.2.4 25 times. Keep a record of the number of times each color is drawn.

> 2. Do you think that 25 times is enough to be able to make a prediction? Why or why not? Would the number of cubes in the bucket make a difference? Why or why not?
> 3. Answer parts A through D.
> 4. Compare your experimental probabilities with the theoretical probabilities from part D. Are there any major discrepancies? What could you do to make your experimental probabilities a better predictor? In other words, what could you do to make it more likely that you would win the game?

◆

5.3 BASIC PROBABILITY RULES

As our probability experiments get more and more complex, it becomes more and more difficult to reasonably run simulations to estimate the experimental probabilities or compute theoretical probabilities without some rules to guide us. We have already introduced some of these rules. Others, though, will take a bit of work before they make sense.

Probability Rule #1: All probabilities must be no less than zero and no greater than one. Using "E" to stand for a particular event, in probability notation we have:

$$0 \le P(E) \le 1$$

Probability Rule #2: The sum of all the probabilities of the individual (and mutually exclusive) outcomes in a sample space must add to 1. Go back and take a look at Figure 5.2.1. Notice that in the frequency table if you add all the theoretical probabilities together (the exact fractional values, not the approximate decimal values) you get 1. Add all the experimental probabilities together for Experiment #1. What do you get as the sum of the probabilities? One, right? The sum should always be 1.

$$\sum P(\text{all outcomes}) = 1$$

Probability Rule #3: The probability of an event not occurring is the probability of the sample space (1) minus the probability of the event occurring. For example, in the *Focus on Understanding* activity following Figure 5.2.2, you were asked to find the probability of the opposite of the events described in the figure. Those "opposite" events are called **complementary** events. Different textbooks will use different symbols to denote complements. We have shown three typical ways below:

$$\text{Probability of the complement of event } A = P(A') = P(\sim A)$$
$$= P(\overline{A}) = 1 - P(A)$$

All of these ways have their pluses and minuses. We are fond of the third way where a bar is placed over the event symbol ($\overline{A}$) to mean "**not A**." This rule is often referred to as the **complement rule**. An example would be helpful here so you can see how the notation works. We have been working with the probability of rolling an even number on one toss of a fair number cube, so let's use that again to illustrate this rule. The complement of rolling an even number would be *not* rolling an even number. We could say this a shorter way by simply saying "rolling an odd number." Since we want to emphasize the "**not**" part, however, we will stick with the complement as "***not*** rolling an even number." We will use E to stand for the event of rolling an even number:

$$P(\text{not rolling an even number}) = 1 - P(\text{rolling an even number})$$
$$P(\overline{E}) = 1 - P(E)$$
$$P(\overline{E}) = 1 - 0.50 = 0.50$$

In this experiment, the probability of the complement of rolling an even number happens to be the same as the probability of rolling an even number. This is not always the case. Suppose E is the event of rolling a 2 on a fair number cube:

$$P(\text{not rolling a 2}) = 1 - P(\text{rolling a 2})$$
$$P(\overline{E}) = 1 - P(E)$$
$$P(\overline{E}) = 1 - \frac{1}{6} = \frac{5}{6}$$

Now we get into the trickier rules. Remember the symbols we introduced for unions (or) and intersections (and) previously? Well, we are going to use them in the next couple of rules.

Probability Rule #4: The General Addition Rule

The probability of event A or event B happening is the probability of event A plus the probability of event B MINUS the probability that both event A and event B will happen at the same time. This is much shorter to write using symbols:

$$P(A \text{ or } B) = P(A \cup B) = P(A) + P(B) - P(A \cap B)$$

While shorter, its meaning is probably (no pun intended) still a mystery, so let's look at a hopefully enlightening example.

EXAMPLE 5.3.1 Suppose we roll a die (calling it a number cube would be a bit misleading since it is not a cube) with 12 faces each marked with a number from 1 to 12. The sample space would consist of the numbers $\{1, 2, 3, 4, 5, 6, 7, 8, 9, 10, 11, 12\}$. These twelve outcomes are all equally likely and are mutually exclusive. Agreed? Okay, let event A be the event that we roll a number divisible by 2 and event B be the event that we roll a multiple of 3. We could write these two events as subsets of our sample space: $A = \{2, 4, 6, 8, 10, 12\}$ and $B = \{3, 6, 9, 12\}$. In a single throw of the number cube, what is the probability of rolling a number divisible by two **or** a number that is a multiple of three? It is very tempting to say it must be just the probability of event A plus the probability of event B. Take a look at what happens if we do it that way:

$$P(A) + P(B) = \frac{6}{12} + \frac{4}{12} = \frac{10}{12}$$

Wow, that's a pretty high probability! Unfortunately, though, we have over-counted some of the numbers. Notice that the outcome of rolling a 6 or a 12 occurs in BOTH set A and set B, so the probability of that outcome has been added in twice. Six and 12 were two of the outcomes in set A and two of the outcomes in set B. Consequently, we have to subtract out the over-count or intersection between the two sets:

$$P(A \cup B) = P(A) + P(B) - P(A \cap B) = \frac{6}{12} + \frac{4}{12} - \frac{2}{12} = \frac{8}{12}$$

By middle school most students should be familiar with Venn diagrams. For students who are highly visual, Venn diagrams are a nice way to represent the preceding experiment. The Venn diagram for this experiment is shown in Figure 5.3.1. Remember that the rectangle is used to represent the sample space and the circles are used to represent the two events. The intersection (rolling a number divisible by 2 and a multiple of 3) is the overlapping area between the two circles. ∎

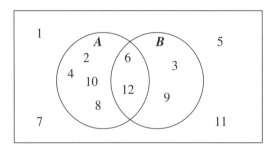

FIGURE 5.3.1 Venn diagram

Another good way to think about the general probability rule is with what is called a **contingency table**. A contingency table is just a way of organizing outcomes that have more than one attribute to identify them. In our table in Figure 5.3.2 we have recorded the number of students in Ms. Majerus's class that bring their lunch to school (or don't) and the number who are girls (or boys). The event that a student brings their lunch to school is designated with the letter L. The complement, $\overline{L}$, means they do not bring their lunch to school. Let G be the event the student is a girl. Let B be the event the student is not a girl (in other words, a boy). Could the event that a student is a boy be represented by $\overline{G}$? Well, sure. It just seemed kind of mean to us, so we went with B instead. The numbers at the end of each row and column represent the total number of students who fit that particular attribute.

From our table we can see that there are 14 girls and 11 boys in Ms. Majerus's class. Twelve of her students bring their lunch, while the remaining 13 do not. We have two attributes to work with here. Students can be classified by their gender or by whether or not they bring their lunch to school. Are the attributes mutually exclusive? Within each individual attribute, yes they are. One cannot be both a boy and a girl. One cannot both bring their lunch and not bring their lunch at the same time. In Ms. Majerus's class there are no special cases. However, across attributes, the

Section 5.4   Conditional Probability and Independence   133

	Bring Lunch (L)	Do Not Bring Lunch ($\overline{L}$)	Totals
Girls (G)	8	6	14
Boys (B)	4	7	11
Totals	12	13	25

FIGURE 5.3.2 Contingency table for students and their lunch choices

events are not mutually exclusive. You can be a girl and bring your lunch to school. The first cell in the table represents the intersection between the event of being a girl **and** bringing lunch to school. We can see that there are 8 girls in this class that bring their lunch to school. Find the cell that would represent boys who do not bring their lunch to school. How many boys in her class do not bring their lunch to school? You should get 7! The following *Focus on Understanding* will help you to read and use the notation of probability and apply the rules we have introduced thus far.

Focus on Understanding

1. Find each of the following probabilities using the contingency table in Figure 5.3.2.

    a. $P(G)$ **b.** $P(B)$ **c.** $P(L)$ **d.** $P(\overline{L})$ **e.** $P(G \cap L)$
    f. $P(B \cap \overline{L})$ **g.** $P(G \cap \overline{L})$ **h.** $P(B \cap L)$

2. Write an English statement describing each of the probabilities in part 1.
3. Add the probabilities you found for 1a and 1b. What do you notice? Why must this be the case?
4. Add the probabilities you found for 1c and 1d. What do you notice? Why must this be the case?
5. Can a student be tallied in both cell #1 ($G \cap L$) and cell #4 ($B \cap \overline{L}$)? Why or why not?
6. Find the sum of the probabilities for the four cells representing the intersections of the attributes. What do you notice?
7. What does the notation $P(G \cup L)$ mean? How is it different from $P(G \cap L)$? Find $P(G \cup L)$.
8. What does the notation $P(B \cup \overline{L})$ mean? How is it different from $P(B \cap \overline{L})$? Find $P(B \cup \overline{L})$.

5.4 CONDITIONAL PROBABILITY AND INDEPENDENCE

Everything we have explored so far has been under the guise of simple probability and more or less pretty straightforward—we hope! In this section we are going to develop the concept of conditional probabilities and independent events. We will begin with conditional probability.

Conditional probability is exactly what the name implies. There is some sort of condition associated with the outcome of an experiment. Let's go back and take a look at the example using a contingency table from the preceding section. In Section 5.3 we presented a rather boring, but hopefully enlightening contingency table. We reintroduce the same table in Figure 5.4.1.

	Bring Lunch (L)	Do Not Bring Lunch ($\bar{L}$)	Totals
Girls (G)	8	6	14
Boys (B)	4	7	11
Totals	12	13	25

FIGURE 5.4.1 Contingency table for students and their Lunch choices

In the *Focus on Understanding* that followed the table, you worked on using notation and finding probabilities for various complements, unions, and intersections based on the information in the table. Suppose we wanted to know the probability that a student did not bring their lunch ($\bar{L}$), given that the student chosen was a girl. The phrase "given that the student chosen was a girl" is the conditional part of the probability. In conditional probability, the sample space is reduced or restricted somehow because you are given or told to restrict the number of possible outcomes to a particular subset of the sample space. In this situation, we are only concerned with the girls as we are told that the student selected was a girl. Of the girls in the sample space (14 altogether), how many do not bring their lunches to school? From the table we see that out of those 14 girls, 6 do not bring their lunches to school. The probability that a student did not bring their lunch to school, given that she is a girl, is then $\frac{6}{14}$.

EXAMPLE 5.4.1 What would be the probability of a student being a boy, given that they brought their lunch to school? The condition that is being met is that of bringing lunch to school, so our sample space is reduced to just those 12 students who brought their lunch to school. Of those 12 students, how many were boys (favorable)? The probability of a student being a boy, given that they brought their lunch to school is therefore $\frac{4}{12}$.

Have you noticed how awkward it is to keep saying that whole conditional phrase bit over and over? You must be thinking that surely there is a shorter way to write that whole sentence and—guess what?—you're right! In the first example, we could write the sentence "the probability that a student did not bring their lunch ($\bar{L}$) given that the student chosen was a girl" as follows, using G for the event that the student is a girl:

$$P(\bar{L}|G) = \frac{6}{14}$$

In this notation, the **condition** always follows the straight up and down bar. Do not be fooled by a "key word" strategy. There are many ways to word conditional probability. The road to success lies in first identifying what the particular subset is from the sample space that you are considering and then looking at the favorable outcomes from that group only. ∎

EXAMPLE 5.4.2 Try this one! Using the same contingency table, if the student selected at random was a boy, what is the probability that he brought his lunch to school (or $P(L|B)$ using the notation).

Since we are restricting the sample space to only the boys, the total number of outcomes possible is 11. From that group, we are looking for the number who brought their lunches to school, which in this case is 4. So:

$$P(L|B) = \frac{4}{11}$$
∎

Go back to Example 5.4.1—"What would be the probability of a student being a boy, given that they brought their lunch to school?" How could we write this using probability notation and how is it different from the example we just did?

Well, in probability notation we would write this question as $P(B|L)$. Notice that in this case the conditional part is that we knew the student selected at random brought their lunch. We wanted to find the probability that from that subset of students he/she is a boy.

In general then, to find $P(A|B)$ means to find the probability of A knowing the outcomes are restricted to subset B.

Students often want to come up with a handy formula to compute conditional probabilities. The one that they naturally try to use as the rule for conditional probability is also unfortunately incorrect. It is very tempting to say (and students will do this, so be prepared as a future teacher) that:

$$P(A|B) \stackrel{?}{=} \frac{P(A)}{P(B)}$$

It would be lovely if the straight up and down bar actually meant "divided by," but it does not. Check it out. Compute $P(L|B)$ using the about "formula." From our table we get:

$$P(L|B) \stackrel{?}{=} \frac{P(L)}{P(B)} = \frac{\frac{12}{25}}{\frac{11}{25}} = \frac{12}{11}$$

We hope this seems just plain wrong to you! Probability Rule #1 stated that all probabilities had to be greater than or equal to zero, but less than or equal to one. Clearly $\frac{12}{11}$ is greater than one! What went wrong? Well, take a look at the numerator. $P(L)$ means that from the entire sample space, what is the probability that a student brought their lunch. We don't want the *entire* sample space in this case, just the boys. How do we fix our "formula?" We consider what we really want to know. We want to know the probability that the student selected at random is a

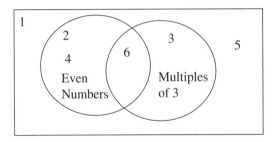

FIGURE 5.4.2 Venn diagram

boy AND brought their lunch. Both of those have to be true. What we need is the intersection of those two attributes! Let's see if that will fix the problem:

$$P(L|B) = \frac{P(L \cap B)}{P(B)} = \frac{\frac{4}{25}}{\frac{11}{25}} = \frac{4}{11}$$

This agrees with the answer we got earlier just by reasoning through the problem.

Probability Rule #5: Conditional Probability Rule

$$P(A|B) = \frac{P(A \cap B)}{P(B)}$$

The **probability of A given B** is the probability of **A intersected with B** divided by the **probability of B**, the given condition, as long as $P(B)$ is not equal to zero.

EXAMPLE 5.4.3 We can also think about conditional probabilities from Venn diagrams. Consider the Venn diagram in Figure 5.4.2 (which is just a simpler version of the diagram in Figure 5.3.1). The sample space consists of all the possible outcomes when we roll a fair number cube. Let event E be the ways that we can roll an even number with a fair number cube and let M be the event that the number we roll is a multiple of 3. Notice that 1 and 5, while in the sample space, are not included in the two circles as they do not have the attributes (even or a multiple of 3) to be included in the event circles. What is the probability that if we roll a multiple of 3, it will be an even number? We will approach this problem both ways—by reasoning from the diagram and from using the conditional probability rule.

To find the probability of an even number given that we rolled a multiple of 3 or $P(E|M)$ we first restrict ourselves to the outcomes in the circle representing the multiples of 3. We have only two numbers in that circle, so the total number of ways we can roll of multiple of 3 (the given) is two. From that circle, how many outcomes are even? We see that 6 is the only outcome that is even from those two outcomes. The probability would then be $\frac{1}{2}$.

Using the conditional probability rule we would have:

$$P(E|M) = \frac{P(E \cap M)}{P(M)} = \frac{\frac{1}{6}}{\frac{2}{6}} = \frac{1}{2}$$

Would you rather be given that it was a multiple of 3 or told it was an even number?

Find $P(M|E)$:

$$P(M|E) = \frac{P(M \cap E)}{P(E)} = \frac{\frac{1}{6}}{\frac{3}{6}} = \frac{1}{3}$$ ∎

What happens if knowing a given condition does not have any impact on the probability that you are trying to find? In that case we say the two events A and B are **independent**. For example, suppose you roll a fair number cube and you know (because you looked) that the number you rolled is a 4. What is the probability that if you roll the number cube again that the number will be a 4? Believe it or not, inanimate objects have no memory of what happened before so it is irrelevant that the first roll was a 4. The probability that the second roll is a 4 is still just $\frac{1}{6}$. The probability neither increases nor decreases because the two rolls are independent of one another.

A word of caution is necessary here. The probability that the second roll is a 4 is not the same as asking what is the probability that you will roll two 4's in a row. These are different probability problems! We will say more about the second question in a bit.

Two events A and B are said to be independent if:

$$\boldsymbol{P(A|B) = P(A)}$$

Classic examples include the aforementioned number cube problem and such experiments as flipping a coin and rolling a number cube. For example, given that you rolled an even number on a fair number cube, what is the probability you will get a head on one toss of a fair coin? It does not matter what came up on the number cube. When we flip the coin, the probability that it will turn up heads is always $\frac{1}{2}$.

Suppose you flipped a (fair) coin nine times and all nine times it came up heads. What is the probability that on the tenth flip it will turn up heads? It does not matter what happened in all the previous trials, the probability that on the next flip the coin turns up heads is still $\frac{1}{2}$.

5.5 MULTIPLICATION RULES

One of the handy things about formulas is that we can manipulate them to suit our needs. For example, **Probability Rule #3** back in Section 5.3 was stated as follows:

$$P(\overline{A}) = 1 - P(A)$$

We could have written equivalently:

$$P(A) + P(\overline{A}) = 1$$

The same is true for the general probability rule (**Probability Rule #4**). Instead of writing:

$$P(A \cup B) = P(A) + P(B) - P(A \cap B)$$

We could have written:

$$P(A \cap B) = P(A) + P(B) - P(A \cup B)$$

Chapter 5 Probability

Using the properties of arithmetic and algebra we can rearrange the formulas to suit ourselves. Generally, when we rearrange these formulas we do not give them a different name, but in the case of conditional probability we do. If we take

Probability Rule #5,

$$P(A|B) = \frac{P(A \cap B)}{P(B)}$$

and rearrange it by multiplying both sides by $P(B)$ we get something like this:

Probability Rule #6: General Multiplication Rule

$$P(B) \cdot P(A|B) = P(A \cap B)$$

The general multiplication rule comes in handy for finding intersections from multistage events. Consider the following:

EXAMPLE 5.5.1 A jar contains 10 gumballs—6 are red and the rest are yellow. What is the probability that if 2 gumballs are selected at random, one after the other with no replacement, that they will both be yellow? At first glance you may think this has nothing to do with the last five pages, but it does. Here is an alternate way of stating the problem: What is the probability that the first gumball is yellow **and** the second gumball is yellow? Or, using probability notation:

$$P(1^{st} \text{ yellow} \cap 2^{nd} \text{ yellow}) = ?$$

Aha! This appears to be an intersection problem, but not as simple as the intersection problems we dealt with back in 5.3. Let's break this down into stages. What is the probability that the first gumball that we take out of the jar is yellow? Easy enough:

$$P(1^{st} \text{ yellow}) = \frac{4 \text{ yellow}}{10 \text{ total}} \text{ or } \frac{4}{10}$$

Here is the tricky part, since we KNOW that the first one is yellow (it has to be in order to meet the conditions of the question) what is the probability that the second gumball is yellow? This is a conditional probability again. We would write this using the notation as:

$$P(2^{nd} \text{ yellow} \mid 1^{st} \text{ yellow}) = ?$$

Given that the first gumball was yellow, and we did not replace the gumball, we now have only 3 yellow gumballs out of 9 total gumballs in the jar so the probability that the second gumball is also yellow would be:

$$P(2^{nd} \text{ yellow} \mid 1^{st} \text{ yellow}) = \frac{3}{9}$$

By the general multiplication rule, the probability that the first one would be yellow and the second one would be yellow is:

$$P(1^{st} \text{ yellow} \cap 2^{nd} \text{ yellow}) = P(1^{st} \text{ yellow}) \cdot P(2^{nd} \text{ yellow} \mid 1^{st} \text{ yellow})$$

$$= \frac{4}{10} \cdot \frac{3}{9} \text{ or } \frac{12}{90} \blacksquare$$

EXAMPLE 5.5.2 Remember the problem we avoided just a bit ago about the probability of rolling two 4's in a row or $P(1^{st}\, 4 \cap 2^{nd}\, 4)$? Well, we can now deal with it. The probability of rolling the first 4 or $P(1^{st}\, 4) = \frac{1}{6}$. The probability that the second roll is a 4, given that the first roll was a 4 or $P(2^{nd}\, 4 | 1^{st}\, 4)$ is also because we established before that the two rolls were independent of one another. So, the probability of rolling two 4's in a row (or rolling a 4 and then rolling another 4) would be:

$$P(1^{st}\, 4 \cap 2^{nd}\, 4) = P(1^{st}\, 4) \cdot P(2^{nd}\, 4 | 1^{st}\, 4) = \frac{1}{6} \cdot \frac{1}{6} \text{ or } \frac{1}{36}$$ ■

The general multiplication rule works whether the events are independent or dependent!

It has been a while since we have done a Classroom Connection and you are probably wondering about that. Most middle school curricula deal with simple probability. Not all, however, address conditional probability and the general multiplication rule—at least not in a direct way. It is in there, it is just not blatantly obvious. Let's take a look at a Classroom Connection from *Connected Mathematics: What Do You Expect?* (page 41)

Classroom Connection

The investigation shown in Figure 5.5.1 uses the ideas of conditional probability and the multiplication rule without really naming them as such.

The player must first decide whether to put the treasure in room A or room B. Based on the size of the room alone it looks like room B might be a good choice—or is it? Work through Parts A through D in Problem 4.1 shown in Figure 5.5.2.

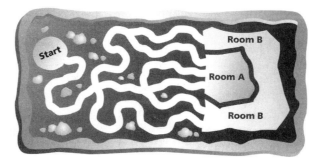

FIGURE 5.5.1 Deep in the dungeon

> **Problem 4.1**
>
> **A.** If you were playing Deep in the Dungeon, in which room would you put the treasure in order to have the best chance of beating Zark? Explain your choice.
>
> **B.** Work with a partner to find a way to simulate Deep in the Dungeon so it can be played without a computer. Your simulation should be a two-person game. One person should hide the treasure, and the other should play the role of Zark. You will need to figure out a way for Zark to make a random selection at each fork.
>
> **C.** Play your simulation of Deep in the Dungeon 20 times with your partner. Take turns hiding the treasure and playing Zark. For each game, record the room that Zark ends in.
>
> **D.** Based on your results from part C, what is the experimental probability that Zark will end in room A? What is the experimental probability that Zark will end in room B?

FIGURE 5.5.2 Simulating multistage events

This *Classroom Connection* is a very interesting one that you can really spend a lot of time exploring. A nice follow-up activity is to create your own dungeon and then work through the prompts (parts A through D) again. Some people will create very elaborate dungeons that require a lot of thought in order to simulate the game and find the corresponding probabilities for the various rooms. ◆

Focus on Understanding

The *Classroom Connection* in Figure 5.5.2 focused on finding the experimental probability that a player would end up either in room A or room B. We can use the general multiplication rule to find the theoretical probability of ending up in either room A or room B.

1. Determine the theoretical probability of taking each of the paths at each fork. (You have probably already worked this out when you created your simulation.)
2. Use the general multiplication rule to determine the theoretical probability that a player would end up in room A.
3. Use the general multiplication rule to determine the theoretical probability that a player would end up in room B.
4. Using the complement rule, find the probability that a player would NOT end up in room A. Is this probability the same as the probability that a player would end up in room B? Why or why not?
5. Carmen believes that she has a shortcut to find the theoretical probability that a player will end up in room A. She says that since there are six paths total that end up at room A or B, and since three of them lead to room A, then the probability that a player would end up in room A must be $\frac{3}{6}$. Is Carmen correct? Why or why not?

5.6 GEOMETRIC PROBABILITY

More and more middle school and secondary school mathematics textbooks are paying attention to what are sometimes called **geometric probabilities**. Geometric probabilities are typically represented through some type of geometric figure like a circle (spinners) or a rectangular grid. Different areas in the figure represent the probabilities associated with particular outcomes of an experiment. Geometric probabilities may also be represented by lengths or by volumes, though area models are the most common in most curricula. We defined simple probability earlier as:

$$\text{Probability} = \frac{\text{number of favorable outcomes}}{\text{number of possible outcomes}}$$

The idea is the same for geometric probability, but instead of talking about the number of outcomes we use the length, area, or volume associated with a particular outcome. We will focus on area models since those are more typical:

$$\text{Geometric Probability} = \frac{\text{area associated with a favorable outcome}}{\text{area for all the outcomes}}$$

EXAMPLE 5.6.1 One of the classic examples of geometric probability is that of finding the probability that a dart will hit a particular ring on a target. Such a target is shown below with various measurements given:

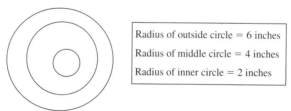

Radius of outside circle = 6 inches
Radius of middle circle = 4 inches
Radius of inner circle = 2 inches

What is the probability of a dart landing in the inner circle? In order to find this probability we need to compare the favorable area (that of the inner circle) to the total area of the circle. In this case, since the radius of the inner circle is 2 inches, we know that its area is $\pi \cdot (2 \text{ inches})^2$ or 4π squared inches. The area of the whole circle would be $\pi \cdot (6 \text{ inches})^2$ or 36π squared inches. Hold on a second before you hyperventilate about π showing up in our geometric probability. Take a look at what happens here:

$$\text{Geometric Probability} = \frac{\text{area associated with a favorable outcome}}{\text{area of all the outcomes}}$$

So in this case we get:

$$\text{Probability of a dart landing in the inner circle} = \frac{\text{area of inner circle}}{\text{area of entire circle}}$$

$$= \frac{4\pi \text{ square inches}}{36\pi \text{ square inches}} = \frac{1}{9}$$

By simplifying this ratio we not only get rid of the π, but our units of measurement (the square inches) "cancel" each other out. Probability is **unit-less**. ∎

EXAMPLE 5.6.2 See if you can find the probability that a dart would land in the middle ring. This is a little trickier, since the middle ring is really a ring and not a complete circle. Take a minute and see what you get before reading our solution. We'll wait for you.

Okay, to find the area of the middle ring we must first find the area of a circle of radius 4 inches (the complete circle) and then subtract out the area of the inner circle. You should get something like this:

Area of middle ring = 16π square inches − 4π square inches = 12π square inches
The probability, then, of a dart landing in the middle ring would be:

$$\text{Probability of a dart landing in the middle ring} = \frac{\textit{area of middle ring}}{\textit{area of entire circle}}$$

$$= \frac{12\pi \text{ square inches}}{36\pi \text{ square inches}} = \frac{1}{3}$$

You may be intensely bothered by the fact that our circles were not concentric. Does that really matter? Nope. The principle is the same whether our circles are concentric or not! ∎

Classroom Connection

Generally, middle school students start out exploring geometric probabilities in a less measurement-intensive situation using spinners and grids to illustrate the various outcomes. Figure 5.6.1 is from *Mathscape: Chance Encounters* (page 14). Spinners and grids are used in this case to have students explore experimental probabilities and the connection between the areas on the spinner and the areas on the grid.

In this activity, middle school students are asked to analyze the results of their spins and design a new game card that minimizes the number of spins required to fill all the boxes. Students use the experimental probabilities from their spins to design a new game card. ◆

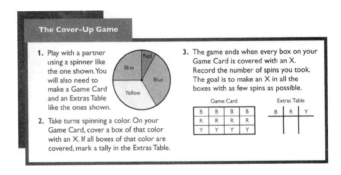

FIGURE 5.6.1 The Cover-Up Game

Section 5.6   Geometric Probability   143

Focus on Understanding

Take a careful look at the spinner in Figure 5.6.1 and the game card shown.

1. Based on the spinner, which color do you think will come up most often? Why? Which color will come up the least? Why?
2. What do you think will happen if the students spin the spinner 12 times? 24 times? Why?
3. How would you design the game card so that in order to completely cover the card you would take fewer spins?
4. How would you change the spinner, if you had to keep the game card the same?

This excerpt is an interesting one because students are really asked to connect the idea of areas on a spinner or circular model to that of areas on a grid. Suppose a student drew a new game card as follows:

Game Card

Blue	Red	Yellow	Red
Yellow	Yellow	Red	Blue
Red	Blue	Blue	Yellow

Is this the same as the game card in Figure 5.6.1? Why or why not? (As you have probably already determined, this new game card is equivalent to the one in Figure 5.6.1.) The "position" of the colors on the card does not matter in this case. It is the area (probability) they represent on both the game card and the spinner that matters.

Classroom Connection

Our next *Classroom Connection* shown in Figure 5.6.2 is from *Connected Mathematics: How Likely Is It?* (page 46). It, too, has middle school students work with spinners to determine the likelihood of certain outcomes. ◆

Focus on Understanding

Use the two spinners pictured in Figure 5.6.2 for the following:

1. Answer the questions posed to the students in Figure 5.6.2. Be sure to explain your reasoning!
2. Are there other ways that you could draw the two spinners but maintain the same probabilities? Draw at least one spinner that is equivalent to Spinner A and at least one spinner that is equivalent to Spinner B.

144 Chapter 5 Probability

11. The cooks at Kyla's school made the spinners shown below to help them determine the lunch menu. They let the students take turns spinning to determine the daily menu. In a–c, decide which spinner you would choose, and explain your reasoning.

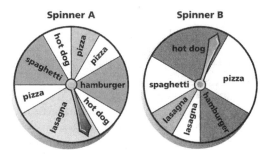

a. Your favorite lunch is pizza.

b. Your favorite lunch is lasagna.

c. Your favorite lunch is hot dogs.

FIGURE 5.6.2 Lunch spinners

3. How would you go about representing Spinner A and Spinner B using a grid or a game card as in the previous Classroom Connection? Create grids that would approximate Spinner A and Spinner B.

Classroom Connection

While spinners and grids are pretty common ways to illustrate geometric probabilities, there are also more geometry/measurement "intensive" ways. Take a look at the practice and application exercise in Figure 5.6.3 from *MathThematics, Book 1* (page 569).

Each of these problems in the portion of the exercise set depicted here expects the student to compute various areas and find the probabilities of certain outcomes based on those areas. Though it's more arithmetic intensive perhaps than the spinners and grids of the previous two examples, the idea is the same. ◆

Focus on Understanding

Time to brush up your measurement skills from geometry! Use Figure 5.6.3 for the following:

1. Read through each of the practice and application exercises in the figure. What prior knowledge do you think students (or you) need in order to answer the questions?

2. Which exercises do you think students would find the most difficult to do? Why? Which ones do you think students would find the easiest to do? Why?

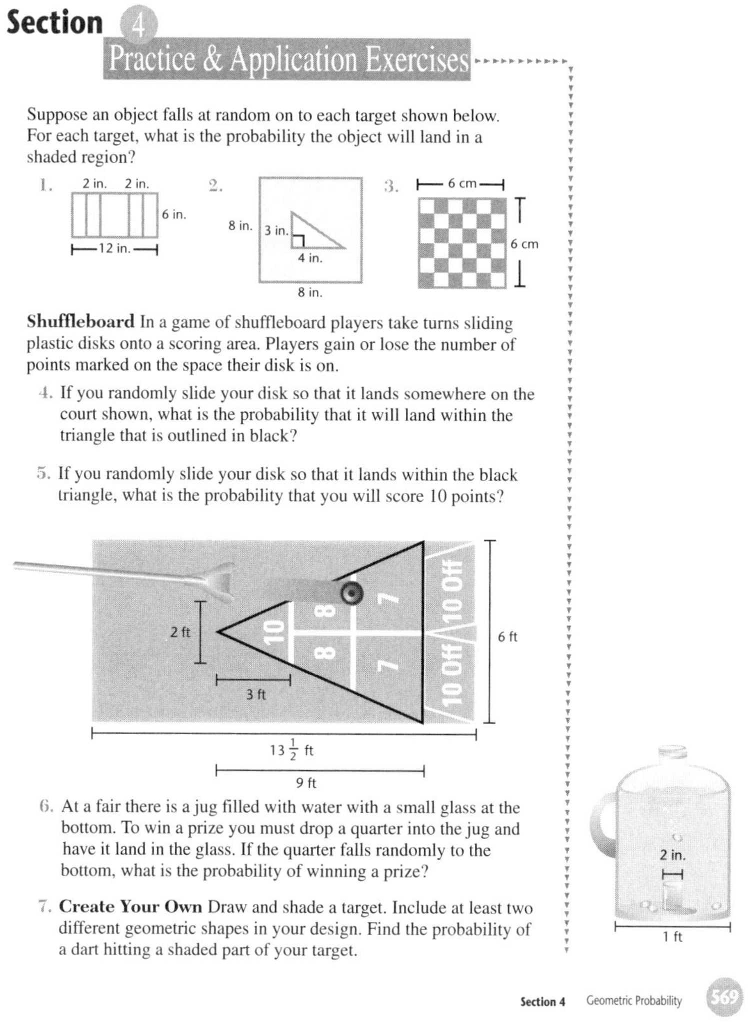

Section 4 Practice & Application Exercises

Suppose an object falls at random on to each target shown below. For each target, what is the probability the object will land in a shaded region?

1. [rectangle 12 in. × 6 in. with vertical bars of width 2 in.]
2. [8 in. square containing a right triangle with legs 3 in. and 4 in.]
3. [6 cm × 6 cm checkered square]

Shuffleboard In a game of shuffleboard players take turns sliding plastic disks onto a scoring area. Players gain or lose the number of points marked on the space their disk is on.

4. If you randomly slide your disk so that it lands somewhere on the court shown, what is the probability that it will land within the triangle that is outlined in black?

5. If you randomly slide your disk so that it lands within the black triangle, what is the probability that you will score 10 points?

[Shuffleboard court: 2 ft, 6 ft, 3 ft, $13\frac{1}{2}$ ft, 9 ft]

6. At a fair there is a jug filled with water with a small glass at the bottom. To win a prize you must drop a quarter into the jug and have it land in the glass. If the quarter falls randomly to the bottom, what is the probability of winning a prize?

7. **Create Your Own** Draw and shade a target. Include at least two different geometric shapes in your design. Find the probability of a dart hitting a shaded part of your target.

[Jug: 2 in. glass opening, 1 ft jug diameter]

Section 4   Geometric Probability   S69

FIGURE 5.6.3 More geometric probabilities

3. Complete the exercises shown in the figure. In the case of exercise 1, the shaded region is the area of the two vertical bars within the rectangle that are marked as having widths of 2 inches. In exercise 2, the shaded region is the area of the square outside of the triangle.

4. Which exercises did you find to be the most difficult? Why? Which ones were the easiest for you? Why?

Chapter 5 Summary

This chapter has introduced you to the fundamental concepts of simulations and simple probability—both theoretical and experimental. While there are many,

many different ways of representing the outcomes of probability experiments, such as frequency tables, contingency tables, spinners, grids, or areas, the idea behind the probability of a particular outcome is still the same:

$$\text{Probability} = \frac{favorable}{possible}$$

At the end of Chapter 6 we will explore more complex probabilities that involve more sophisticated ways of determining the number of possible outcomes. The essential idea, however, will remain *exactly* the same. You guessed it!

$$\text{Probability} = \frac{favorable}{possible}$$

The key terms and ideas from this chapter are listed below:

experiment   120
outcomes   120
sample space   120
event(s)   120
theoretical probability   122
experimental probability   122
Law of Large Numbers   124
complementary events   130

mutually exclusive events   127
general addition rule   131
Venn diagram   132
contingency table   132
conditional probability   134
independent events   137
geometric probability   141

Assessment is an integral part of every curriculum from the elementary school all the way through college. The question always arises—*what is it that students should be able to do after completing this lesson/unit/chapter*? We have included here our intended learning goals for Chapter 5. Students who have a good grasp of the concepts developed in Chapter 5 should be successful in responding to these items:

- Explain or describe what is meant by each of the terms in the vocabulary list.
- Identify outcomes of an experiment and determine the sample space for the experiment.
- Determine the probabilities of events associated with an experiment.
- Determine probabilities of events using frequency tables, Venn diagrams, or contingency tables.
- Use the basic probability rules to deduce other probabilities or to analyze probability situations.
- Determine probabilities associated with conditional events.
- Determine probabilities associated with independent events.
- Determine geometric probabilities associated with spinners, grids, or area models.

Your course instructor may have additional or different assessable outcomes for your class. As teachers (or future teachers) you should think about the assessment outcomes and learning goals for each chapter as you work through them.

EXERCISES FOR CHAPTER 5

1. Using the *Data Analysis and Probability Standards for Grades 6–8* from the NCTM's *Principles and Standards for School Mathematics* found at www.nctm.org, identify the middle school objectives that are found in Chapter 5.
2. Using the state standards for mathematics for the content area of data analysis and probability for your state, identify the middle school objectives that are found in Chapter 5. The following website may be useful: www.doe.state.in.us This website will allow you to access the web pages of the state departments of education for the 50 states. From the state web pages you should be able to find your state's mathematical standards.
3. What is the difference (if any) between a theoretical probability and an experimental probability? Can these two ever be the same? Create an example to illustrate your explanation.
4. George insists that if he plays the same five lottery numbers long enough (out of 50 choices), he eventually has to win the lottery. He claims that the Law of Large Numbers supports his reasoning. Do you believe that the Law of Large Numbers supports George's claim? Why or why not?
5. Mr. Ramone conducted an experiment with his seventh grade class. In this experiment he put some colored blocks into a bucket and students took turns drawing a block out of the bucket. After each draw, the block was returned to the bucket. The students were not told the total number of blocks in the bucket. The results of their draws are recorded in the frequency table below:

Color of Block	Frequency
Blue	10
Red	6
Green	3
White	1
Total	20

Based on the data in the table, what is the probability that if a block is drawn at random it will be:

a. red? **b.** white? **c.** green? **d.** blue?
e. red or white? **f.** green or blue? **g.** blue and white?

6. Cheyenne and her friend rolled two, six-sided number cubes 25 times. They kept track of the sum of the numbers on each roll and then determined the experimental probability for the possible sums based on the data they collected. Their table is shown below:

Sum of Numbers	Experimental Probability Based on 25 Rolls
2	0.01
3	0.02
4	0.05
5	0.08

Sum of Numbers	Experimental Probability Based on 25 Rolls
6	0.21
7	0.28
8	0.14
9	0.06
10	0.03
11	0.03
12	0.01

When they showed their table to their friend, Charlie, he said that they must have made a mistake somewhere along the line when finding the experimental probabilities. Why do you think Charlie said that? What evidence is there in the table that a mistake was made?

7. A local pizza parlor likes to keep data on whether customers like thick or thin crust and pepperoni or no pepperoni. The owner decided to use a contingency table to record his survey results.

	Thick Crust (C)	Thin Crust ($\overline{C}$)	Totals
Pepperoni (P)	53	32	85
No Pepperoni ($\overline{P}$)	14	21	35
Totals	67	53	120

Using the table, find each of the following probabilities:

a. $P(P)$ b. $P(\overline{C})$ c. $P(P \cup C)$ d. $P(P \cap \overline{C})$
e. $P(C|P)$ f. $P(\overline{P}|\overline{C})$ g. $P(P|C)$ h. $P(\overline{P}|C)$

8. A biology class is raising money for a project. They have 30 mice altogether. Fourteen of them are males and eight of the male mice are white. There are a total of 20 white mice.

   a. A contingency table has been started at the right. Fill in the rest of the table.

	Male	Female	Total
White	8		20
Not White			
Total	14		30

   If a mouse is selected at random:
   b. Find the probability that the selected mouse is female.
   c. Find the probability that the selected mouse is a white female (in other words, white and female).
   d. Are the events "white" and "female" mutually exclusive? Explain.
   e. Are the events "white" and "female" independent? Explain.
   f. Find the probability that the selected mouse is either male or white.
   g. Given that the selected mouse is not white, find the probability that it is male.

9. Given the maze below, find the probability that a path chosen at random will lead to Room A.

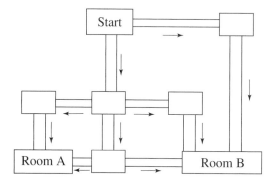

10. A box contains 8 red blocks, 3 white blocks, and 1 blue block.
    a. A block is drawn at random. Find the probability that it will be a blue block.
    b. A block is drawn at random. Find the probability that it will be a red block.
    c. A block is drawn at random. Find the probability that it will be a blue or white block.
    d. Create an arca diagram that represents this probability experiment. Is there more than one way to create such a diagram? Explain why or why not.
    e. A block is drawn at random and is NOT put back into the box. A second block is then drawn at random from the box. Given that the first block was red, what is the probability that the second block will also be red?
    f. Suppose two blocks are drawn at random, one after another, without replacement. What is the probability of drawing a blue block and then a red block? Is this the same probability as drawing first a red block and then a blue block? Why or why not?
    g. Suppose three blocks are drawn at random, one after another, without replacement. What is the probability of getting first a red block, then a white block, and then a blue block?
    h. If the blocks were replaced in the box after each draw in parts e, f, and g, would it change the probabilities you were asked to find? Why or why not?
11. The area diagram below is made up of sets of equilateral triangles. Find the probability that if a dart is thrown at this triangular target it will land in a white space.

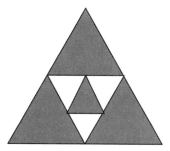

12. Bailey and Max were designing grids to simulate a game they played in class drawing colored blocks out of a bucket. There were 16 blocks in the bucket.

Eight were green, six were purple, and two were orange. Their grids are shown below:

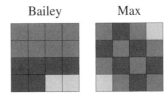

Bailey   Max

Max says that Bailey's grid is not correct because the different colors of blocks in the grid have to be "spread out" in order to accurately represent the probabilities of drawing the blocks out of the box. Bailey argues that it doesn't matter where she puts the colors on the grid as long as she has the correct number of blocks represented. Who is right and why?

13. Mr. and Mrs. Jones currently have three girls ages 3, 5, and 8. They are expecting another child. What is the probability that their fourth child will also be a girl? Does it matter that the other three children are girls? Why or why not?

14. Lindsey made up a new game to play in her sixth grade class. In her game, players take turns rolling two tetrahedral dice (four sided) numbered 1 through 4 on the sides and find the sum of the two dice. If an even number sum is rolled, then Player A scores a point. If an odd number sum is rolled, then Player B gets a point.
   a. Construct the sample space for Lindsey's game.
   b. Determine the theoretical probability for each of the outcomes in the sample space.
   c. Is Lindsey's game a fair game? Why or why not?

15. Consider the following Venn diagram:

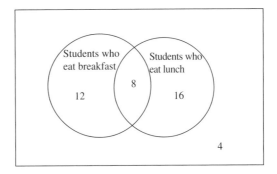

   a. What is the probability that a student selected at random from this group eats lunch?
   b. What is the probability that a student selected at random from this group eats lunch or breakfast?
   c. What is the probability that a student selected at random from this group does not eat breakfast?
   d. What is the probability that a student selected at random from this group eats lunch given that they eat breakfast?

e. What is the probability that a student selected at random from this group eats breakfast given that they do not eat lunch?

16. Mrs. Sullivan's seventh grade class conducted a survey to see how many students were involved in extracurricular activities. Fifty students were surveyed. Here's what her class found:

   - 28 students were involved in musical activities like band or choir
   - 19 students were involved in sports
   - 21 students were involved in theatre
   - 9 students were involved in music and sports
   - 13 students were involved in music and theatre
   - 8 students were involved in sports and theatre
   - 6 students were involved in all three activities

   a. Draw a Venn diagram to represent the results of this survey. There were some students who were not involved in any of these three activities!
   b. What is the probability that a student selected at random from this survey is involved with sports?
   c. What is the probability that a student selected at random from this survey is involved with music and theatre, but not sports? (You may have to think about this one a bit.)
   d. What is the probability that a student selected at random from this survey is not involved in any of these three activities?
   e. What is the probability that a student selected at random from this survey is involved with sports or music?
   f. What is the probability that a student selected at random from this survey is involved in music given that they are involved in theatre?

CHAPTER 6

Counting Techniques

6.1 THE MULTIPLICATION PRINCIPLE OR THE FUNDAMENTAL COUNTING PRINCIPLE
6.2 PERMUTATIONS
6.3 COMBINATIONS
6.4 MIXED COUNTING PROBLEMS

You are probably wondering why a college course in statistics needs a chapter on counting. After all, some college statistics courses do not include a chapter on counting, so why should we bother? We have a couple of reasons for including counting techniques. First of all, although some college courses do not cover counting techniques, it is very prevalent in standards-based middle school curricula and our goal is to connect to what you will be teaching in the middle school. Secondly, back in Chapter 5 we began a discussion of simple probability. As it turns out, being able to properly count all the outcomes in a sample space without driving oneself nuts in the process is a handy thing to know!

Classroom Exploration 6.1

Take a moment and read the letter to Dr. Math in Figure 6.0.1. It is from *Mathscape: Looking Behind the Numbers* (page 43).

1. Define what you think the owner of the Unique Diner means by a "unique combination." Can you think of an "un-unique combination?"
2. Describe how you would go about solving the owner's dilemma. It might be helpful to think about how you would go about creating a smaller number of lunch specials first—say 12.

Writing

10. Answer the letter to Dr. Math.

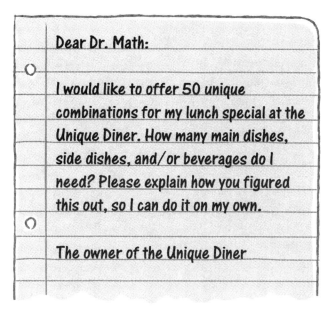

FIGURE 6.0.1 The Unique Diner

Most people learn to count at least to 10 before they ever start kindergarten. So what is the big deal about counting? Students in elementary statistics courses inevitably get confused between questions that ask "How many ways can you ...?" versus "What is the probability of ...?" Whenever the question asks "How many ways ...?" the solution MUST be a whole number!! Whoever heard of being able to do something four and a half ways?! Be careful though, sometimes your calculator may give you an answer that does not look like a whole number even though it is. When the question asks, "What is the probability of ...?" the solution must be between 0 and 1. Why? Remember that probability is always bounded between 0 and 1, so a negative answer or an answer bigger than one simply doesn't make sense in a probability context.

There are essentially three basic ways that we count in statistics. The three basic counting techniques we will explore are the multiplication of choices, permutations, and combinations. Sometimes those three ways are mixed together to create more sophisticated counting and probability problems like those found at the end of this chapter.

154 Chapter 6 Counting Techniques

6.1 THE MULTIPLICATION PRINCIPLE OR THE FUNDAMENTAL COUNTING PRINCIPLE

Back in elementary school, once you had a pretty good handle on counting at least to 100, you probably explored several different models of multiplication. There are area models and array models and all sorts of models for multiplying two numbers together. One such model that you might have studied was called the *Cartesian Product* model. Since it's been a while since elementary school, let's look at a simple example to help recall how it goes.

Classroom Connection

In the picture shown in Figure 6.1.1 from *Mathematics in Context: Take a Chance* (page 28), Robert has three shirts and two pairs of pants. The students are directed to find all the outfits that Robert can wear to school. Elementary school children

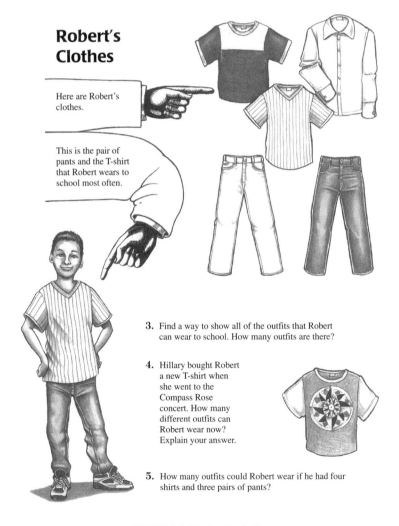

FIGURE 6.1.1 Robert's clothes

Section 6.1   The Multiplication Principle or the Fundamental Counting Principle   155

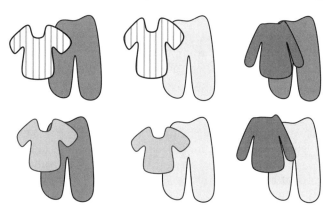

FIGURE 6.1.2 Matching Robert's shirts and pants

might actually use cutouts of the three shirts and the two pairs of pants to construct their solution to the problem. Pants and shirts would be matched until all possible pairs had been discovered. A picture of their solution might look something like Figure 6.1.2. ◆

Physically matching pants with shirts is one way to answer the question of how many outfits, though it gets a little cumbersome if you have a lot of shirts and a lot of pants. In this case, we can see that there are six possible outfits consisting of one shirt and one pair of pants.

Others students may match shirts and pants by drawing arrows, as in Figure 6.1.3. Drawing arrows has the advantage that your finished picture looks

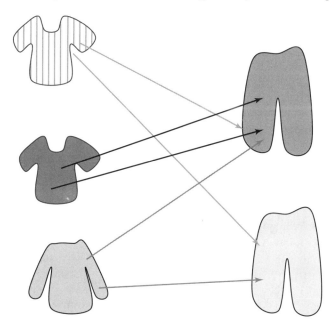

FIGURE 6.1.3 Arrows matching shirts and pants

a little less "messy" as it has fewer items in the picture. We have color coded the lines. Doing so would make it easier for elementary students to identify which items have been matched with what. Notice that once again we get six possible outfits consisting of one shirt and one pair of pants. Let's see. We have three shirts and two pairs of pants. We have generated two different ways of looking at the problem, but we got six possible outfits both times. A conjecture must be forming!

Yet another way to approach the problem is to draw a **counting tree** or a **tree diagram**. We discussed this technique back in Chapter 5, so hopefully you remember how it goes! We will let the first branch of our tree diagram represent the shirts and the second branches represent the pants. The resulting tree diagram is shown in Figure 6.1.4. Once again we get six possible outfits using a tree. We hope you weren't really surprised.

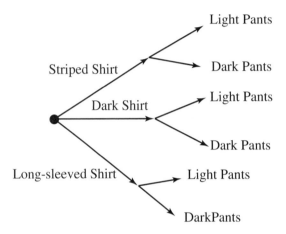

FIGURE 6.1.4 Tree diagram of shirt and pants outfits

Okay, so what does this have to do with Cartesian products or with counting for that matter? Well, notice that we had three different shirts and two different pairs of pants. Our total number of possible outcomes happened to be six no matter which way we approached the problem. Coincidence? Not likely! To find the Cartesian product back in elementary school, you ended up taking the product of the two factors (3 and 2) to get the total number of ways one could pick one choice from one category and one choice from another. In this case $3 \times 2 = 6$, which agrees with the three ways we solved the problem. In college-level statistics, we call the good old Cartesian product that you learned in elementary school the **Multiplication of Choices Principle**. Fancier name, same idea. In middle school it is often referred to as the **Fundamental Counting Principle**.

Why do we need a principle for counting that just involves multiplying the number of choices for one thing together with the number of choices for another when we can clearly solve the problem in at least three other ways? Good question. The answer is that there are times when tree diagrams and drawing diagrams are not

Section 6.1   The Multiplication Principle or the Fundamental Counting Principle   157

the most terribly efficient ways (i.e., quickest) to find how many possible outcomes there are for some sample space.

What if there are more than two groups from which to choose? The same principle applies. Simply take the number of choices for the first group multiplied by the number of choices for the second group multiplied by the number of choices for the third group and so on.

> **Definition 6.1.1 The Multiplication of Choices or the Fundamental Counting Principle**
>
> If there are m possibilities for the first choice, n possibilities for the second choice, p possibilities for the third choice, and so on, then the total number of possible outcomes by selecting one from each group is given by $m \times n \times p$ and so on.

Classroom Connection

In the scenario shown in Figure 6.1.5 from *Connected Mathematics: Clever Counting* (page 7), a robbery has occurred and the witness has been asked to work with a police sketch artist to develop a composite photo of the robber. The witness has to make choices from three different categories in this problem situation.

A tree diagram for this situation would simply be too unruly. Making cut-outs and matching sets of three doesn't sound like much fun either. This is a good time to apply the Multiplication of Choices Principle!   ◆

Focus on Understanding

1. Answer the question in Part A in Figure 6.1.5. Does it matter what order you pick the attributes? In other words, could you pick eyes, then the nose, and then hair rather than hair, eyes, nose? Why or why not?
2. Answer the question in Part B in Figure 6.1.5. How would the answer change if there were five choices for mouth rather than three?
3. Sometimes the suspect has distinguishing marks such as tattoos, moles, scars, and so on. How many possible sketches are there if the witness can choose from tattoos, moles, scars, or no distinguishing marks in addition to the four other attributes described in the problem?
4. In the Problem 1.1 Follow-Up, middle school students are asked to determine how many of the facial descriptions have droopy eyes. What kind of strategy do you think they would use to answer the question? Is there more than one way to answer the question?
5. How is the second question in the Problem 1.1 Follow-Up different from the first? How would you determine the number of facial descriptions that have a bald head and a thin mean mouth? How is it possible to use the Multiplication of Choices Principle to answer this question?

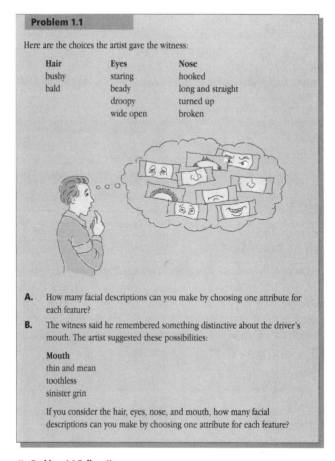

FIGURE 6.1.5 Making faces

Classroom Connection

Standards-based curricula generally introduce the Fundamental Counting Principle through situations in which a tree diagram is first used to determine the number of ways in which something can be done. Take a look at the following example from *Mathscape: Looking Behind the Numbers* (pages 28 and 29), shown in Figure 6.1.6.

In the introduction to the lesson, students are asked to count the bottom branches of the tree diagram to see how many different lunch combinations there are. In the final question of this section, the Fundamental Counting Principle is introduced. ◆

Section 6.1   The Multiplication Principle or the Fundamental Counting Principle   159

Focus on Understanding

1. In some situations, the Fundamental Counting Principle provides the same information as counting the branches of a tree diagram. Why is this so?
2. Draw two tree diagrams, based on the Unique Diner's Lunch Special in Figure 6.1.6. Let one diagram have "Main Dish Choice" at the top, and let the other have "Beverage Choice" at the top. What is the difference between the two diagrams?
3. Exploration 6.1.1 at the beginning of this chapter asked you to help Dr. Math answer a letter from the owner of the Unique Diner. Please go back and reread the letter and then answer the following questions:

   a. How many unique solutions do you think there are to the problem posed? Explain your reasoning.
   b. How many unique solutions are there if the owner of the Unique Diner wanted to create 54 unique lunch combinations instead of 50? Are there more solutions or fewer solutions? Why do you think that is? (Hint: How many factors or divisors are there of 54 versus 50?)

FIGURE 6.1.6 Lunch specials

160 Chapter 6 Counting Techniques

Similar types of problems to Exploration 6.1.1 are found in other middle school curricula. The exercise in Figure 6.1.7 is from page 53 of *Connected Mathematics: Clever Counting*.

This type of problem is a nice way to work writing into a middle school classroom. It will also give you, as a teacher, a good idea of whether or not your students understand the ideas behind the Fundamental Counting Principle.

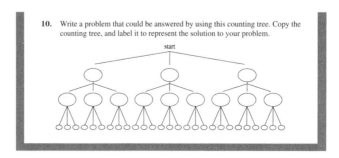

FIGURE 6.1.7 Creating a new counting problem from a tree diagram

6.2 PERMUTATIONS

We think it only fair to warn you that these next two sections have very similar sounding names and tend to be confusing to many students. Section 6.3 will explore combinations, while this section will focus on permutations. Although these two counting techniques sound similar and can appear similar at times, they are different in very special ways.

Permutations count the total number of different arrangements you can make from a given number of items. **The order of the items in the arrangements matters!** This means that an arrangement of blue-green-red is very different from green-red-blue.

It is perhaps easiest to think of permutations to begin with as a special case of the Multiplication of Choices Principle. After all, we understand that method pretty well now. (Or so we hope!) What is special about permutations? Let's look at an example and see.

EXAMPLE 6.2.1 **Nick's Car Rally**

Nick has seven miniature cars. The cars are blue, green, red, purple, white, gold, and silver. How many ways can he line his cars up to make a parade for his dad?

At first glance this problem seems like it would be an ideal problem to give elementary or middle school children to model either with actual miniature cars or with colored blocks. That method would work, but it soon becomes a major hassle trying to keep track of all the different outcomes. It might also seem that this problem has absolutely nothing to do with multiplication of choices, since in the multiplication of choices we always had at least two groups from which we were choosing selections. Here we only have one group of seven cars. How can that possibly be related to the previous section? The connection is that we are going to pick one car at a time from the group and then go back and pick from the group again!

In teaching statistics for years and years, we think it helps to think of what we call a "slot" model. Nick wants to arrange or order all seven of his cars in this parade. So we begin with seven "slots."

_____ _____ _____ _____ _____ _____ _____

The first slot represents all the ways he can pick the first car. The second slot represents all the ways he can pick the second car. We continue picking cars until we have them all arranged for the parade.

The first time when Nick goes to choose a car, he has seven from which to pick so we record a 7 in the first slot. There are seven ways to pick the first car.

$\dfrac{7}{1^{st}}$ $\dfrac{}{2^{nd}}$ $\dfrac{}{3^{rd}}$ $\dfrac{}{4^{th}}$ $\dfrac{}{5^{th}}$ $\dfrac{}{6^{th}}$ $\dfrac{}{7^{th}}$

Now we go back to the car lot to pick the second car! Nick only has six cars remaining from which he can pick. So we record a 6 in the second slot.

Key Idea!! Nick is picking the cars **without replacement**. This means that once he selects a car, it is gone from the pool from which he can choose.

$\dfrac{7}{1^{st}}$ $\dfrac{6}{2^{nd}}$ $\dfrac{}{3^{rd}}$ $\dfrac{}{4^{th}}$ $\dfrac{}{5^{th}}$ $\dfrac{}{6^{th}}$ $\dfrac{}{7^{th}}$

For the third position in the parade, Nick now has five cars from which to choose. You get the idea, right? Our slot model ends up looking like this:

$\dfrac{7}{1^{st}}$ $\dfrac{6}{2^{nd}}$ $\dfrac{5}{3^{rd}}$ $\dfrac{4}{4^{th}}$ $\dfrac{3}{5^{th}}$ $\dfrac{2}{6^{th}}$ $\dfrac{1}{7^{th}}$

Okay, but what does this have to do with the Fundamental Counting Principle? Well, Nick has 7 choices for the first position, 6 choices for the second position, 5 choices for the third position, and so on. Using the multiplication of choices principle he has $7 \times 6 \times 5 \times 4 \times 3 \times 2 \times 1$ or 5,040 ways he can arrange his cards for the parade. ∎

Thank goodness Nick only had seven cars to arrange. If he had 36 cars, the multiplication would have gotten a bit tedious! And thus, we introduce the need for some kind of mathematical notation to deal with arranging large numbers of items from a large group.

There are times in mathematics when we need to multiply a whole string of descending numbers (like 10, 9, 8, 7, 6, 5, 4, 3, 2, 1) together. Counting using permutations is one such time. To indicate a string of descending numbers all multiplied together from the first number down to 1, we use what is called **factorial notation**. The factorial symbol looks exactly the same as an exclamation point. In fact, it is an exclamation point. It just means something different in mathematics. So 6! means $6 \times 5 \times 4 \times 3 \times 2 \times 1$ or 720 and 56! means $56 \times 55 \times 54 \times 53 \times 52 \times \ldots \times 3 \times 2 \times 1$ or ???. By this point in time you are probably saying to yourself that there must be an easier way to do this rather than multiplying a string of 56 numbers out on your calculator. Yes, indeed there is!! Notice that we cagily did not give you the numerical result of 56 factorial (56!). We are going to get the calculator to compute this for us and save us the trouble of all that keystroking. We are using the TI-83 plus for this, but other calculators have similar keys.

To find 56!, first type 56 on the home screen. Next, go to the **MATH** menu. You should see something like this on your screen:

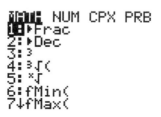

We are interested in the PRB menu, so use the arrow keys (left or right, it doesn't matter) to toggle over to PRB (short for probability). Once PRB is highlighted you should see the menu that corresponds to PRB. Notice that the fourth choice on the list is "!". That's the one we want. Either use the down arrow key to toggle to choice 4 or simply type 4, it doesn't matter. Your home screen should now be displaying 56!, so press the ENTER key. Something amazing happens. Actually, two amazing things happen. First of all, your calculator multiplies 56 times 55 times etc... all the way down to 1 for you. That's pretty amazing. Secondly, it probably appears as if it has gone nuts because the answer looks strangely like a decimal. Remember at the beginning of this chapter we said that the solution to a question that begins "How many ways can you ...?" is always, without fail, a whole number. So what's the deal? In order to handle extremely large or extremely small numbers your calculator will automatically go into scientific notation. The number you see on your screen is a really, really big whole number written in scientific notation. You should see something like this:

7.109985878E74

The number before the "E" represents the decimal portion of scientific notation, while the number following the "E" indicates the power of 10. So, 7.109985878E74 really means $7.109985878 \times 10^{74}$. If we were to take this number out of scientific notation we would "move the decimal point" 74 places to the right! That is a really, really big whole number. Do you have to do that? No, but please be kind to your instructor and write your numbers in real scientific notation $7.109985878 \times 10^{74}$ and not the calculator-speak way of 7.109985878E74. Why? If you write your answers using real scientific notation your instructor will have a better idea if you actually understand what the calculator is telling you than if you just blindly copy down what your home screen says. So is $7.109985878 \times 10^{74}$ a whole number? You betcha! It's a huge whole number!

Definition 6.2.1 Factorials

The number of different arrangements of a set or group of *n* objects is ***n* factorial** (written n!), where

$$n! = n \cdot (n-1) \cdot (n-2) \cdot (n-3) \cdot \ldots \cdot 3 \cdot 2 \cdot 1$$

This definition holds when an outcome is only allowed to occur once, like in our example with Nick's cars. As mentioned earlier, this idea is referred to as an experiment **without replacement**. If the same object can be reused multiple times, then the experiment is done **with replacement**. We will come back to the idea of with replacement in a bit.

Back to permutations. Try 7! using the factorial key on your calculator. You should get the same answer as we did by multiplying it out the long way—5,040. Anytime you need to know how many ways to arrange **all** the members of a particular group you can either multiply each of the numbers individually using the multiplication key or you can use the factorial key. The choice is yours.

Classroom Connection

Figure 6.2.1 from *Mathscape: Looking Behind the Numbers* (page 30) shows three different itineraries for a band. While all three itineraries contain the same three cities, the order is different in each case. ◆

Focus on Understanding

In the *Classroom Connection* shown in Figure 6.2.1, students are asked to conduct an experiment to find out how many times their band is selected to perform in a particular order assuming there are only four bands competing. To make things easier, call the bands A, B, C, and D.

1. How many ways can the four bands perform?
2. In the questions for the students in this *Classroom Connection*, they were asked how many times their band performed first, second, third, or fourth. Out of all the ways you found for the four bands to perform, how many of those do you think had Band A performing first? Explain how you know.
3. List all the possible ways (the sample space) for the four bands to perform. Is this a difficult task or a hard task? Why?
4. How many ways out of all the possible ways you found will Band D perform last?
5. Draw a tree diagram to represent all the possible ways that Band B can perform first. Would tree diagrams for the other bands performing first look similar or different? Why?

164 Chapter 6   Counting Techniques

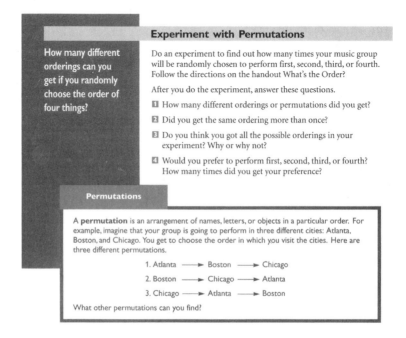

FIGURE 6.2.1 The Battle of the Bands

As students progress through this lesson from *Mathscape*, they are asked to complete a tree diagram to find the number of different permutations for the four musical groups. By observing the patterns that occur among the branches of the tree diagrams and using the Fundamental Counting Principle, students can develop a formula much like we did in Example 6.2.1.

There are times when we may not want to arrange all the objects in a set or group of objects. We don't have to abandon what we've developed so far, we just need to adjust it a bit. The same ideas and principles will still be applied!

EXAMPLE 6.2.2 **Nick's Car Rally, Part Two**

Okay, so what if Nick only wanted to pick five cars from his group of seven to be in his parade rather than arranging all seven? Excellent question! Using our tried and true "slot" model again we see that Nick has seven choices for this first car, six for the second, five for the third, four for the fourth, and three for the fifth and final position in the parade.

$$\frac{7}{1^{st}} \quad \frac{6}{2^{nd}} \quad \frac{5}{3^{rd}} \quad \frac{4}{4^{th}} \quad \frac{3}{5^{th}}$$

Once again we can use the Fundamental Counting Principle to multiply our choices at each position together. This time we get $7 \times 6 \times 5 \times 4 \times 3 = 2{,}520$ ways to arrange five cars out of seven in a parade. You are probably thinking that there must be an easier way to do this using the factorial key again or using the definition we developed earlier. Let's see if we can think our way through this. ■

In our original problem we arranged all the cars to get 7! or 5,040. Now we arranged only five of the original seven, so our answer must be smaller than 7!. In fact, we don't need to multiply by the 2 or the 1 representing the sixth and seventh positions. Consequently, if we take 7! and divide by 2 times 1 (which is just 2) we should get 2,520. In a sense, we divided out the number of ways we could fill the unused "slots" in our problem.

Will this always work? Let's try another situation and see. Suppose Nick wants to arrange four of his seven cars in the parade. Using our original model we know that he can do this $7 \times 6 \times 5 \times 4$ ways or 840 ways. Using 7! we would have to divide out the number of ways to fill the fifth, sixth, and seventh positions, which would be $3 \times 2 \times 1$ or 6. Taking 7! and dividing by 6 we get 840 once again. Amazing, isn't it? It probably crossed your mind that the part we are dividing by is beginning to look suspiciously like a factorial itself. You're right, it is. In this last instance we divided by 3!. In the previous situation we divided by 2!. Can we make a generalization? Of course!

Definition 6.2.2 General Permutations

The number of **permutations** of size k that can be formed from a set of n distinct objects is denoted by $_nP_k$, and can be calculated using:

$$_nP_k = \frac{n!}{(n-k)!}$$

When Nick wanted to arrange five out of seven cars we really took:

$$_7P_5 = \frac{7!}{(7-5)!} = \frac{7!}{2!} = \frac{5{,}040}{2} = 2{,}520$$

Remember a couple of pages or so back when we introduced you to the PRB menu on a graphing calculator to use the factorial key? There were several other choices under that same menu. One of them, in fact, was a permutation of k objects taken from a set of n objects. Return to the PRB menu by pressing MATH. Use the arrow key to toggle over to the PRB menu. On a TI-83 plus, the second choice reads nPr, which is the same formula as nPk. To use this key you must first enter in the total number of objects, n, then select nPr from the PRB menu, enter the number of objects to be selected, k, and then press ENTER. Using our example of Nick's Car

166 Chapter 6   Counting Techniques

Opening Locks

The manager of Fail-Safe suspects that the security guard stole Rodney's CD players. He suggested to Detective Curious that the guard had tried different combinations until the lock opened. The detective wondered how long it would take to try every possible combination.

 Pushing Buttons

Some of the lockers at Fail-Safe have push-button locks that consist of five lettered buttons.

To open a push-button lock, the letters must be pressed in the correct sequence.

> **Problem 2.1**
>
> A lock sequence may have two, three, four, or five letters. A letter may not occur more than once in a sequence.
> **A.** How many two-letter sequences are possible?
> **B.** How many three-letter sequences are possible?
> **C.** How many four-letter sequences are possible?
> **D.** How many five-letter sequences are possible?
> **E.** The security guard would not have known whether the sequence that would open Rodney's lock consisted of two, three, four, or five letters. How many possible lock sequences might she have had to try?

FIGURE 6.2.2 Opening locks

Rally, Part Two, we would enter 7, select *nPr* from the PRB menu, enter 5, and then press ENTER. Once again we get 2,520. Like the Fundamental Counting Principle from 6.1, there is more than one way to arrive at the same answer. The choice is up to you and which method best fits the problem at hand!

Classroom Connection

The Classroom Connection shown in Figure 6.2.2 is from *Connected Mathematics: Clever Counting* (page 15). This example highlights one of the reasons that students sometimes get confused between permutations and combinations. Even though the problem refers to a "*combination*" lock and the number of "*combinations*," it is really a permutations problem. Ask any middle school student who has ever struggled with their locker and they will tell you that order matters!! ◆

Focus on Understanding

After reading through the excerpt from *Connected Mathematics* shown in Figure 6.2.2:

1. Answer questions A through D using the Fundamental Counting Principle.
2. Answer questions A through D using Definition 6.2.2.
3. Which of the questions A through D are easier to answer using the Fundamental Counting Principle? Which are easier to answer using Definition 6.2.2?
4. Suppose that instead of 5 letters to work with the guard had to use all 26 letters of the alphabet. Do you think one method might be easier than another in this case? Why or why not?
5. Answer questions A through D using all 26 letters of the alphabet instead of just the 5 shown in the original problem. Use whichever method you prefer.

While middle school curricula do contain permutations (and combinations), they typically DO NOT use Definition 6.2.2 to solve permutation problems. Students in the middle school are generally asked to make lists, tree diagrams, or use the Fundamental Counting Principle to solve such problems. As a college student, however, you should be familiar with the definition and the notation associated with the definition.

The *Classroom Connection* you just explored was another example of an experiment **without replacement**. Once a letter was pressed on the lock, it could not be pressed again. We promised that we would come back to experiments **with replacement**. Suppose that in the *Battle of the Bands* Classroom Connection (Figure 6.2.1) from *Mathscape*, each band could play more than once. This situation would be an example of an experiment with replacement. The same would be true if we allowed the guard in *Opening Locks* (Figure 6.2.2) from *Connected Mathematics* to press each letter more than once.

Focus on Understanding

1. Develop a formula for the number of different arrangements of a set of n objects with replacement.
2. Which number do you think is larger, the number of arrangements of a set of objects with replacement, or the number of objects of the same set without replacement? Explain your reasoning.

3. Answer questions A through D from Figure 6.2.2 if the guard is allowed to press each letter more than once. Are your answers larger (or smaller) than your previous answer when the guard was not allowed to press the same letter more than once?

Again, most middle school curricula do not worry about finding or using a formula to deal with permutation problems involving "with replacement" situations. In reality, these types of situations are really just versions of the Fundamental Counting Principle.

6.3 COMBINATIONS

Unlike permutations, the order of the items taken from a set of objects does not matter for **combinations**. Blue-green-red and green-red-blue are not considered different outcomes when using combinations. For example, in the lesson from *MathThematics: Book 2* (page 568) shown in Figure 6.3.1, notice that some permutations have been crossed out on the tree diagram. The permutations that are crossed out simply provide different arrangements of items, not any new items in the group.

Listing all the possible permutations for 4 horses grouped in pairs is messy even with a tree diagram. We know from our experience with permutations in Section 6.2 that the number of permutations of 2 horses selected from a group of 4 horses has to be $4 \cdot 3 = 12$ or $_4P_2 = 12$. Since the order that the horses were selected in doesn't matter, six arrangements were crossed off the list. Is there an easier way to find the combinations without listing all of the permutations first? As **Exploration 2** from *MathThematics: Book 2* (page 557) continues in Figure 6.3.2, it suggests that students make a more refined list.

Notice that Wendell and Judith have started an organized list beginning by matching Elmer with each of the other 9 horses. Remember that with combinations the group {Elmer, Winston} is the same as the group {Winston, Elmer}—order doesn't matter. To start the list of horses we can match with Winston, we can skip Elmer because the group {Elmer, Winston} has already been counted.

Focus on Understanding

1. Complete the list of horses that can be paired with Winston in Figure 6.3.2. How many new pairs did you form?
2. Continue the list for the 8 remaining horses. Do you notice a pattern in the number of new pairs that are formed each time? Explain the pattern you see.
3. How many pairs (combinations) of 2 horses did you find altogether using 10 different horses? Is there an easy way to add up all of those numbers?
4. Suppose there were 16 horses instead of 10. Make a conjecture about how many combinations of 2 horses can be formed. Explain. (If you feel like you do not have enough information to make a conjecture, start an organized

Section 6.3 Combinations 169

GOAL

LEARN HOW TO...
* list and find all the possible combinations
* distinguish a combination from a permutation

As You...
* consider ways to group

KEY TERM
* combination

Exploration 2

Combinations

Wendell and Judith plan to free ten horses: Elmer, Winston, Molly, Archibald, Quentin, Angelina, Samantha, Cromwell, Chloe, and Old Barnaby. Each name can be represented by its first two letters: El, Wi, Mo, Ar, Qu, An, Sa, Cr, Ch, and Ol.

9 a. Which of the following selections represents the same group of ten horses listed above?

Selection 1:	Wi, Mo, Qu, An, Ol, Ch, Cr, Sa, El, Ar
Selection 2:	Mo, Qu, An, Sa, Cr, Ch, Ol, Ch, Mo, Ar
Selection 3:	Sa, Ch, Ol, Ar, Mo, El, Qu, Wi, Cr

b. Does changing the order of the horses change the group itself?

▶ A selection of items in which order is not important is called a **combination.** One way to find all possible combinations in a situation is to first list all the possible permutations.

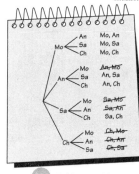

10 The diagram show all the ways Wendell and Judith could have paired up the mares Molly (Mo), Angelina (An), Samantha (Sa), and Chloe (Ch).
   a. How many permutations of two of these four mares are there?
   b. Discussion Why are some of the permutations crossed out?
   c. How many different combinations are possible in this situation?

11 Try This as a Class How is a combination different from a permutation?

556 Module 8 Heart of the City

FIGURE 6.3.1 Combinations of horses

▶ Sometimes it is not practical to list all the possible permutations and then cross out repeats to find the combinations. Instead, you can make an organized list just of the combinations.

12 Suppose Judith and Wendell want to divide all ten horses into groups of two and lead each group on a different route to the Park. Copy and extend the table as you answer the questions.

First horse picked	Possible second horses to pick	Combinations of two horses	Number of combinations
El	Wi Mo Ar Qu An Sa Cr Ch Ol	El, Wi El, Mo El, Ar El, Qu El, An El, Sa El, Cr El, Ch El, Ol	?
Wi	?	?	?

FIGURE 6.3.2 Combinations of horses (*Continued*)

list again! We're betting you don't have to go very far before you notice a pattern again.)

5. Suppose there were 5 horses instead of 10 and this time you want combinations of 3 horses rather than 2. Do you think your conjecture from Question 4 will still work? Check it out by making an organized list of all the possible combinations of 3 horses from a group of 5.

Looking for patterns is an excellent way to form conjectures and generalizations about ways of attacking all types of problems. In the preceding *Focus on Understanding*, you probably had a pretty nice pattern going as long as you were just pairing horses up. For example, if there were 6 horses we're guessing (hoping, actually) you would tell us that there would be $5 + 4 + 3 + 2 + 1 = 15$ different **paired** combinations. When you had to list combinations of 3 horses, however, the pattern was probably not as clear, though there is a pattern. Take a look at the following example.

EXAMPLE 6.3.1 Six students—Amy, Bill, Cathy, David, Eliza, and Frank—are going to a math contest. Since none of them can drive, two parents volunteered to take three students in each of their cars. How many combinations of three students can be made from this group of students?

Making an organized list is still a great strategy to use, but a tree diagram would work as well. Notice that we started with Amy and did all of the combinations with Bill as the second student before moving on to the next student in our list.

First Student	Second Student	Third Student	Combination	Number of Combinations
Amy	Bill	Cathy	ABC	4 using Amy, Bill
	Bill	David	ABD	
	Bill	Eliza	ABE	
	Bill	Frank	ABF	
Amy	Cathy	David	ACD	3 using Amy, Cathy
	Cathy	Eliza	ACE	
	Cathy	Frank	ACF	
Amy	David	Eliza	ADE	2 using Amy, David
	David	Frank	ADF	
Amy	Eliza	Frank	AEF	1 using Amy, Eliza
Bill	Cathy	David	BCD	3 using Bill, Cathy
	Cathy	Eliza	BCE	
	Cathy	Frank	BCF	
Bill	David	Eliza	BDE	2 using Bill, David
	David	Frank	BDF	

First Student	Second Student	Third Student	Combination	Number of Combinations
Bill	Eliza	Frank	BEF	1 using Bill, Eliza
Cathy	David	Eliza	CDE	2 using Cathy, David
	David	Frank	CDF	
Cathy	Eliza	Frank	CEF	1 using Cathy, Eliza
David	Eliza	Frank	DEF	1 using David, Eliza

So altogether we end up with 20 possible combinations of three students from our group of six students. That's a lot to list, but there is a pattern! The whole point of looking for patterns is to aid in making generalizations and, from a generalization, discovering a shortcut to find the number of combinations without listing them all. In middle school curricula, students generally stick to making lists and tree diagrams when finding all the possible combinations. Do not despair. The patterns you explored in the previous paragraphs give rise to the next definition. ■

Definition 6.3.1 Combinations

The number of ways to form **combinations** of size k from a set of n distinct objects is denoted by $_nC_k$, which can be calculated using:

$$_nC_k = \frac{n!}{k!(n-k)!}$$

You may see some textbooks use an alternate notation $\binom{n}{k}$ for combinations instead of $_nC_k$. These two notations mean exactly the same thing. **Combinations** are used for experiments done without replacement where **order does not matter**.

When working with combinations, we essentially need some handy way of getting rid of the groups that have the same members, just in a different order, since for combinations order does not matter. Remember that ABC is the same as CAB in combinations.

Why the extra factor in the denominator for combinations if we are choosing k items from a set of n objects? Essentially, we are dividing out the number of "repeats" with the $k!$ factor. If the order does not matter, there will be $k!$ groups which have exactly the same objects in them. Since we don't want to count groups that are essentially the same, only the ones with different objects, we divide out the number of repeated groups, which is $k!$.

Just like you used your calculator to compute $_nP_r$ for permutations, we can find $_nC_r$ for combinations as well. We will use the calculator to check to see that we really found all of the possible combinations in Example 6.3.1. To find $_nP_r$ we had to use the MATH menu, and the same is true for $_nC_r$. Remember that we have to tell the calculator how many total items we have to choose from first, so enter 6 as it is the *n* in this calculation. Now, press the MATH key, toggle over to PRB on the MATH menu, arrow down to choice 3, which is $_nC_r$, and press ENTER. Your home screen should now look like:

$$6 \text{ nCr } \blacksquare$$

The cursor should be flashing behind this as it is waiting for you to tell it how many objects you want to choose from six. Type "3", press ENTER again, and it should tell you there are 20 ways to choose three objects from a collection of six objects. This is much faster than listing all 20 choices as we did in Example 6.3.1. However, there are advantages to making a list. For example, if we wanted to know how many of our groups of size three contained Frank, we could very easily go through the list and count all the groups with Frank. Just performing the computation on the calculator does not give us this kind of information.

Focus on Understanding

1. How does the combination formula in Definition 6.3.1 relate to the permutation formula in Definition 6.2.2?
2. How many ways can a group of 4 objects be selected from a group of 12 if order does not matter? If order matters, how many ways can the 4 objects be selected? (Use your calculator to find these values.)
3. How many ways can a group of 12 objects be selected from a group of 30 objects if order does not matter? If order matters? (Use your calculator to find these values.)
4. Which will be larger for a given set of *n* objects: the number of permutations of size *k*, or the number of combinations of size *k*? Why?
5. Which number is larger: the number of ways to choose a committee of 2 from a group of 7 or a committee of 5 from a group of 7? (Use your calculator to find these values.)
6. Which number is larger: the number of ways to choose a committee of 3 from a group of 8 or the number of ways to choose 5 from a group of 8? (Use your calculator to find these values.)
7. Which number is larger: the number of ways to choose a committee of 6 from a group of 10 or the number of ways to choose 4? (Use your calculator to find these values.)
8. What did you notice about your answers to parts 3, 4, and 5? Explain what you think is happening. (Hint: Is $\frac{n!}{k!(n-k)!}$ the same as $\frac{n!}{(n-k)!k!}$? Why?)

Again, permutations and combinations are very confusing to some students, as you can see in Figure 6.3.3 from *Mathscape: Looking Behind the Numbers* (page 45).

Writing

16. Answer the letter to Dr. Math.

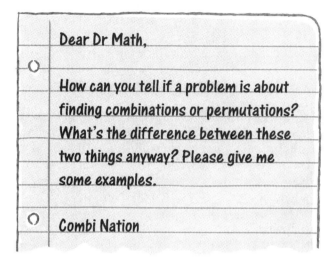

FIGURE 6.3.3 Confused student

Focus on Understanding

1. Write a response to Combi Nation using examples to illustrate your explanation.
2. Write a brief description of the differences and similarities among the three counting techniques you have studied thus far: The Fundamental Counting Principle, Permutations, and Combinations.

6.4 MIXED COUNTING PROBLEMS

Sometimes counting problems require the use of more than one of the three basic techniques in order to determine how many ways there are to do something. Let's look at some classic examples of these types of problems.

EXAMPLE 6.4.1 **The Deli Count-er**

The local deli offers a lunch special every Wednesday. For $2.99 you can get your choice of bread, filling, and cheese. You may only select one from each group. Choices are displayed in the table below. How many possible sandwiches could you make at this deli?

174   Chapter 6   Counting Techniques

Bread	Filling	Cheese
White	Turkey	Provolone
Wheat	Tuna	Mozzarella
Rye	Meatballs	Cheddar
Oat Nut	Salami	Swiss
Pumpernickel	Chicken	American
Sourdough	Steak	
	Ham	

    This problem should look very familiar. Since we only get to pick one of each—bread, filling, and cheese—we can directly apply the Fundamental Counting Principle. The number of possible sandwiches would be 6 × 7 × 5 = 210 different sandwiches. No big deal.

    Suppose, however, that instead of allowing only one choice of filling, the deli lets the customer choose two different fillings for their sandwich. Now how many possible sandwiches are there?

    It seems like this problem should be very similar to the one we just solved. You are probably guessing that there are more than 210 ways, since we now get to pick two different kinds of fillings. Since we only get one choice of bread, we know that the number of ways to pick the bread is still 6. By the same reasoning, our choices for cheese remain at 5. The only thing that has changed in the problem is that we now get to choose 2 fillings from a list of 7. Does the order of the fillings matter? No, so this is a job for combinations. The number of ways we can choose 2 fillings from a list of 7 is $_7C_2 = 21$ different combinations of 2 fillings. How many possible sandwiches are there? Using the Fundamental Counting Principle again we get:

Number of sandwiches = Number of ways to choose the bread

                   × Number of ways to choose the fillings

                   × Number of ways to choose the cheese

or

                  Number of sandwiches = 6 × 21 × 5 = 630

That's a lot of different sandwiches!   ∎

    The basic idea is that if you get more than one choice in a situation like this you can use permutations (if the order matters) or combinations (if the order doesn't matter) to find that subpart of the overall counting problem. In a sense, you are using the problem-solving strategy of making a simpler problem before addressing the larger question.

    Let's extend this problem further. How many possible sandwiches can be made if the customer is allowed to choose 1 bread, 3 fillings, and 2 cheeses? We already know that the number of ways to choose the bread is still just 6. The number of ways to choose the fillings would be $_7C_3 = 35$. The number of ways to choose the cheeses would be $_5C_2 = 10$. We've used combinations to find the number of ways to choose particular items like fillings and cheese and now we will apply the Fundamental Counting Principle to determine the total number of possible sandwiches:

$$\text{Number of possible sandwiches} = 6 \times 35 \times 10 = 2100$$

If a customer was really picky and insisted that the order the fillings and cheeses went on the sandwich were important, then permutations would be used. Most of the time though in problems like this the order does not matter, so combinations are used.

EXAMPLE 6.4.2 **The Caterer**

A large school district is planning on having an end-of-the-year party in appreciation for all the hard work the teachers have done. The local caterer offers 10 different types of desserts. The committee planning the party would like to offer people a choice of more than one dessert, so they decide to pick 5 desserts from the caterer's list of 10.

Question 1: How many ways can the committee choose five desserts from the list of 10?

This question should be very straightforward. There are 10 desserts, they get to pick 5, order doesn't matter, so it is combinations. $_{10}C_5 = 252$ ways to choose 5 desserts from the list of 10.

Suppose the caterer's list of 10 desserts is split into chocolate desserts and non-chocolate desserts. There are 6 chocolate desserts and 4 non-chocolate desserts on the list.

Question 2: How many ways can the committee choose 5 desserts from the list of 10 and get 3 chocolate desserts and 2 non-chocolate desserts?

Aha! This is like the deli problem from Example 6.4.1 only we have chocolate and non-chocolate desserts rather than fillings and cheeses. (Plus, we don't have to worry about picking bread!) Does the order matter? No! The number of ways the committee can choose 3 chocolate desserts from a list of 6 is $_6C_3 = 20$. The number of ways they can choose 2 non-chocolate desserts from 4 is $_4C_2 = 6$. So, the total number of ways the committee could choose 5 desserts and get 3 chocolate desserts and 2 non-chocolate desserts would be $20 \times 6 = 120$.

Question 3: How many ways can the committee choose 5 desserts from the list of 10 (6 chocolate and 4 non-chocolate) and get 4 chocolate desserts and 1 non-chocolate dessert?

This question is exactly like the previous one, except the number chosen from each subgroup (chocolate and non-chocolate) is different. The number of ways to choose 4 chocolate desserts from 6 is $_6C_4 = 15$. The number of ways to choose 1 non-chocolate dessert from 4 is $_4C_1 = 4$. Once again we apply the Fundamental Counting Principle. The total number of ways the committee can choose 5 desserts and get 4 chocolate desserts and 1 non-chocolate dessert is $15 \times 4 = 60$.

Question 4: What is the *probability* that if the committee chooses 5 desserts at random, 3 will be chocolate and 2 will be non-chocolate?

To answer this question we must use what we learned about probability from Chapter 5 along with the counting techniques from Chapter 6. Recall that:

$$\text{Probability} = \frac{\text{total number of favorable ways}}{\text{total number of possible ways}}$$

The numerator in this case would be all the ways we could get the desired outcome of 3 chocolate desserts and 2 non-chocolate desserts. Do we know this? Sure, we found the number of ways for this particular outcome in Question 2. So the numerator must be 120. What about the denominator? Really what the denominator is in this case is the total number of possible ways to pick *any* 5 desserts, chocolate or not. We already know this value as well since it was the solution to Question 1. The denominator is therefore 252. The probability that the committee chooses 5 desserts at random and gets 3 chocolate and 2 non-chocolate would be:

Probability of three chocolate & two non-chocolate

$$= \frac{\text{Number of ways to get 3 chocolate \& 2 non-chocolate}}{\text{Number of ways to get 5 desserts}}$$

$$= \frac{120}{252} \qquad \blacksquare$$

Both of these ideas—counting and probability—are used in later chapters to explore sampling distributions. Any time we want to take a sample from a population we have to consider how many possible samples of a given size exist from that population and we would like to have some idea of what the probability is of getting a particular sample. Isn't it nice when ideas come together like that?

Focus on Understanding

The big fish tank at the local pet shop has 12 gold-colored fish and 8 calico-colored fish. A customer comes into the store to buy 6 of the fish from this tank.

1. Do you think the order the fish are selected from the tank matters? Why or why not?
2. How many ways can the customer get 6 gold fish if order doesn't matter?
3. How many ways can the customer get 6 calico fish if order doesn't matter?
4. How many ways can the customer get 4 gold fish and 2 calico fish if order doesn't matter?
5. How many ways can the customer get 3 of each if order doesn't matter?
6. How many ways can the customer get 6 fish regardless of their color? (Hint: If the color is unimportant then all of the fish can be treated as if they were the same color.)
7. Is this a situation with replacement or without replacement? How can you tell?

8. What is the probability that if 6 fish are selected at random 3 will be gold and 3 will be calico?
9. What is the probability that if 6 fish are selected at random 4 will be gold and 2 will be calico?
10. What is the probability that if 6 fish are selected at random all 6 will be calico?

Other types of odd counting problems exist. Another standard counting problem is the problem of arranging non-unique items. Typically, you will see this type of problem using the letters of particular words, but it could be arrangements of books on a shelf or television commercials during a two-hour movie.

EXAMPLE 6.4.3 **You Say "Potato"** How many ways can you arrange the letters in the word "potato?"

The first thing we need to decide in order to answer this question is whether or not the order matters. In other words, is "potato" different from "tapoto"? Those are clearly different words, so the order does make a difference. Since "potato" has six letters, the order matters, and since we get to use all six letters it seems that this is just an application of permutations. So, $_6P_6 = 720$ different arrangements of the letters in "potato."

Upon closer inspection, however, this doesn't quite work. The trouble is we can't tell one "o" from another or one "t" from another. Take a look.

$$\text{potato} \quad \text{potato} \quad \text{potato} \quad \text{potato}$$

The first "potato" is the original word. In the second "potato" we interchanged the two "o's." Couldn't you tell by looking? In the third "potato" we left the "o's" alone and interchanged the two "t's." Again, hard to tell, isn't it? In the last "potato" we interchanged the "o's" and the "t's." They all look exactly the same don't they? The only way to see if the "o's" and "t's" have been interchanged is to somehow distinguish one from another. In mathematics, we often use subscripts to distinguish one variable for another, so we'll use the same idea here to tell these repeated letters apart:

$$po_1t_1at_2o_2 \quad po_2t_1at_2o_1 \quad po_1t_2at_1o_2 \quad po_2t_2at_1o_1$$

Notice that there are two ways to arrange the "t's" in the word "potato" and two ways to arrange the "o's" in the word "potato." The same thing would happen if we looked at a different arrangement like "tatoop." If the repeated letters are given subscripts then there are two ways we can arrange the "t's" and two ways we can arrange the "o's." This give us four more words that for all intents and purposes we cannot tell apart unless we put subscripts on the repeated letters. The real question becomes how can we get rid of these extra arrangements that do not look any different without the subscripts?

178 Chapter 6   Counting Techniques

Definition 6.4.1 The number of ways to arrange n items where not all of the items are unique is:

$$\frac{n!}{k_1! k_2! k_3! \ldots} \text{ where } k_1 + k_2 + k_3 + \ldots = n$$

Essentially, this computation takes the number of permutations of all of the items taken together and then divides out all of the possible arrangements of the non-unique items in the group.

If we apply the formula in Definition 6.4.1 to our "potato" problem we get:

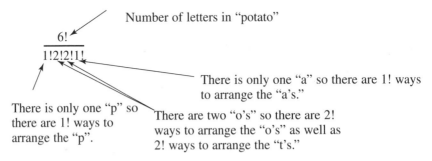

Notice that if we add up the numbers in the denominator (not the factorial values, just the numbers themselves) we get 6, which is the number of letters in the word. This is one of the advantages of using this method of solving these types of problems. You can readily check to see if you have accounted for all of the letters in the problem. The denominator uses both permutations and the Fundamental Counting Principle. The number of ways each letter can be arranged or ordered (permutations) is multiplied by all the possible arrangements of the other letters (Fundamental Counting Principle). Performing the computation results in 180 different arrangements of the letters in the word "potato." ∎

Other people may approach this problem in a different way. We think looking at it from a permutations standpoint probably makes the most sense to most students.

Chapter 6 Summary

This chapter focused on the three basic counting techniques of the multiplication of choices (Fundamental Counting Principle), permutations, and combinations, along with a couple of variations on those ideas. To distinguish between permutations and combinations you should ask yourself "Does the order make a difference?" If the answer is "yes," then permutations are needed. If the answer is "no," then combinations should be used. The three key formulas for this chapter are:

Fundamental Counting Principle or Multiplication of Choices Principle:

Number of total ways = number of possibilities for the first choice *times* number of possibilities for the second choice *times* number of the possibilities for the third choice

Permutations:

$$_nP_k = \frac{n!}{(n-k)!}$$

Combinations:

$$_nC_k = \frac{n!}{k!(n-k)!}$$

These counting techniques, coupled with the concepts of probability from Chapter 5, will help drive the remaining chapters in this book as we explore various statistical distributions and statistical inference.

The key terms and ideas from this chapter are listed below:

Multiplication of Choices Principle   156

Fundamental Counting Principle   156

with replacement   163

without replacement   163

permutations   160

combinations   168

Assessment is an integral part of every curriculum from the elementary school all the way through college. The question always arises—*what is it that students should be able to do after completing this lesson/unit/chapter?* We have included here our intended learning goals for Chapter 6. Students who have a good grasp of the concepts developed in this chapter should be successful in responding to these items:

- Explain, describe, or give an example of what is meant by each of the terms in the vocabulary list.
- Distinguish among permutations, combinations, and the Fundamental Counting Principle in problem situations.
- Construct a tree diagram to portray all the possible outcomes in a counting situation.
- Construct an organized list or table to portray all the possible outcomes in a counting situation.
- Distinguish between experiments with replacement and experiments without replacement.
- Appropriately apply the Fundamental Counting Principle, permutations, or combinations in a counting situation.
- Determine probabilities associated with counting situations.
- Determine the number of arrangements possible of a given number of items not all of which are unique.

Your course instructor may have additional or different assessable outcomes for your class. As teachers (or future teachers) you should think about the assessment outcomes and learning goals for each chapter as you work through them.

180 Chapter 6 Counting Techniques

EXERCISES FOR CHAPTER 6

1. Using the *Data Analysis and Probability Standards for Grades 6–8* from the NCTM's *Principles and Standards for School Mathematics* found at www.nctm.org, identify the middle school objectives that are found in Chapter 6.
2. Using the state standards for mathematics for the content area of data analysis and probability for your state, identify the middle school objectives that are found in Chapter 6. The following website may be useful: www.doe.state.in.us This website will allow you to access the web pages of the state departments of education for the 50 states. From the state web pages you should be able to find the state's mathematical standards.
3. Katie is making ice cream sundaes for her friends. She has three flavors of ice cream (vanilla, strawberry, and chocolate) and four different toppings (caramel, pineapple, strawberry, and hot fudge). If each sundae is made with one flavor of ice cream and one topping, how many different sundaes can Katie make?
4. Elizabeth has 54 pairs of shoes and 36 hats. How many ways can she select one hat and one pair of shoes to wear?
5. How many different permutations are there of all of the letters in the following words?
   a. banana   b. armadillo   c. statistics   d. anaconda
6. The local PTA has hired Krazy Katerers to cater the next awards dinner. The menu from Krazy Katerers is shown below:

Salads	Entrees	Vegetables	Desserts
Garden	Roasted Turkey	Green Beans	Chocolate Mousse
Potato	Smoked Pork	Broccoli	Velvet Crumb Cake
Carrot	Barbeque Beef	Creamed Spinach	Rhubarb Pie
Jello	Chicken Kiev	Corn-on-the-cob	Banana Cream Pie
Three-bean	Honey Salmon	Harvard Beets	Assorted Cookies
	Shrimp Scampi	Glazed Carrots	Coconut Cake
		Sliced Tomatoes	
		Baby Peas	

   a. How many possible dinners are there if the caterers allow the PTA to pick two salads, two entrees, three vegetables, and three desserts from the above menu?
   b. How many possible dinners are there if the caterers allow the PTA to pick two salads, three entrees, four vegetables, and two desserts?
   c. The PTA knows that it wants to serve Chicken Kiev and Shrimp Scampi as the entrees. How many possible dinners are there if they still get to pick two salads, three vegetables, and three desserts?
7. A box contains 20 blocks. Fifteen of the blocks are red, the rest are green. If 6 blocks are selected at random:
   a. How many ways can 6 red blocks be selected?
   b. How many ways can 4 red blocks and 2 green blocks be selected?
   c. How many ways can 1 red block and 5 green blocks be selected?
   d. How many ways can 3 red blocks and 3 green blocks be selected?
   e. How many ways can any 6 blocks be selected regardless of color?

f. What is the probability of drawing 4 red blocks and 2 green blocks from the box?

   g. What is the probability of drawing all red blocks from the box?

8. Ted has a cat carrier which will hold three cats. Unfortunately, Ted has five cats: Muffy, Nero, Oprah, Penny, and Quade.

   a. Make an organized list to show all the ways Ted can chose three cats at random to put in the cat carrier.

   b. How many of the ways in part (a) contain Oprah?

   c. How many of the ways in part (b) contain both Muffy and Nero?

   d. Based on your answers to parts (a) and (b), what is the probability that Oprah will end up in the cat carrier?

   e. Based on your answers to parts (a) and (c), what is the probability that Muffy and Nero will end up in the cat carrier?

9. There are seven cars entered in a local stock car race. How many different ways can these seven cars be lined up for the race? Does order make a difference? Why or why not?

10. Suppose a "combination" lock has the numbers 0 to 39 on it. In order to open the lock, a person must dial the correct three number sequence like 12 left, 24 right, 6 left.

    a. Why is the name "combination" lock misleading?

    b. How many different three-number sequences are possible assuming that numbers can be repeated?

    c. How many different three-number sequences are possible if numbers may not be repeated?

    d. Sometimes with these types of locks once the first two numbers are determined, the lock will go ahead and open. How many different sequences are possible if this is the case?

    e. How many different three-number sequences are possible if all three numbers are even numbers?

    f. How many different three-number sequences are possible if only the middle number has to be even?

11. Theron has 10 coins in his pocket. He has four pennies and six nickels. Three coins are selected at random without replacement.

    The following frequency distribution table uses the number of nickels drawn out of the three coins selected. For example, the first cell in the column denoted "Number of Nickels" means that all three coins were nickels and none were pennies.

Number of Nickels Selected (Event)	Probability of the Event
3	
2	
1	
0	

    Complete the table by finding the corresponding probabilities associated with the events in the table. In other words, find the probabilities that all three coins would be nickels, two would be nickels, one would be a nickel, and there would be no nickels.

12. In the Wacko Lottery, players pay $5.00 and get to pick six numbers ranging from 0 to 60.
   a. Does it matter whether or not you can pick the same number more than once? Why or why not?
   b. Does order matter here? Why or why not?
   c. Assuming that the same number can be picked more than once, how many possible pick-six lottery numbers are there?
   d. Assuming that the order matters and the same number cannot be picked more than once, how many possible pick-six lottery numbers are there?
   e. If the order doesn't matter and the same number cannot be picked more than once, how many possible pick-six lottery numbers are there?
   f. What is the probability of any one particular lottery number—say 12 23 10 09 45 60—coming up if order doesn't matter and the same number cannot be picked more than once?

Random Variables and Probability Distributions

CHAPTER 7

7.1 WHAT IS A RANDOM VARIABLE?
7.2 THE MEAN OF A RANDOM VARIABLE
7.3 VARIANCE AND STANDARD DEVIATION
7.4 BINOMIAL RANDOM VARIABLES
7.5 THE NORMAL CURVE
7.6 NORMAL APPROXIMATIONS

7.1 WHAT IS A RANDOM VARIABLE?

Middle school children, like all children, enjoy playing games. Games involve situations where there is more than one possible outcome and where the score of the game is determined by which of these outcomes occurs. Look at the example in Figure 7.1.1, taken from *Connected Mathematics: What Do You Expect?* (page 50).

5.1 Shooting the One-and-One

Nicky is playing basketball on her school team this year. In the district finals, the team is 1 point behind with 2 seconds left in the game. Nicky has just been fouled, and she is in a one-and-one free-throw situation. This means that Nicky will try one shot. If she makes the first shot, she gets to try a second shot. If she misses the first shot, she is done and does not get to try a second shot. Nicky's free-throw average is 60%.

FIGURE 7.1.1 A one-and-one situation

In this case, there are three possible outcomes:

- She misses her first shot and scores 0.
- She hits her first shot and misses her second, scoring 1 point.
- She makes both her shots, scoring 2 points.

The one-and-one situation is an example of a **chance experiment**, a situation with more than one possible outcome. A **random variable** is a quantitative variable whose value is determined by the outcome of a chance experiment. Nicky's score is a random variable with three possible values: 0, 1, and 2. Which of these values her score will have depends on the outcome of the one-and-one situation.

Students using *Connected Mathematics* go on to explore both experimental and theoretical probabilities for this one-and-one situation. Figure 7.1.2 is from *What Do You Expect* (page 51).

There are several different ways that students could simulate a one-and-one situation (parts B and C in Figure 7.1.2). The main idea is for each stage of the simulation to have the same probabilities as the stages in the real situation. To simulate a single stage (one of Nicky's free throws), a student might use 10 colored counters (blocks, chips, etc.), 6 of one color and 4 of another; let's say 6 blue chips and 4 red chips. The student could put the chips in a container of some kind, mix

---

Problem 5.1

A. Which of the following do you think is most likely to happen?
- Nicky will score 0 points. That is, she will miss the first shot.
- Nicky will score 1 point. That is, she will make the first shot and miss the second shot.
- Nicky will score 2 points. That is, she will make two shots.

Record what you think before you analyze the situation.

B. Plan a way to simulate this situation. Describe your plan.

C. Use your plan from part B to simulate Nicky's one-and-one situation 20 times. Record the result of each trial.

D. Based on your results, what is the experimental probability that Nicky will score 0 points? That she will score 1 point? That she will score 2 points?

E. Make an area model for this situation, using a 10 by 10 grid. What is the theoretical probability that Nicky will score 0 points? 1 point? 2 points?

F. How do the three theoretical probabilities compare with the three experimental probabilities?

■ **Problem 5.1 Follow-Up**

1. Suppose Nicky is in a two-shot free-throw situation. This means that she will get a second shot even if she misses the first shot. What is the theoretical probability that Nicky will score 0 points? That she will score 1 point? That she will score 2 points? Explain your reasoning.

2. How do the theoretical probabilities for the one-and-one situation compare to the theoretical probabilities for the two-shot situation?

FIGURE 7.1.2 A one-and-one situation (*Continued*)

them, and choose one without looking. (Teacher tip: Cloth bags are a lot quieter than plastic buckets.) If a red chip is chosen, Nicky missed her first shot and scores 0. If a blue chip is chosen, Nicky made her first shot and gets a second one. The chip is replaced and another is chosen to simulate her second shot. She scores 1 if the second chip is red and 2 if it's blue.

The TI-83 plus can be a useful tool for conducting simulations. Hit the MATH button and use the right arrow key to go to PRB (for "probability").

```
MATH NUM CPX PRB
1:rand
2:nPr
3:nCr
4:!
5:randInt(
6:randNorm(
7:randBin(
```

Select 5: randInt(and you should see randInt(on your screen; this stands for "random integer." The command randInt(0,9) gives you a random integer from 0 to 9, just like a table of random digits. The calculator is more flexible, however. Entering randInt(1,6) gives a random integer from 1 to 6, as if you rolled a six-sided number cube. Even better, you can get random integers in bunches. The command randInt(1,6,5) is like rolling 5 dice.

Classroom Exploration 7.1

	Nicky's Score	Experimental Probability
1. Suppose you were using 6 blue chips and 4 red chips to simulate Nicky's free throws as described previously. Would it make a difference if the first chip was not replaced before the second chip was drawn? Explain.		
2. Can you use fewer than 10 chips to simulate one of Nicky's free throws? Can you use more than 10 chips? Explain.	0	
3. Explain how you could use a table of random digits (0 through 9) to simulate a one-and-one situation.	1	
4. Do parts B, C, and D in Figure 7.1.2. Fill in the table at the right.		
5. What is the total of all of the probabilities at the right? Is this a coincidence or should this always happen?	2	

Part E in Figure 7.1.2 asks students to make an area model for this situation, using a 10 by 10 grid. What do they mean by that? Look at the diagram in Figure 7.1.3.

186 Chapter 7 Random Variables and Probability Distributions

	2nd shot hits Probability = 0.6	2nd shot misses Probability = 0.4
1st shot hits Probability = 0.6	2 2 2 2 2 2 2 2 2 2 2 2 2 2 2 2 2 2 2 2 2 2 2 2 2 2 2 2 2 2 2 2 2 2 2 2	1 1 1 1 1 1 1 1 1 1 1 1 1 1 1 1 1 1 1 1 1 1 1 1
1st shot misses Probability = 0.4	0 0 0 0 0 0 0 0 0 0 0 0 0 0 0 0 0 0 0 0 0 0 0 0	0 0 0 0 0 0 0 0 0 0 0 0 0 0 0 0

FIGURE 7.1.3 Area model for one-and-one situation

Focus on Understanding

1. How many rows are in the grid? Why are 6 rows selected to go with "1st shot hits?"
2. How many columns are in the grid? Why are 4 of them selected to go with "2nd shot misses?"
3. The grid shows the outcomes for 100 one-and-one situations if they occurred in the same proportion as their theoretical probabilities. Find their theoretical probabilities and fill in the table at the right.
4. Explain what the probabilities have to do with areas in the grid. Hint: Think of the length and width of the grid as 1 unit. What are the length and width of each section of the grid?
5. Compare the theoretical probabilities to the experimental probabilities you got earlier. What do you think would happen if you had simulated Nicky's situation 100 times, instead of just 20?

Nicky's Score	Theoretical Probability
0	
1	
2	

You have made two tables listing Nicky's possible scores and the probability of each, one with experimental probabilities and the other with theoretical probabilities. Both of these illustrate the idea of a probability distribution. The **probability distribution** of a random variable is a rule or table that allows us to find probabilities for values of the variable. Formally, a probability distribution for a random

x	$P(x)$
0	$P(0)$
1	$P(1)$
2	$P(2)$

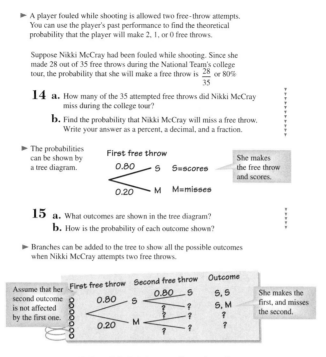

▶ A player fouled while shooting is allowed two free-throw attempts. You can use the player's past performance to find the theoretical probability that the player will make 2, 1, or 0 free throws.

Suppose Nikki McCray had been fouled while shooting. Since she made 28 out of 35 free throws during the National Team's college tour, the probability that she will make a free throw is $\frac{28}{35}$ or 80%

14 a. How many of the 35 attempted free throws did Nikki McCray miss during the college tour?

b. Find the probability that Nikki McCray will miss a free throw. Write your answer as a percent, a decimal, and a fraction.

15 a. What outcomes are shown in the tree diagram?

b. How is the probability of each outcome shown?

▶ Branches can be added to the tree to show all the possible outcomes when Nikki McCray attempts two free throws.

FIGURE 7.1.4 A two-shot situation

variable is a function $P(x)$ that gives the probability that the variable has value x. When we make a table listing Nicky's possible scores and the probability of each, we are letting x = *Nicky's score* and making a table for the function $P(x)$ like the one illustrated on the previous page.

Free throw shooting seems to be very popular. Figure 7.1.4 shows an excerpt from *MathThematics 2* (page 363) that also deals with free throw shooting. The previous example from *Connected Mathematics* dealt with a basketball player named Nicky; this one is about a basketball player named Nikki. What's the probability of that?! Never mind.

Unlike Nicky, Nikki is in a two-shot situation. The example deals with theoretical probability, rather than experimental probability, and the authors of *MathThematics* have chosen to illustrate the situation with a tree diagram, rather than an area model. Nikki is also a better shooter than Nicky.

Focus on Understanding

1. Use a tree diagram to find the probabilities for Nikki's two-shot situation, and give a table for the probability distribution of Nikki's score.
2. Use an area model to find probabilities for Nikki's two-shot situation. Compare your results to part 1.

188 Chapter 7 Random Variables and Probability Distributions

> **3.** Go back to Nicky's one-and-one situation. Use a tree diagram to find the probabilities and compare them to the results you got with the area model.
>
> **4.** Compare the area model and the tree diagram as approaches to finding probabilities. List at least one advantage of area models over tree diagrams and vice versa.

For both situations we've looked at so far, the random variable had only three possible values: 0, 1, and 2. In general, there is no limit to the number of possible values a random variable can have. Look at the situation in Figure 7.1.5. It comes from *Mathscape: Chance Encounters* (page 22).

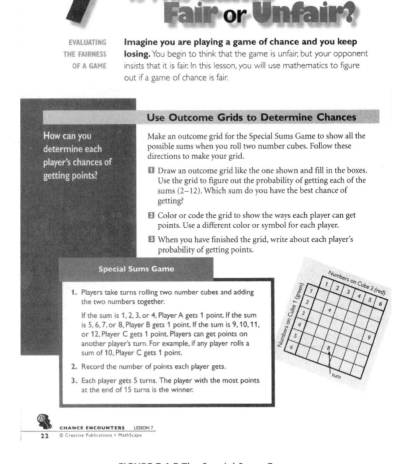

FIGURE 7.1.5 The Special Sums Game

Focus on Understanding

1. Make a grid for the possible sums from rolling two number cubes, as shown in Figure 7.1.5.
2. Make a table for the probability distribution of $x = $ *sum from rolling two number cubes*.
3. Refer to Figure 7.1.5. Use the table from #2 above to find the probability that A wins a point when the two number cubes are rolled once. What is the probability that B wins a point? What about C?

7.2 THE MEAN OF A RANDOM VARIABLE

In Chapter 3, we introduced the mean, variance, and standard deviation as ways of describing data. In this section and the next, we will return to these ideas in a different context, as a way of describing random variables. We'll start with the mean of a random variable, also called the **expected value**.

Connected Mathematics introduces expected value in a follow-up to Nicky's one-and-one situation from the previous section. Figure 7.2.1 is from *Connected Mathematics: What Do You Expect?* (page 52).

The grid that we made for this situation is the perfect tool for thinking about the problem in Figure 7.2.1. There are 100 cells in the 10 by 10 grid, illustrating the outcomes we would expect from 100 of Nicky's one-and-one situations. The 100 outcomes consist of 40 zeros, 24 ones, and 36 twos. Thus, for part A, we would expect the total number of points to be:

$$0(40) + 1(24) + 2(36) = 96$$

For part B, to figure the average number of points per trip, we would divide the total number of points by the total number of trips:

$$\frac{96 \text{ points}}{100 \text{ trips}} = 0.96 \text{ points per trip}$$

Problem 5.2

Suppose Nicky has a 60% free-throw average and is in a one-and-one free-throw situation 100 times during the season.

A. What total number of points would you expect Nicky to score in these 100 trips to the free-throw line?

B. What would Nicky's average number of points per trip be? This is the **expected value** for this situation.

FIGURE 7.2.1 Expected value

190 Chapter 7 Random Variables and Probability Distributions

This is the expected value for the number of points that Nicky scores in a one-and-one situation. Does she ever actually score 0.96 points when she goes to the free throw line? No; of course not. The expected value is an average. It is a long-term average value for Nicky's score if the situation is repeated many times. To understand the general formula for expected value, let's look at the calculations we just did and do a little algebra.

$$\frac{0(40) + 1(24) + 2(36)}{100} =$$

$$\frac{0(40)}{100} + \frac{1(24)}{100} + \frac{2(36)}{100} =$$

$$0\left(\frac{40}{100}\right) + 1\left(\frac{24}{100}\right) + 2\left(\frac{36}{100}\right) =$$

$$0 \cdot P(0) + 1 \cdot P(1) + 2 \cdot P(2)$$

In other words, we could find the expected value for Nicky's score by taking each of the possible values times the probability for that value and adding the results. This is the general formula for the expected value or mean of a random variable.

Expected Value or Mean of a Random Variable

$$E(x) = \mu = \sum [x \cdot P(x)] = x_1 \cdot P(x_1) + x_2 \cdot P(x_2) + \ldots + x_n \cdot P(x_n)$$

The **expected value** or **mean of a random variable** is a long-term average value for the variable.

Notice that there are two different terms (and two different symbols) for the same idea. The expected value of x, symbolized by $E(x)$, and the mean of the probability distribution, symbolized by the Greek letter μ (pronounced *mu*), are exactly the same thing. The symbols $E(x)$ and μ (or sometimes μ_x) are used interchangeably.

Notice that when we say that the mean is an "average value" for the random variable, we are not just adding the values and dividing by how many values we have. The mean of a random variable is a *weighted* average. Each value of the random variable is weighted according to its probability. The value 0 receives more weight than the value 1 because it occurs more often; it has greater probability (40% versus 24%).

Focus on Understanding

In the previous section, you made an area model grid for Nikki, an 80% free throw shooter in a two-shot situation.

1. Use your grid to determine the total number of points you would expect an 80% free throw shooter to get in 100 two-shot situations.

2. Find the average number of points per trip for the 100 two-shot trips to the free throw line.
3. You also determined the probability distribution for x = *the number of points Nikki makes in a two-shot situation*. Use $\mu = \sum[x \cdot P(x)]$ to find the average number of points Nikki scores in two-shot situations. Is your answer the same as #2?
4. A middle school student doesn't like your answer for #3. He says that the average number of points in a two-shot situation should be: $\frac{0+1+2}{3} = \frac{3}{3} = 1$. What should you tell him?

Middle school students use the idea of expected value to make decisions in a variety of situations. One typical type of question asks students to decide whether or not a game is "fair." The example on the Special Sums Game in the previous section contained just this type of question. In that case, each player received the same number of points (1) if one of their sums came up. Therefore, it was reasonable to just compare probabilities. The game would be fair if each player had the same probability of getting a point. The situation is more complicated if the number of points a player scores is not the same on every outcome. Now the question becomes, "Is the expected number of points the same for each player?" Look at Figure 7.2.2, taken from *Connected Mathematics: What Do You Expect* (page 25).

Focus on Understanding

See Figure 7.2.2. Suppose we are playing the Prime Number Multiplication Game and we look at just a single roll of two number cubes. Let x = *the number of points scored by Player A in one roll* and y = *the number of points scored by Player B in one roll*.

1. Make a table for the probability distribution of x and find the expected score for Player A.

7. Raymundo invented the Prime Number Multiplication game. In this game, two number cubes are rolled. Player A scores 10 points if the product is prime, and Player B scores 1 point if the product is not prime. Raymundo thinks this scoring system is reasonable because there are many more ways to roll a nonprime product than a prime product.

  a. If the cubes are rolled 100 times, how many points would you expect Player A to score? How many points would you expect Player B to score?

  b. Is Raymundo's game a fair game? Explain why or why not.

FIGURE 7.2.2 The Prime Number Multiplication Game

192 Chapter 7 Random Variables and Probability Distributions

2. Make a table for the probability distribution of *y* and find the expected score for Player B.
3. Is it a fair game? Explain.
4. Instead of looking at a single roll, as we did above, *Connected Mathematics* has students look at 100 rolls. What would be the expected number of points for each player in 100 rolls of two number cubes?
5. Looking at 100 rolls would be the obvious thing to do if we were going to model this situation with a 10 by 10 grid. But since we are rolling dice, it would seem more natural to use a different grid where the numbers would work out nicer. Rewrite the problem (you're a teacher, so you can do that) so that it would work out nicely on an outcome grid, then solve the new problem that way.
6. Suppose that we wanted to change the rules to make the game fair. If we keep the rule for Player B the same, but changed the number of points that Player A gets for rolling a prime product, how many points should Player A get?

Another type of problem has middle school students use expected value to decide between different options. Figure 7.2.3 is from *Connected Mathematics: What Do You Expect* (page 65).

Focus on Understanding

1. Answer the question in Figure 7.2.3.
2. Based on Tasa's explanation, Mr. Fujita decides to change Option 2 so that Tasa would still earn, on average, $10 each time she mows the lawn. Call this Option 2A. He will still pay her $30 if she rolls a sum of 7. How much should he pay her if she does not roll a 7?
3. Mr. Fujita has another idea. Call this one Option 2B. If Tasa chooses Option 2B, he will still pay her $3 if she does not roll a 7. How much should he pay her when she does roll a 7, if he still wants the average payment to be $10?

---

3. Mr. Fujita hires Tasa to mow his lawn for the summer. When Tasa asks him how much he will pay her, he offers her two options:

Option 1: Mr. Fujita will pay Tasa $10 each time she mows his lawn.

Option 2: Each time Tasa mows Mr. Fujita's lawn, she will roll a pair of number cubes. If the sum on the cubes is 7, Mr. Fujita will pay her $30. If the sum is not 7, he will pay her only $3.

Which option should Tasa choose? Give mathematical reasons to support your answer.

FIGURE 7.2.3 Tasa's choice

7.3 VARIANCE AND STANDARD DEVIATION

In Chapter 3, we used variance and standard deviation to measure the spread or variability in a data set. In a similar way, statisticians use variance and standard deviation to measure the spread or variability in the values of a random variable. Recall that variance is the average (or predicted average) of the squared deviations. Let's revisit this idea in the context of random variables.

Squared Deviations

Refer back to Tasa's choice in the example at the end of Section 7.2. Let x = *Tasa's earnings for mowing the lawn once* if she chooses Option 2A. The probability distribution for x is given at the right, together with the deviation and squared deviation for each value of x. Remember from before that $\mu = \$10$.

x	$P(x)$	Deviation $= (x - \mu)$	(Deviation)^2 $= (x - \mu)^2$
30	$\frac{1}{6}$	$30 - 10 = 20$	$20^2 = 400$
6	$\frac{5}{6}$	$6 - 10 = -4$	$(-4)^2 = 16$

Figure 7.3.1 shows a geometric view of the deviations and squared deviations.

The **deviation** $(x - \mu)$ for each value indicates the relationship between that value and the mean. For example, the value 6 has a deviation of -4, indicating that it is 4 units ($\$4$) below the mean ($\$10$). Each deviation is "squared"; that is, a square is built using the deviation as its side. The area of one of these squares $(x - \mu)^2$ is called a **squared deviation**. Notice that the farther a value is from the mean, the larger the squared deviation. Values that are the same distance from the mean will have the same squared deviation, regardless of whether they are above or below the mean.

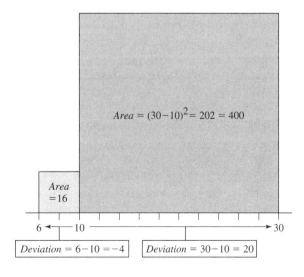

FIGURE 7.3.1 Deviations and squared deviations

194 Chapter 7 Random Variables and Probability Distributions

		\multicolumn{6}{c}{Second Cube}
		1	2	3	4	5	6
First Cube	1	6	6	6	6	6	30
	2	6	6	6	6	30	6
	3	6	6	6	30	6	6
	4	6	6	30	6	6	6
	5	6	30	6	6	6	6
	6	30	6	6	6	6	6

FIGURE 7.3.2 Area model for option 2A

Figure 7.3.2 shows an area model for Tasa's earnings using Option 2A. The entries show Tasa's earnings for 36 rolls, if they occur exactly in proportion to their theoretical probabilities.

Classroom Exploration 7.3

1. Make a new grid similar to Figure 7.3.2, but replace each value for Tasa's earnings with its squared deviation.
2. Use your grid to find the average of the squared deviations for the 36 rolls.
3. In computing the average squared deviation in #2, you should have done something like the computation shown below:

$$\frac{400(6) + 16(30)}{36}$$

Show that this is equivalent to:

$$(30 - 10)^2 \cdot \left(\frac{1}{6}\right) + (6 - 10)^2 \cdot \left(\frac{5}{6}\right)$$

Variance

The average squared deviation you found in Exploration 7.3 is the **variance** of Tasa's earnings. Part 3 of the exploration shows how the computations you did relate to the formula usually given for variance.

$$(30 - 10)^2 \cdot \left(\frac{1}{6}\right) + (6 - 10)^2 \cdot \left(\frac{5}{6}\right)$$
$$= (30 - \mu)^2 \cdot P(30) + (6 - \mu)^2 \cdot P(6)$$

The variance is the weighted average of the squared deviations. Each squared deviation is weighted according to its probability. The general formula for variance is given in the following box.

Section 7.3 Variance and Standard Deviation

Variance of a Random Variable x

$$Var(x) = \sigma^2 = \sigma_x^2 = \sum(x - \mu)^2 \cdot P(x)$$
$$= (x_1 - \mu)^2 \cdot P(x_1) + (x_2 - \mu)^2 \cdot P(x_2) + \ldots + (x_n - \mu)^2 \cdot P(x_n)$$

The **variance** is a weighted average of the squared deviations, with each being weighted according to its probability.

Notice that, just as there are different symbols commonly used for the mean or expected value (μ, μ_x, and $E(x)$), there are different symbols used for variance ($Var(x)$, σ^2, and σ_x^2). The symbol σ is the Greek letter sigma. For those who know $\sum$ as sigma, yes, it is a capital sigma, while σ is a lowercase sigma.

For Tasa's earnings using Option 2A, there are only two different values for x: $x_1 = 30$ and $x_2 = 6$. Computing the variance using the preceding formula would go something like this:

$$\sigma^2 = \sum(x - \mu)^2 \cdot P(x)$$
$$= (x_1 - \mu)^2 \cdot P(x_1) + (x_2 - \mu)^2 \cdot P(x_2)$$
$$= (30 - 10)^2 \cdot \left(\frac{1}{6}\right) + (6 - 10)^2 \cdot \left(\frac{5}{6}\right)$$
$$= 80$$

This should be the same as the average squared deviation you got in Exploration 7.3.

There is also an alternate formula for variance which gives the same values with (usually) easier computations.

Computation Formula for Variance

$$\sigma^2 = \sum[x^2 \cdot P(x)] - \mu^2$$

Let's compute the variance for Tasa's earnings using Option 2A again, this time with the computation formula.

$$\sigma^2 = \sum[x^2 \cdot P(x)] - \mu^2$$
$$= \left[30^2 \cdot \left(\frac{1}{6}\right) + 6^2 \cdot \left(\frac{5}{6}\right)\right] - 10^2$$
$$= [150 + 30] - 100$$
$$= 80, \text{ the same value we got from the other formula}$$

Suppose Tasa chose Option 2B instead. The probability distribution for y = *Tasa's earnings using Option 2B* is given at the right. Remember that the mean is again $10.

y	P(y)
45	$\frac{1}{6}$
3	$\frac{5}{6}$

Focus on Understanding

1. Compare the probability distributions for Tasa's earnings using Option 2A and Option 2B. For which option are the values more spread out? Based on this, which should have the larger variance?
2. Compute the variance for Option 2B and compare it to the variance for Option 2A. Does the result confirm your answer for #1?

Let's think about the units that variance is measured in. In the example of Tasa's earnings, x is in dollars. Her average earnings, μ, is also in dollars. When we compute a deviation $(x - \mu)$ for one of the values, we are subtracting dollars from dollars, so we still have dollars. However, once we square the deviations, we no longer have dollars, but dollars times dollars or square dollars (whatever those are). The variance is a weighted average of these squared deviations, so it's also in square dollars. In general, variance is never in the same units that the values of the variable are in, but rather the square of those units. Thus, the variance is not directly comparable to values of the variable. We saw the same thing when we looked at variance in Chapter 3. For this reason, statisticians often use the other measure of spread, standard deviation.

The Standard Deviation and z-Scores

Standard Deviation of a Random Variable

$$\sigma = \sqrt{\sigma^2}$$
$$= \sqrt{\sum (x - \mu)^2 \cdot P(x)}$$

The **standard deviation** can be interpreted as a typical or standard amount that values of the variable deviate from the mean.

For example, the variance of Tasa's earnings using Option 2A is 80 (dollars)^2. So, the standard deviation is:

$$\sigma = \sqrt{\sigma^2} = \sqrt{80} = 8.94$$

Section 7.3 Variance and Standard Deviation

Since we've taken the square root of (dollars)², we're back in dollars again, so the standard deviation is $8.94.

In Chapter 3, we used the standard deviation to compute z-scores. In the same way, we can use the standard deviation to find z-scores for values of a random variable.

The z-Score for a Value of a Random Variable

$$z = \frac{x - \mu}{\sigma}$$

- The sign of the z-score tells us whether the x-value is above or below the mean. If the z-score is positive, the x-value is above the mean; if the z-score is negative, the x-value is below the mean.
- The magnitude of the z-score tells us how far the x-value is from the mean, measured in standard deviations.

For example, a z-score of -1.5 indicates that the value is 1.5 standard deviations below the mean.

Looking back once again at Tasa's earnings using Option 2A, let's compute the z-scores for the two possible values in that option: $x = 6$ and $x = 30$. For Option 2A, $\mu = 10$, and $\sigma = \sqrt{80} \approx 8.94$.

For $x = 6$: $\quad z = \dfrac{x - \mu}{\sigma} = \dfrac{6 - 10}{\sqrt{80}} = -0.447$

For $x = 30$: $\quad z = \dfrac{x - \mu}{\sigma} = \dfrac{30 - 10}{\sqrt{80}} = 2.236$

Thus, if she earns $6, that would be 0.447 standard deviations below her expected average, while $30 is 2.236 standard deviations above her expected average.

Focus on Understanding

Let $y = $ Tasa's earnings using Option 2B.

1. Find the standard deviation for y.
2. Find the z-scores for the two possible values in Option 2B: $y = 3$ and $y = 45$. Interpret your results.
3. Compare your answers to the z-scores we got for $x = 6$ and $x = 30$ in Option 2A. Do you find this surprising? Explain.

Using the TI-83 plus

We can use the TI-83 plus to find the mean, variance, and standard deviation of a random variable. We'll use Option 2A as an example one more time.

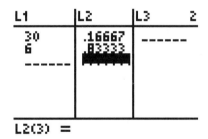

FIGURE 7.3.3 A probability distribution in the TI-83 plus

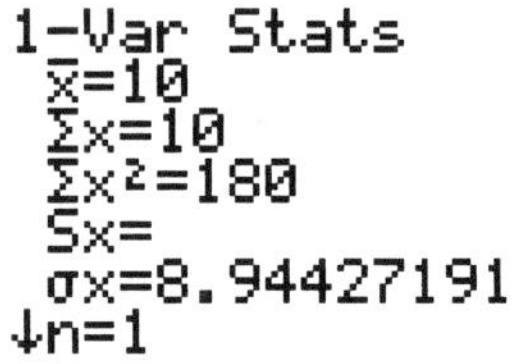

```
1-Var Stats
 x̄=10
 Σx=10
 Σx²=180
 Sx=
 σx=8.94427191
↓n=1
```

FIGURE 7.3.4 One-variable statistics from the TI-83 plus

Hit STAT, and select 1: Edit ...

We are going to enter the *x*-values in L1 and their probabilities in L2. If necessary, use the arrow keys to move to the top of each of those columns, and hit CLEAR and ENTER to empty those columns.

Enter the values 30 and 6 in L1. Enter 1/6 and 5/6 in L2. You should see something like Figure 7.3.3.

Hit STAT again, and use the right arrow key to go to CALC. Select 1: 1-Var Stats. You should see 1-Var Stats on the screen. We want 1-Var Stats(L1,L2), so type in the rest, using 2nd 1 to get L1 and 2nd 2 to get L2. Hit ENTER. You should see the information in Figure 7.3.4.

The calculator is really set up to deal with frequency distributions, rather than probability distributions, so we need to do a little interpretation. We could interpret each of the calculator's values, but let's just stick to the mean, variance, and standard deviation for now.

- The mean μ is what the calculator calls $\bar{x}$ (as if we were dealing with a sample, rather than a random variable).
- The standard deviation we want is σ_x, not S_x (which is blank anyway).
- The calculator does not list a value for the variance, but we can just square the standard deviation to get it.

In this case, the values for the mean, standard deviation, and variance agree with those we had figured previously.

The TI-83 plus is particularly useful if a random variable has lots of different possible values, as you can see in the following *Focus on Understanding*.

Focus on Understanding

At the end of Section 7.1, you found the probability distribution for x = *the sum showing on two number cubes*. This distribution is shown at the right.

1. Use the calculator to find the mean, standard deviation, and variance for this random variable.
2. You are also going to use the calculator to simulate 100 rolls of two number cubes.

   a. First hit MATH and use the right arrow key to go to PRB.
   b. Select 5: randInt(and use it to enter 5: randInt(1,6,100). You've just made 100 rolls with one number cube.
   c. Hit STO 2nd L1 ENTER. You've just stored your 100 rolls as the list L1.
   d. Repeat steps b and c, but store the results in L2 instead. You now have 100 rolls of one cube stored in L1 and 100 rolls for a second cube stored in L2.
   e. Now we need to use L1 and L2 to get 100 sums. Hit 2nd L1 + 2nd L2 STO 2nd L3.
   f. Hit STAT and EDIT to see the results of what we've done so far. You should see 100 numbers in each column. The numbers in L1 and L2 should look like the results from rolling a single number cube. The numbers in L3 should be the sums of the numbers from L1 and L2.
   g. Hit STAT, and go to CALC. Select 1: 1-Var Stats and enter 1-Var Stats(L3). This is a sample, so we should be using S_x, rather than σ_x. What are the mean, standard deviation, and variance of your sample?

3. Compare the results from #2 to the results you got in #1. Should they be exactly the same? Should they be close? Are they? Explain.
4. Compare your results with another student. Should your answers to #1 be the same? What about #2? Explain.
5. A student thinks he has a quicker way to do #2. Instead of making L1 and L2 first and adding the results, he enters 5: randInt(2,12,100) STO L3 to get his sums directly into L3 without doing L1 and L2 first. Will that work? Explain.

x	P(x)
2	1/36
3	2/36
4	3/36
5	4/36
6	5/36
7	6/36
8	5/36
9	4/36
10	3/36
11	2/36
12	1/36

7.4 BINOMIAL RANDOM VARIABLES

A common situation that shows up in middle school curricula and many real-world applications involves a multistage process in which every stage is essentially the same—what we call **repeated identical trials**. We have already seen two-stage versions of this: rolling two number cubes and shooting two free throws, for example. What happens if we extend these to more than two stages?

Classroom Exploration 7.4

Suppose that our old friend Nicky, the 60% free throw shooter, is going to shoot 5 shots. Let $x =$ *the number of shots she hits in the 5 attempts*. We'd like to make a table for the probability distribution of x.

1. Choose a simulation method that would have the same probabilities as one of Nicky's shots. Describe your choice.
2. Use your method to simulate 5 shots. Count the number of hits, and tally the result in Table 1 below.
3. Repeat your 5-shot simulation 19 more times (95 more shots). Tally the number of hits in each 5-shot group in Table 1.
4. Fill in the frequencies and experimental probabilities to complete Table 1.
5. Combine your data with data from other students and complete Table 2 below.
6. Compare the experimental probabilities in your two tables. Which would you expect to be closer to the theoretical probabilities? Why?

Table 1: Data from 20 Five-Shot Simulations

x	Tally	Frequency	Experimental Probability
0			
1			
2			
3			
4			
5			

Table 2: Combined Data

x	Combined Frequency	Experimental Probability
0		
1		
2		
3		
4		
5		

Well, that simulation was a lot of fun, but it's not practical to conduct a simulation whenever we need to find a probability. We need a method for finding the theoretical probability in this type of situation.

Suppose we want to find $P(3)$, the probability that she hits exactly 3 of the 5 shots. We could do this with a tree diagram, but the tree is a bit large (and even larger if she takes more shots). Maybe we can do it without drawing the whole tree.

One outcome with exactly 3 hits would be HHHMM, hitting the first 3 and missing the last 2. The path through the tree leading to this outcome looks something like this:

$$\xrightarrow{0.6} H \xrightarrow{0.6} H \xrightarrow{0.6} H \xrightarrow{0.4} M \xrightarrow{0.4} M$$

The probability of this particular outcome is:

$$P(HHHMM) = (0.6)(0.6)(0.6)(0.4)(0.4) = (0.6)^3(0.4)^2$$

Unfortunately, this is not the only outcome with exactly 3 hits. Another is HHMHM. Its path looks like this:

$$\xrightarrow{0.6} H \xrightarrow{0.6} H \xrightarrow{0.4} M \xrightarrow{0.6} H \xrightarrow{0.4} M$$

Its probability is: $P(HHMHM) = (0.6)(0.6)(0.4)(0.6)(0.4) = (0.6)^3(0.4)^2$, the same as before.

Will we get the same probability for every path with 3 hits? Yes. In this tree, any branch leading to an H has probability 0.6, and any branch leading to an M has probability 0.4. If we have exactly 3 H's we'll have exactly 2 M's, and we'll always end up with $(0.6)^3(0.4)^2$.

So now the question becomes, "How many paths have 3 H's and 2 M's?" If we knew the answer to this question, we could take the number of paths times the probability of each path to get the total probability of getting 3 hits.

Think of this question as a problem about choosing blanks to put letters in. We have 5 blanks representing the 5 shots.

$$\underline{}_{\text{1st}} \quad \underline{}_{\text{2nd}} \quad \underline{}_{\text{3rd}} \quad \underline{}_{\text{4th}} \quad \underline{}_{\text{5th}}$$

We have to fill the 5 blanks using 3 H's and 2 M's. We can think of this as a two-stage process:

Stage 1: Choose 3 of the 5 blanks for H's.
Stage 2: Choose 2 of the remaining blanks for M's.

When we choose 3 blanks for H's, are repetitions allowed? No, since we can't put two letters in the same blank. Does the order in which the blanks are chosen matter? No. For example, choosing the 1st, 3rd, and 4th blanks gives the same outcome as choosing the 4th, 3rd, and 1st blanks. So, what counting method should we use? Combinations!

For Stage 1 (choosing 3 of the 5 blanks for H's), we are choosing 3 blanks from 5 blanks. The number of possibilities is:

$$_5C_3 = \binom{5}{3} = \frac{5!}{3!2!} = \frac{5 \cdot 4 \cdot 3 \cdot 2 \cdot 1}{3 \cdot 2 \cdot 1 \cdot 2 \cdot 1} = 10$$

Chapter 7 Random Variables and Probability Distributions

For Stage 2 (choosing 2 of the remaining blanks for M's), there are only 2 blanks remaining. The number of possibilities is:

$$_2C_2 = \binom{2}{2} = \frac{2!}{2!0!} = \frac{2 \cdot 1}{2 \cdot 1 \cdot 1} = 1$$

There's really not a choice to be made for the second stage. Once we choose blanks for the H's, there is only one way to fill in the M's.

By our reasoning, there should be $_5C_3$ or 10 different outcomes with exactly 3 H's. See if you can list them all without peeking at the list below.

HHHMM HHMHM HHMMH HMHHM HMHMH
HMMHH MHHHM MHHMH MHMHH MMHHH

So, we now know that there are 10 different outcomes with exactly 3 hits and that each of these outcomes has a probability of $(0.6)^3(0.4)^2$. Therefore, the total probability of Nicky getting exactly 3 hits in 5 free throws is:

$$P(3) = \binom{5}{3}(0.6)^3(0.4)^2 = 0.3456$$

Compare this theoretical probability to the experimental probabilities for getting 3 hits from Exploration 7.4. The experimental probabilities should be close to the theoretical probability, and the one from the combined data is probably closer.

Focus on Understanding

	x	P(x)
1. We have just found the probability that Nicky, the 60% free throw shooter, hits exactly 3 shots in 5 attempts: $P(3) = 0.3456$. Use the same method to fill in the table at the right, where $x =$ *the number of free throws Nicky will hit in 5 attempts.* 2. Compare your results to the experimental probabilities in Exploration 7.4. Which experimental probabilities are closer to the theoretical probabilities, those from your simulation or those from the combined data?	0	
	1	
	2	
	3	0.3456
	4	
	5	

The type of situation we've been dealing with in this section is called a **binomial experiment**. The following box describes the kind of situations that qualify as binomial experiments.

Binomial Experiments

A binomial experiment consists of a series of repeated identical trials satisfying these conditions:

- There must be a fixed number of trials. Call the number of trials n.
- Each trial must have two possible outcomes. Call one outcome *success* and the other *failure*.
- The probability of *success* must be the same on every trial; call it p. The failure probability is $q = 1 - p$.
- The trials must be independent. In other words, the outcome of one trial cannot affect the probability on another trial.
- The random variable x is the number of successes in the n trials. This is called a **binomial random variable**.

Binomial Probabilities

Probabilities for a binomial random variable are given by the formula:

$$P(x) = \binom{n}{x} p^x q^{n-x}$$

This formula is called the **binomial probability density function (binomial pdf for short)**.

Nicky's 5-shot situation is a binomial experiment.

- Each shot is a trial. There are a fixed number of trials ($n = 5$).
- Each trial has two possible outcomes (*success* if she hits, *failure* if she misses).
- The probability of *success* is the same on each trial ($p = 0.6; q = 0.4$).
- The trials are independent. This is probably not quite true in a real situation. If she hits her first shot, it may affect the probability that she hits her next shot. However, since we have no way of knowing exactly what this new probability would be, we are assuming that the probability stays the same. In other words, we are assuming that her shots are independent.
- The random variable we are concerned about is the number of successes in the n trials (the number of hits in her 5 shots).

We can use the binomial pdf $P(x) = \binom{n}{x} p^x q^{n-x}$ to find probabilities for any binomial random variable. This is essentially what we did for Nicky's situation. One shortcoming of the binomial pdf, however, is that it can only be used to find probabilities for one value of x at a time. For example, suppose we wanted to find the probability that Nicky makes at most 3 of her 5 shots. There are two different approaches we could take. The direct approach would be to find the probabilities for each value separately and add the results together.

$$P(x \leq 3) = P(x = 0, 1, 2, \text{or } 3)$$
$$= P(0) + P(1) + P(2) + P(3)$$

$$= \binom{5}{0}(0.6)^0(0.4)^5 + \binom{5}{1}(0.6)^1(0.4)^4 + \binom{5}{2}(0.6)^2(0.4)^3$$
$$+ \binom{5}{3}(0.6)^3(0.4)^2$$
$$= 0.0102 + 0.0768 + 0.2304 + 0.3456$$
$$= 0.6630$$

Alternatively, we could use the indirect approach, focusing on the values of x we don't want, rather than those we want. Using the complement rule:

$$P(x \leq 3) = 1 - P(x = 4 \text{ or } 5)$$
$$= 1 - \left[\binom{5}{4}(0.6)^4(0.4)^1 + \binom{5}{5}(0.6)^5(0.4)^0\right]$$
$$= 1 - [0.2592 + 0.0778]$$
$$= 0.663$$

Look at the example in Figure 7.4.1. It comes from *Connected Mathematics: What Do You Expect* (page 77).

> **11.** Mindy is taking a ten-question true-false test. She forgot to study, so she is guessing at the answers.
>
>   **a.** What is the probability that Mindy will get all the answers correct?
>
>   **b.** What is the probability that Mindy will get at least nine answers correct?

FIGURE 7.4.1 Mindy's test

Focus on Understanding

1. Is Mindy's Test a binomial experiment? Explain.
2. Answer the questions in Figure 7.4.1.
3. Find the probability that Mindy gets at least one answer correct.

Mean, Variance, and Standard Deviation for a Binomial Random Variable

Let's return to Nicky's 5-shot situation to find the mean, variance, and standard deviation for the number of shots she hits. For reference, the probability distribution is given at the right. Let's start with the mean. Using the formula from Section 7.2:

$$\mu = \sum[x \cdot P(x)]$$
$$= 0(0.01024) + 1(0.0768) + 2(0.2304) + 3(0.3456)$$
$$+ 4(0.2592) + 5(0.07776)$$
$$= 3$$

x	$P(x)$
0	0.01024
1	0.0768
2	0.2304
3	0.3456
4	0.2592
5	0.07776

Now, wait a minute! There was an easier way to figure that out. The mean is another word for expected value. Nicky is a 60% free throw shooter. When we find the mean for the number of hits that Nicky gets, we are asking, "How many shots would we expect Nicky to get?" Well, 60% of them, of course! Sixty percent of 5 shots is:

$$5(0.6) = 3 \quad \text{Much easier!}$$

Notice that, in computing the mean this short way, we took the number of trials (5) times the success probability (0.6). In general, for a binomial random variable: $\mu = np$.

To compute the variance, we'll use the computation formula:

$$\sigma^2 = \sum[x^2 \cdot P(x)] - \mu^2$$
$$= [0^2(0.01024) + 1^2(0.0768) + 2^2(0.2304) + 3^2(0.3456) + 4^2(0.2592)$$
$$+ 5^2(0.07776)] - 3^2$$
$$= 1.2$$

It's not as obvious as the mean, but there is a shortcut for this, too. For a binomial random variable: $\sigma^2 = npq$. In this case:

$$\sigma^2 = npq = 5(0.6)(0.4)$$
$$= 1.2, \text{the same answer we got the long way.}$$

Either way we compute the variance, the standard deviation is:

$$\sigma = \sqrt{\sigma^2} = \sqrt{1.2} = 1.095$$

The short formulas we've just presented are summarized in the following box.

Mean, Variance, and Standard Deviation for Binomial Random Variables

$$\mu = np \qquad \sigma^2 = npq \qquad \sigma = \sqrt{npq}$$

Caution:

Now that you have seen the short formulas in the box, it is tempting to want to use them all the time. Notice that these short formulas apply for *binomial random variables only*. For other random variables, we must use the formulas in Sections 7.2 and 7.3.

> ## Focus on Understanding
>
> Find the mean, variance, and standard deviation for the number of correct answers Mindy gets. (See Figure 7.4.1.)

Using the TI-83 plus

Earlier, we found the probability that Nicky, the 60% free throw shooter, would hit exactly 3 shots in 5 attempts. Let's use a calculator shortcut to get the same result.

1. Notice that the second function for the VARS key is DISTR (for "distributions"). Hit 2nd DISTR.
2. Scroll down until you see 0: binompdf(. This stands for "binomial probability density function." Select this option, and binompdf(should appear on the screen.
3. In order to use the binomial pdf, we need to know the number of trials, n, the success probability, p, and the number of successes, x. In this case, $n = 5$, $p = 0.6$, and $x = 3$. We want the calculator screen to show binompdf(5, 0.6, 3) in that order. Type in the rest of this command and hit ENTER.
4. You should get 0.3456. This is the same answer we got earlier.

In general, the command binompdf(n, p, x) computes $P(x)$, the probability of getting exactly x successes in a binomial experiment with n trials and success probability p on each trial.

The TI-83 plus also has an additional feature called a cumulative distribution function (cdf, rather than pdf). A **cumulative distribution function** gives the total probability for all values of a random variable that are less than or equal to a fixed value. Earlier, we also found the probability that Nicky would get *at most* 3 hits in 5 attempts. We can do this on the calculator using the cumulative distribution function.

1. Hit 2nd DISTR as we did before.
2. Scroll down past 0: binompdf(to A: binomcdf.
3. We want the calculator screen to show binomcdf(5, 0.6, 3). Type in the rest of this command and hit ENTER.
4. You should get 0.66304, which is the same as we got earlier (with one more decimal place than we showed).

In general, the command binomcdf(n, p, k) computes $P(x \leq k)$, the total probability of all values of x less than or equal to k, where x is a binomial random variable. What we just found was $P(x \leq 3)$. Suppose that we wanted to use the cumulative distribution function to find $P(x \geq 3)$; that is, the probability that Nicky gets at least

3 hits in 5 shots. We do it indirectly, by using the complement rule:

$$P(x \geq 3) = P(x = 3, 4, \text{ or } 5)$$
$$= 1 - P(x = 0, 1, \text{ or } 2)$$
$$= 1 - P(x \leq 2)$$

Using the calculator, enter:

$$1-\text{binomcdf}(5, 0.6, 2)$$

You should get 0.68256.

As usual, with the indirect approach we subtract the probability for the values we do not want (0, 1, and 2). We can use a similar approach to find a probability like $P(2 \leq x \leq 4)$. Start with:

$$P(x \leq 4) = P(x = 0, 1, 2, 3, \text{ or } 4)$$

This probability is easy to get from the calculator, but unfortunately, it includes some values we don't want (0 and 1). We need to subtract the probability of these values. Thus, the probability we want is:

$$P(2 \leq x \leq 4) = P(x = 2, 3, \text{ or } 4)$$
$$= P(x = 0, 1, 2, 3, \text{ or } 4) - P(x = 0 \text{ or } 1)$$
$$= P(x \leq 4) - P(x \leq 1)$$

Using the calculator, enter:

$$\text{binomcdf}(5, 0.6, 4) - \text{binomcdf}(5, 0.6, 1)$$

You should get 0.8352.

Focus on Understanding

Look back at *Mindy's Test* in Figure 7.4.1. Let $x =$ *the number of questions she gets correct*. Use the binomial pdf and cdf features of your calculator to answer these questions.

1. What is the probability that Mindy gets exactly 6 correct answers?
2. What is the probability that she gets 6 or fewer correct answers?
3. What is the probability that she gets at least 6 correct answers?
4. Find $P(4 \leq x \leq 8)$.

7.5 THE NORMAL CURVE

The Standard Normal Curve

Imagine we are tossing 100 coins. Let $x =$ *the number of heads showing in the 100 tosses*. Suppose we wanted to make a bar graph of the probability distribution of x.

What is the smallest number x can be? What is the largest? In theory, we could use the binomial pdf from the previous section to find the probability for each value of x from 0 to 100. Doing this task by hand would be a mammoth undertaking. It's a good thing we have computers. Figure 7.5.1 shows this bar graph, at least for values of x between 30 and 70. We didn't bother showing values below 30 or above 70, since the bars are so short they don't even show up.

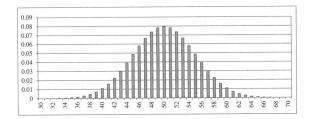

FIGURE 7.5.1 The number of heads in 100 tosses

What we want you to notice about this graph is that its shape seems to follow a smooth bell-shaped curve called a **normal curve**, the subject of this section. Figure 7.5.2 shows the **standard normal curve**.

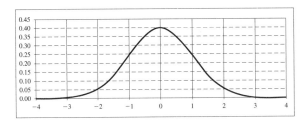

FIGURE 7.5.2 The standard normal curve

The standard normal curve is not just a picture; it has an equation. You don't need to know the equation for the kinds of problems you are going to do, but just so that you've seen it, here it is:

$$f(z) = \frac{1}{\sqrt{2\pi}} \cdot e^{-\left(\frac{z^2}{2}\right)}$$

This is the probability density function (pdf) or density curve for a standard normal random variable z. (The letter z is usually used to stand for the standard normal random variable.) The z in the previous equation is a different type of variable than those we have studied so far in this chapter. It is *continuous*, rather than *discrete*. Recall from Chapter 1 that a quantitative variable is discrete if the possible values for the variable are isolated points on the number line, while the possible values for a continuous variable form a continuous interval on the number line. The possible values for $x =$ *the number of heads in 100 coin tosses* are isolated points: the whole numbers 0, 1, 2, . . ., 100. The value of z in the standard normal pdf does not have to be a whole number, it can be any real number; the possible values form a continuous interval on the number line.

Continuous random variables are quite different from discrete random variables. The most important difference for us is that, for a continuous random variable, probabilities are given by areas under the density curve. Those of you who have had calculus found areas under curves by using integration. Don't worry about that here. Statisticians typically make tables for the areas they need or use calculators (or computers) to find them. That's what we'll be doing also.

Focus on Understanding

1. Look at Figure 7.5.2. Recall from Chapter 3 that the mean is a "balancing point" for the distribution. Based on the graph, find the mean of the standard normal random variable.
2. Since areas under the curve represent probabilities, what is the total area under the entire curve?
3. What is the area to the left of the line $z = 0$? What is the area to the right of the line $z = 0$?

Some Facts about the Standard Normal Distribution

- $\mu = 0$
- $\sigma = 1$
- The density curve $f(z)$ is symmetric about $z = 0$.
- Probabilities correspond to areas under the density curve. See *Table 1: Standard Normal Curve Areas*, located inside the front cover of this text.

Table 1: Standard Normal Curve Areas, located inside the front cover of this book, gives areas under the normal curve to the left of z-values. For example, suppose we wanted to find $P(z \leq 1.28)$. This probability would be the same as the area below the standard normal curve to the left of the vertical line $z = 1.28$. This area is shown in Figure 7.5.3.

To find this area from the standard normal table:

- Find the ones and tenths digits (1.2) in the left column. (See Figure 7.5.4.)
- Find the correct hundredths digit (0.08) in the top row.
- In the row for 1.2 and the column for 0.08, you should find 0.8997. That is the area to the left of $z = 1.28$, so $P(z \leq 1.28) = .8997$.

Suppose we had wanted $P(z < 1.28)$, rather than $P(z \leq 1.28)$? The area for $P(z < 1.28)$ looks exactly like the shaded area in Figure 7.5.3, except it does not

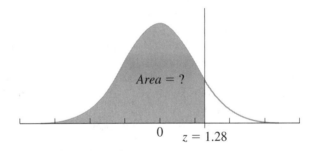

FIGURE 7.5.3 The area represented by $P(z \leq 1.28)$

z_0	0.00	0.01	0.02	0.03	0.04	0.05	0.06	0.07	0.08	0.09
0.5	0.6915	0.6950	0.6985	0.7019	0.7054	0.7088	0.7123	0.7157	0.7190	0.7224
0.6	0.7257	0.7291	0.7324	0.7357	0.7389	0.7422	0.7454	0.7486	0.7517	0.7549
0.7	0.7580	0.7611	0.7642	0.7673	0.7704	0.7734	0.7764	0.7794	0.7823	0.7852
0.8	0.7881	0.7910	0.7939	0.7967	0.7995	0.8023	0.8051	0.8078	0.8106	0.8133
0.9	0.8159	0.8186	0.8212	0.8238	0.8264	0.8289	0.8315	0.8340	0.8365	0.8389
1.0	0.8413	0.8438	0.8461	0.8485	0.8508	0.8531	0.8554	0.8577	0.8599	0.8621
1.1	0.8643	0.8665	0.8686	0.8708	0.8729	0.8749	0.8770	0.8790	0.8810	0.8830
1.2	0.8849	0.8869	0.8888	0.8907	0.8925	0.8944	0.8962	0.8980	0.8997	0.9015
1.3	0.9032	0.9049	0.9066	0.9082	0.9099	0.9115	0.9131	0.9147	0.9162	0.9177
1.4	0.9192	0.9207	0.9222	0.9236	0.9251	0.9265	0.9279	0.9292	0.9306	0.9319

FIGURE 7.5.4 Using Table 1

include the line segment on the edge where $z = 1.28$. What is the area of a line segment? A mathematician would say that the area of a line segment is 0, but students argue with this. They say that if the area was 0, we would not be able to see it. Rather than get into a long philosophical discussion about the difference between a line and a picture of a line, let's just say that the area of a line segment is very small, so small that for practical purposes we'll call it 0. So, for practical purposes, $P(z < 1.28)$ and $P(z \leq 1.28)$ are the same. They are both 0.8997. For the same reason, we would say that $P(z = 1.28) = 0$, since $P(z = 1.28)$ would be the area directly above the single point 1.28 on the z-axis (just a line segment). The point is that, for a continuous random variable, whether the endpoint is included or not makes no difference in the probability; $P(z < z_0) = P(z \leq z_0)$. Note that this is not the case for discrete random variables. If $x = $ *the number of heads showing in 100 coin tosses*, $P(x < 50)$ is *not* the same as $P(x \leq 50)$, and $P(x = 50)$ is not 0.

Suppose we want to find $P(z \geq -0.67)$. This probability is represented by the shaded area in Figure 7.5.5.

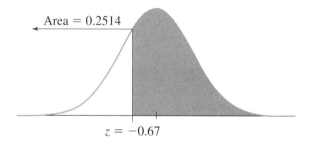

FIGURE 7.5.5 The area for P(z ≥ −0.67)

The problem is that Table 1 gives areas to the left of z-values, and we want the area to the right. There are two approaches we could take:

1. We could use the complement rule. Using Table 1 in the same way as in the previous example, the area to the left of $z = -0.67$ is 0.2514. Since the total area is 1, the area we want is:

$$P(z \geq -0.67) = 1 - P(z < -0.67) = 1 - 0.2514 = 0.7486$$

2. We could use the fact that the standard normal curve is symmetric about $z = 0$. This is faster, but a little harder to picture. Think of flipping the entire picture about the line of symmetry. The area on the right side of $z = -0.67$ is the same as the area on the left side of $z = 0.67$. If we use Table 1 to find the area to the left of $z = 0.67$ we get 0.7486 directly. In other words:

$$P(z \geq -0.67) = P(z \leq 0.67) = 0.7486$$

Both of the methods used are fine. As we go on, we don't want to show both ways every time, so we'll generally just show the first method. (You can use the other method if you want to.)

Suppose we want to find $P(-1.25 < z < -0.50)$. Looking up $z = -1.25$ and $z = -0.50$ in Table 1, we find that $P(z \leq -1.25) = 0.1056$ and $P(z \leq -0.50) = 0.3085$ (see Figure 7.5.6). The area we want is the difference between these two:

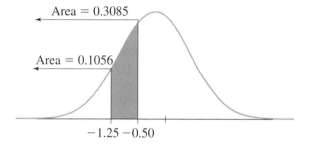

FIGURE 7.5.6 The area for P(−1.25 < z < −0.50)

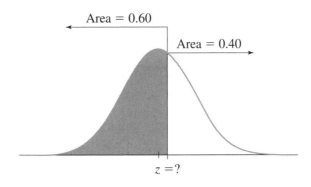

FIGURE 7.5.7 Using a probability to find a z-value

z_0	0.00	0.01	0.02	0.03	0.04	0.05	0.06	0.07	0.08	0.09
0.0	0.5000	0.5040	0.5080	0.5120	0.5160	0.5199	0.5239	0.5279	0.5319	0.5359
0.1	0.5398	0.5438	0.5478	0.5517	0.5557	0.5596	0.5636	0.5675	0.5714	0.5753
0.2	0.5793	0.5832	0.5871	0.5910	0.5948	0.5987	0.6026	0.6064	0.6103	0.6141
0.3	0.6179	0.6217	0.6255	0.6293	0.6331	0.6368	0.6406	0.6443	0.6480	0.6517
0.4	0.6554	0.6591	0.6628	0.6664	0.6700	0.6736	0.6772	0.6808	0.6844	0.6879

FIGURE 7.5.8 Using Table 1 to find a z-value

$$P(-1.25 < z < -0.50) = 0.3085 - 0.1056 = 0.2029$$

So far, we have been starting with z-values and using Table 1 to find areas (probabilities). We can also do the reverse. We can start with a probability (an area) and use it to find a z-value. Suppose that we want to find a value z_0 so that $P(z \geq z_0) = 0.40$. We can use Table 1 again, but in reverse. Instead of starting with z_0 and using it to find the area, we are going to start with the area and use the table to find z_0. We have a slight problem, however. The table gives the area to the left of z_0. The probability we are given is the area to the right. No big deal. If $P(z \geq z_0) = 0.40$, then $P(z < z_0) = 0.60$. (See Figure 7.5.7.)

So we are looking for an area of 0.60 (See Table 1 and Figure 7.5.8). We find two areas in the table that are close to 0.60, both 0.5987 and 0.6026. We'll just use the closer one, 0.5987. By looking at the left column and the top row, we can read off the z-value: $z_0 = 0.25$.

Focus on Understanding

For each of the problems below, draw a normal curve and label it correctly before solving the problem.

1. Find each of the probabilities below for the standard normal random variable z.

a. $P(z \leq 1.75)$
b. $P(z < 1.75)$
c. $P(z = 1.75)$
d. $P(z \geq -0.62)$
e. $P(-1.5 \leq z \leq 2.5)$
f. $P(z < -2 \text{ or } z > 2)$

2. Find a number z_0 so that $P(z \geq z_0) = 0.025$.

So far, we've been working with only one particular normal curve, the <u>standard</u> normal curve. However, most normal curves are <u>not</u> standard. Fortunately, there is a very simple way to adapt what we know about the standard normal curve to non-standard situations.

Non-Standard Normal Curves

Focus on Understanding

Let x = *the speed of a vehicle (in miles per hour) passing a certain checkpoint on I-70*. Suppose that x has a probability distribution that follows a normal curve with a mean of 66.8 and a standard deviation of 2.5.

1. Is the random variable x discrete or continuous?
2. We are given that x has a normal distribution, but is it a *standard* normal distribution? Explain.

Suppose the speed limit at the checkpoint in the preceding *Focus on Understanding* is 65 miles per hour and we want to find the probability that a randomly selected vehicle will be speeding. This probability is represented by the area to the right of 65 in the distribution of x. (See Figure 7.5.9.) To find this area, we are going to use the fact that, if we adjust the scale on the horizontal axis correctly, all normal curves are the same size and shape. If we can find the z-value that is in the same place within the standard normal distribution as the x-value 65, we can use the standard normal table to find the area we want. There is a simple formula for doing this. In fact, you've seen it before. It's our old friend the z-score.

$$\textbf{z-scores}: z = \frac{x - \mu}{\sigma}$$

The z-value we want is:

$$z = \frac{x - \mu}{\sigma} = \frac{65 - 66.8}{2.5} = -0.72$$

So, the area to the right of $x = 65$ is the same as the area to the right of $z = -0.72$. We can find this area using Table 1. Remember, however, that the table gives

214 Chapter 7 Random Variables and Probability Distributions

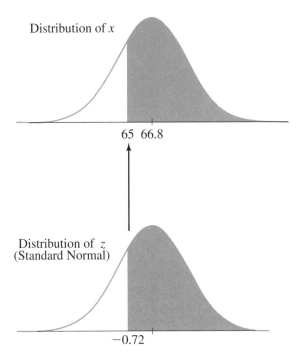

FIGURE 7.5.9 The probability that a vehicle is speeding

the area to the left, rather than the area to the right, so we have to look at the complement.

$$\begin{aligned} P(x > 65) &= P(z > -0.72) \\ &= 1 - P(z \le -0.72) \\ &= 1 - 0.2358 \quad \text{(from Table 1)} \\ &= 0.7642 \end{aligned}$$

So the probability that a randomly selected vehicle will be speeding is 0.7642. In other words, 76.42% of all vehicles are speeding! The police can't stop everyone who's speeding. What if they plan to stop only the fastest 10% of vehicles? At what speed would someone risk getting a ticket? This question is exactly the reverse of the question we just answered. In the first question, we started with an x-value (65 mph) and used it to find an area (0.7642) which represents a probability. In the new problem, we know a probability (10%). We want to use it to find an x-value (the speed at which we'd risk getting a ticket).

In general, there are two types of problems dealing with normal distributions. The first type is outlined in the diagram below:

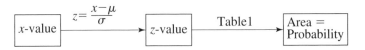

We use the formula $z = \frac{x-\mu}{\sigma}$ to convert the given x-value to a z-value. Then we use Table 1 to find the area representing the probability we want. To reverse this procedure, we need a formula for getting an x-value from a z-value. If we solve the same equation for x instead of z, we get:

$$x = \mu + z\sigma$$

The procedure for the second type of problem is diagrammed below (moving from right to left):

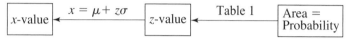

Starting with a given probability representing an area under a normal curve, we use Table 1 to find the z-value that goes with that area. We can then use the formula $x = \mu + z\sigma$ to find the correct x-value.

To find the fastest 10% of vehicles, we are looking for an x-value, call it x_0, so that $P(x \geq x_0) = 0.10$. (See Figure 7.5.10.) Remember that Table 1 gives the area to the left. So, when we look at Table 1, we are looking for an area of 0.90, not 0.10. The closest area in the table is 0.8997, which is the area to the left of $z = 1.28$. Finding the x-value that is in the same place as this z-value:

$$x = \mu + z\sigma = 66.8 + 1.28(2.5) = 70$$

Thus, the fastest 10% of vehicles are those that are traveling at 70 miles per hour or faster.

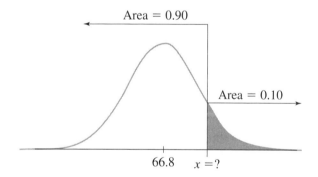

FIGURE 7.5.10 The fastest 10% of vehicles

Focus on Understanding

Assume that we are at the same checkpoint we've been discussing, where speeds are normally distributed with a mean of 66.8 and a standard deviation of 2.5 (in miles per hour). For each of the problems below, draw a normal curve and label it correctly before solving the problem.

1. Find the probability that the next vehicle will be going less than 62 miles per hour.
2. What percent of vehicles travel at speeds between 66 and 68 miles per hour?
3. The slowest 2% of vehicles travel at what speed or slower?

We have been solving problems about normal distributions by using Table 1 to find either areas from z-values or z-values from areas. We can also use the TI-83 plus calculator to accomplish the same result.

Using the TI-83 plus

The TI-83 plus has a built-in cumulative distribution function for the standard normal curve. You get to this function the same way we got to the binomial cdf: hit 2nd DISTR. You should see three functions for normal curves:

```
DISTR DRAW
1:normalpdf(
2:normalcdf(
3:invNorm(
4:tpdf(
5:tcdf(
6:X²pdf(
7↓X²cdf(
```

The first would be useful if we wanted the height of the standard normal curve at any point, but what we usually want is a probability; that is, an area under the curve. For this, we want the second function. Select 2: normalcdf. You should see normalcdf(on your screen. Earlier in this section, we found $P(-1.25 < z < -.50)$, the area between $z = -1.25$ and $z = -0.50$ under the standard normal curve. To find this probability using the TI-83 plus, complete the command to read normalcdf(-1.25, -0.5), and hit ENTER. You should get 0.2028876933, which, rounded to four decimal places, is 0.2029, the same value we got earlier.

We can get even fancier if we want to. We can get the calculator to draw the graph as well. First we have to set the window to see the graph. Hit WINDOW and enter the settings below:

```
WINDOW
  Xmin=-4
  Xmax=4
  Xscl=1
  Ymin=-.2
  Ymax=.5
  Yscl=1
  Xres=
```

Hit 2nd QUIT to get out of the window controls. Now hit 2nd DISTR again, but this time use the right arrow key to move from DISTR to DRAW. Select 1: ShadeNorm(.

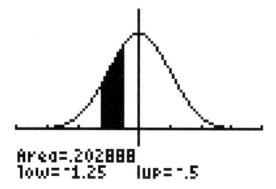

FIGURE 7.5.11 ShadeNorm(−1.25, −0.5) on the TI-83 plus

Complete the command ShadeNorm(−1.25, −0.5), and hit ENTER. Your screen should look something like Figure 7.5.11. Once again, we get the area 0.202888, or 0.2029 to four decimal places.

Note: If you are going to use the ShadeNorm function again, you should hit 2nd DRAW and select 1: ClrDraw first. Otherwise, the calculator will shade the new graph without clearing the previous one, drawing one on top of the other—probably not what you want.

In general, the command normalcdf(z_1, z_2) calculates $P(z_1 \leq z \leq z_2)$, the area between z_1 and z_2 under the standard normal curve. The command ShadeNorm(z_1, z_2) produces a graph shading the same area. In both cases, the TI-83 plus requires both a lower and an upper endpoint for z. What if we wanted to find $P(z \leq 1.28)$, as we did earlier? We have an upper endpoint, but not a lower one. What should we use for the lower endpoint? Notice that in the graphs of the standard normal distribution we have shown, the curve appears to meet the horizontal axis somewhere close to $z = -4$ and $z = 4$. It actually doesn't, but it is very close to 0 there. Looking at Table 1, we find that to 4 decimal places, the area to the left of $z = -4$ and the area to the right of $z = 4$ are both 0. Based on this, it would seem that $z = -4$ could be used in place of a missing left endpoint, and $z = 4$ for a missing right endpoint. It really depends on how many decimal places of accuracy we need. To be on the safe side, we'll use $z = -5$ and $z = 5$. So, to find $P(z \leq 1.28)$, enter normalcdf(−5, 1.28). Your answer, when rounded to 4 decimal places, should be 0.8997, the same value we got earlier.

What about going the other way, starting with an area and finding a z-value? We did this kind of problem earlier when we found a number z_0 so that $P(z \geq z_0) = 0.4$. You might remember the idea of inverse functions from algebra. The function we want is the inverse of the standard normal cdf. Instead of inputting z-values and getting an area, we want to input an area and get a z-value. On the TI-83 plus, this function is invNorm (for inverse normal).

To find a number z_0 so that $P(z \geq z_0) = 0.4$, hit 2nd DISTR. Then select 3: invNorm(, and hit ENTER. Just as in Table 1, the area we need is the area to the left (0.6), not the area to the right (0.4). Complete the command so that it shows invNorm(0.6) and hit ENTER. You should get 0.2533471011, which agrees with the closest value we could find in the table (0.25). In general, the calculator can be much more accurate than the table, since the z-values aren't limited to two decimal places.

What about situations involving non-standard normal distributions? Earlier, for example, we found the probability that a randomly selected vehicle would be traveling over 65 miles per hour. We could do things exactly as we did before, except that we substitute the calculator for Table 1. In other words, we could still use the formula $z = \frac{x-\mu}{\sigma}$ to convert $x = 65$ to $z = -0.72$ and enter in normalcdf(-0.72, 5). We would get 0.7642, just as before. However, the TI-83 plus gives us a shortcut. Rather than converting x-values to z-values, the calculator allows us to enter normalcdf(x_1, x_2, μ, σ) to find the area between $x = x_1$ and $x = x_2$ for a normal distribution with mean μ and standard deviation σ. If we want to find the area to the right of $x = 65$ in a normal distribution with $\mu = 66.8$ and $\sigma = 2.5$ (Figure 7.5.9), the only question is what we should use for the upper endpoint. Based on the idea of using $z = 5$ when using a standard normal distribution, we should be okay if we use a value at least 5 standard deviations to the right of the mean:

$$\mu + 5\sigma = 66.8 + 5(2.5) = 79.3$$

But, rather than calculating this, we could just realize that any rough overestimate is good enough. Let's use $x_2 = 200$. It is highly unlikely that anyone would be traveling over 200 miles per hour. Entering normalcdf(65, 200, 66.8, 2.5) should give the probability that a vehicle is traveling between 65 and 200 miles per hour. You should get 0.7642, as we did before.

Likewise, if we enter invNorm(A, μ, σ), the TI-83 plus will find a number x_0 so that A is the area to the left of x_0 in a normal distribution with mean μ and standard deviation σ. So, to find the speed for the fastest 10% of vehicles, remember that the calculator, like Table 1, uses area to the left. Enter invNorm(0.9, 66.8, 2.5). You should get 70 miles per hour, as before.

Focus on Understanding

Use your calculator to answer the same questions as the last *Focus on Understanding*. Check that you get the same results as before.

You have now seen the two basic types of problems dealing with normal distributions: finding probabilities (areas) for given values of a normal random variable and finding values of the variable corresponding to given areas (probabilities). In the next section, we will use normal curves to approximate probabilities for binomial random variables such as $x =$ *the number of heads in 100 coin tosses*.

7.6 NORMAL APPROXIMATIONS

When Is a Binomial Distribution Approximately Normal?

At the beginning of the previous section (in Figure 7.5.1), we showed a bar graph for the probability distribution of $x = $ *the number of heads showing in 100 coin tosses*. The graph appeared to be very close to a normal curve, but this random variable is not really normal, it is binomial. Each toss is a trial. Every trial has two possible outcomes (heads and tails). The probabilities are the same for each trial. Lastly, the outcome of one toss does not affect the probabilities on the other tosses, so the trials are independent. It satisfies all of the conditions for a binomial random variable, but its distribution is very close to a normal curve, what we would call *approximately* normal.

So, we have seen one case where the distribution of a binomial random variable is approximately normal. It seems reasonable to ask whether this is always true. Is every binomial distribution approximately normal? If not, is there a way we can tell whether the distribution is close to normal without graphing the distribution?

Look at Figure 7.6.1. It shows six different binomial distributions with different values for n (the number of trials) and p (the success probability). Some appear quite close to a normal curve, while others don't look very close at all. In Exploration 7.6, you are going to be arranging the six graphs according to how closely they resemble a normal curve. You won't be getting out your calipers to measure. You're just judging by appearance, so there may be room for some debate, but here are a couple of ideas to consider:

- **Smooth vs. chunky:** These are not technical statistics terms, but we think you'll know what we mean; for example, in Figure 7.6.1, B seems smoother than C. A normal distribution is smooth, not chunky.
- **Symmetric vs. skewed:** In a perfectly **symmetric distribution**, one half of the graph is the exact mirror image of the other. In a **skewed distribution**, one half is more bunched together, and the other is more stretched out horizontally. The stretched out side may appear to have a longer "tail" than the bunched together side, which may appear to have a shorter tail or no tail at all. The distribution is said to be skewed in the direction of the stretched out side, the side with the longer tail. In Figure 7.6.1, E is skewed to the right, F is skewed to the left, and C is symmetric. A normal distribution is symmetric.

Exploration 7.6

Refer to Figure 7.6.1.

1. Arrange the six graphs according to how closely each graph resembles a normal curve. Start with the one that least resembles a normal curve, and end with the one that is closest to normal. In case of doubt, we suggest that you consider smoothness to be a little more important than symmetry.

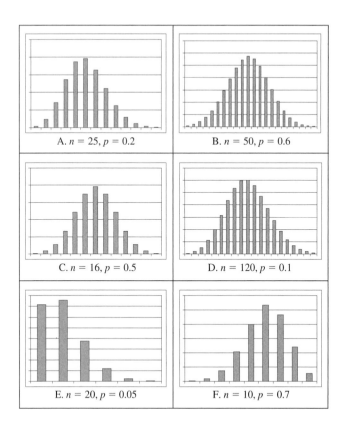

FIGURE 7.6.1 Assorted binomial distributions

2. Is there a direct relationship between the size of *n* and the order you chose? What about the size of *p* and your order? Explain.
3. To find a relationship that makes sense of this order, take the steps listed below:

   a. For each distribution, find the value of *q* (the failure probability) and compute:

   $n \cdot \min(p, q)$ (Note: $\min(p, q)$ is either *p* or *q*, whichever is smaller.)

   b. Now look at the value of $n \cdot \min(p, q)$ for each graph in the order you arranged them in #1.

   c. Complete the following statement: *A binomial distribution gets closer to a normal distribution as the value of* $n \cdot \min(p, q)$ _____.

In Exploration 7.6, you should have found that as the value of $n \cdot \min(p, q)$ increases, the distribution gets closer to normal. But how large does this value need to be for the distribution to be considered approximately normal? The answer really depends

Section 7.6 Normal Approximations 221

on how accurate we need our approximations to be, but we'll follow a common rule of thumb, given in the box below.

When Is a Binomial Distribution Approximately Normal?

A binomial distribution is considered approximately normal if:

$$n \cdot \min(p, q) > 5$$

We now have a method for deciding whether a binomial distribution is approximately normal, but we still have one more issue to deal with before we can use normal distributions to estimate binomial probabilities. This issue has to do with the difference between discrete distributions and continuous distributions.

The Continuity Correction

As an example, suppose we are going to toss 25 coins, and we want to find the probability of getting fewer than 14 heads. Let $x = $ *the number of heads showing in the 25 tosses*. What type of random variable is x? You should say "binomial." Why? Each toss is a trial; there are 25 trials. Each trial has two possible outcomes, heads and tails. A head is a success, since that's what we're counting; a tail is a failure. The success probability is the same on every trial (0.5). Lastly, the trials are independent, since the outcome of one toss does not affect the probabilities on the other tosses.

Since x is binomial, we could use the formula $P(x) = \binom{n}{x} p^x q^{n-x}$ to find the probability of getting fewer than 14 heads, but we would have to do each value of x separately and add the results:

$$P(x < 14) = P(x \leq 13) = P(x = 0, 1, 2, \ldots, 13)$$
$$= P(0) + P(1) + P(2) + \ldots + P(13)$$
$$= \binom{25}{0}(0.5)^0(0.5)^{25} + \binom{25}{1}(0.5)^1(0.5)^{24} + \binom{25}{2}(0.5)^2(0.5)^{23} + \ldots$$
$$+ \binom{25}{13}(0.5)^{13}(0.5)^{12}$$
$$= \ldots \text{ This looks like way too much work! There must be a better way.}$$

There are, in fact, a couple of better ways. One way would be to use a TI-83 plus calculator. We could hit 2nd DISTR and select A: binomcdf(. Notice that for $P(x < 14)$, the highest x-value we are including is 13. So, entering binomcdf(25, 0.5, 13) and rounding to 4 decimal places, we get $P(x < 14) = 0.6550$. Great! But suppose we didn't have a TI-83 plus calculator? In that case, we could use a normal curve to approximate this probability.

In order to use a normal curve to approximate probabilities for a binomial random variable, we need to know that this particular binomial distribution is approximately normal. Using the rule of thumb that we just discussed:

$$n \cdot \min(p, q) = 25(0.5) = 12.5 > 5$$

Since $n \cdot \min(p, q)$ is greater than 5, we know that the distribution of x is approximately normal. The idea will be to use a normal distribution with the same mean and standard deviation as x. So, what are the mean and standard deviation of x? Luckily, since x is binomial, we have nice short formulas for the mean and standard deviation: $\mu = np$ and $\sigma = \sqrt{npq}$.

$$\mu = np = 25(0.5) = 12.5$$

$$\sigma = \sqrt{npq} = \sqrt{25(0.5)(0.5)} = \sqrt{6.25} = 2.5$$

We're going to make up a new random variable (call it y) that has the same mean and standard deviation as x, but has a normal distribution, rather than a binomial distribution. We'll change the x to y, and we'll be all set. But wait! (Don't you just hate when they say that in TV commercials?) We still have a problem.

Notice that when we were finding $P(x < 14)$ above, we switched to $P(x \leq 13)$. This makes sense, since getting fewer than 14 heads is the same as getting 13 heads or fewer. Binomial random variables are discrete; they take on whole number values only. The x in a binomial experiment is the number of successes. You can't have a fractional number of successes. For a binomial random variable x, $P(x < 14) = P(x \leq 13)$. But a normal distribution is not discrete, it is continuous. Since y is a normal random variable, $P(y < 14)$ and $P(y \leq 13)$ are not the same. They differ by the area between $y = 13$ and $y = 14$ under the normal curve. (See Figure 7.6.2.) So which should we use? Let's try each one and see which is closer.

To find $P(y \leq 13)$, we find the z-value that's in the same place as $y = 13$. We use the same formula as we have used before, except we have y instead of x.

$$z = \frac{y - \mu}{\sigma} = \frac{13 - 12.5}{2.5} = 0.20$$

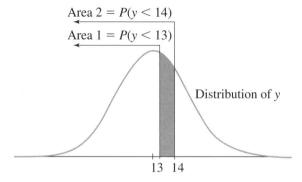

FIGURE 7.6.2 Estimating a binomial probability with a normal curve

Now, using either Table 1 or the TI-83 plus, we get:

$$\text{Area 1} = P(y \leq 13) = P(z \leq 0.20) = 0.5793$$

In the same way, we find $P(y < 14)$:

$$z = \frac{y - \mu}{\sigma} = \frac{14 - 12.5}{2.5} = 0.60$$

$$\text{Area 2} = P(y < 14) = P(z \leq 0.60) = 0.7257$$

Now, let's compare both answers to the value we got earlier. Using **binomcdf**, we found that $P(x \leq 13) = 0.6550$. Using a normal random variable y with the same mean and standard deviation as the binomial random variable x, we found that $P(y \leq 13) = 0.5793$. This is too small! It is off by $0.6550 - 0.5793 = 0.0757$, more than 7% too low. Likewise, we found $P(y < 14) = 0.7257$. This one is too big! It is off by $0.7257 - 0.6550 = 0.0707$, more than 7% too high. As they say in commercials, "But wait!" What if we split the difference between $y = 13$ and $y = 14$, and we used $y = 13.5$? Using the same method:

$$z = \frac{y - \mu}{\sigma} = \frac{13.5 - 12.5}{2.5} = 0.40$$

$$P(y \leq 13.5) = P(z \leq 0.40) = 0.6554$$

Comparing this to the value we got earlier, we are just a little teeny bit too high. We're only off by $0.6554 - 0.6550 = 0.0004$, just four hundredths of 1%. Maybe Goldilocks wouldn't call it "just right," but it's pretty darn close.

What we've just discovered is called the **continuity correction**, a way of adjusting values when we go from a binomial random variable to a normal random variable so that the probabilities work out almost the same. The best approximation for $P(x \leq 13)$ is $P(y \leq 13.5) = P(y \leq 13 + 0.5)$, while the best estimate for $P(x < 14)$ is also $P(y \leq 13.5) = P(y \leq 14 - 0.5)$. Notice that in one case we add 0.5, while in the other we subtract 0.5. We could actually develop four different rules, one for each of the following probabilities: $P(x \leq k)$, $P(x < k)$, $P(x \geq k)$, and $P(x > k)$. Let's try to keep things simple, however, and just concentrate on the first case. As you will see, we can always express any probability we need in terms of the first type, $P(x \leq k)$.

The Continuity Correction

Suppose x is a binomial random variable with $n \cdot \min(p, q) > 5$, so that its distribution is approximately normal. Let y be a normal random variable with the same mean and standard deviation as x. Then, if k is any integer between 0 and n:

$$P(x \leq k) \approx P(y \leq k + 0.5)$$

The symbol $\approx$ means "is approximately equal to."

To demonstrate the versatility of the continuity correction in the box above, let's stick with our 25 coin tosses and use a normal approximation to estimate the probabilities below. We'll compare our estimates to the values we would get from the binomial cdf.

a. $P(x > 10)$
b. $P(10 \leq x \leq 15)$
c. $P(x = 12)$

a. To find $P(x > 10)$, we use the complement rule to rewrite it in terms of $P(x \leq k)$ then use the continuity correction to switch to a probability about a normal random variable y.

$$P(x > 10) = P(x = 11, 12, 13, \ldots, 25)$$
$$= 1 - P(x = 0, 1, 2, \ldots, 10)$$
$$= 1 - P(x \leq 10)$$
$$\approx 1 - P(y \leq 10.5)$$

Now we can follow the usual procedure for finding probabilities in a normal distribution. Remember that we made up the normal random variable y to have the same mean and standard deviation as x: $\mu = 12.5$ and $\sigma = 2.5$

$$z = \frac{y - \mu}{\sigma} = \frac{10.5 - 12.5}{2.5} = -0.80$$

$$P(x > 10) \approx 1 - P(y \leq 10.5)$$
$$= 1 - P(z \leq -0.80)$$
$$= 1 - 0.2119 \text{ (from Table 1 or TI-83 plus)}$$
$$= 0.7881$$

So, $P(x > 10) \approx 0.7881$. For comparison, the actual value of $P(x > 10)$, rounded to 4 decimal places, is 0.7878. Our estimate from the normal approximation is only off by 0.0003.

b.
$$P(10 \leq x \leq 15) = P(x = 10, 11, 12, 13, 14, 15)$$
$$= P(x = 0, 1, \ldots, 15) - P(x = 0, 1, \ldots, 9)$$
$$= P(x \leq 15) - P(x \leq 9)$$
$$\approx P(y \leq 15.5) - P(y \leq 9.5)$$

Now we have two z-values to find; call them z_1 and z_2. (See Figure 7.6.3.)

$$z_1 = \frac{y_1 - \mu}{\sigma} = \frac{9.5 - 12.5}{2.5} = -1.2$$

$$z_2 = \frac{y_2 - \mu}{\sigma} = \frac{15.5 - 12.5}{2.5} = 1.2$$

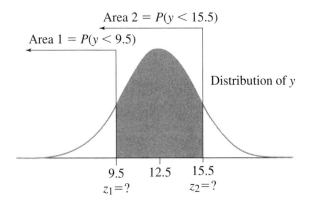

FIGURE 7.6.3 Estimating $P(10 \leq x \leq 15)$

$$P(10 \leq x \leq 15) \approx P(y \leq 15.5) - P(y \leq 9.5)$$
$$= P(z \leq z_2) - P(z \leq z_1)$$
$$= P(z \leq 1.2) - P(z \leq -1.2)$$
$$= 0.8849 - 0.1151$$
$$= 0.7698$$

For comparison, using the binomial cdf, $P(10 \leq x \leq 15) = 0.7705$. Our estimate is only off by 0.0007.

c. This one really shows the need for the continuity correction. If we simply changed x to y, we would have $P(y = 12)$, which represents the area above a single point, in other words a line segment. We would get a probability of zero, clearly incorrect. Instead, we have to be creative to represent $P(x = 12)$ in terms of probabilities in the form $P(x \leq k)$.

$$P(x = 12) = P(x = 0, 1, \ldots, 12) - P(x = 0, 1, \ldots, 11)$$
$$= P(x \leq 12) - P(x \leq 11)$$
$$\approx P(y \leq 12.5) - P(y \leq 11.5)$$

Again, we have two z-values to find; call them z_1 and z_2. (See Figure 7.6.4.)

$$z_1 = \frac{y_1 - \mu}{\sigma} = \frac{11.5 - 12.5}{2.5} = -0.40$$

$$z_2 = \frac{y_2 - \mu}{\sigma} = \frac{12.5 - 12.5}{2.5} = 0$$

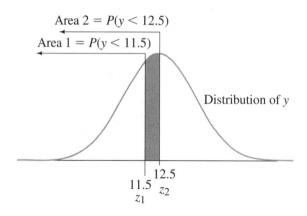

FIGURE 7.6.4 Estimating $P(x = 12)$

$$P(x = 12) \approx P(y \le 12.5) - P(y \le 11.5)$$
$$= P(z \le 0) - P(z \le -0.40)$$
$$= 0.5000 - 0.3446$$
$$= 0.1554$$

The corresponding probability from the binomial pdf is 0.1550, so again our estimate is very good, only off by 0.0004.

Part (c) shows what's really going on here. When we use the continuous random variable y to estimate probabilities for a discrete random variable x, we have to account for the fact that the values of x must be whole numbers, but the values of y include all of those fractional values between the whole numbers. When we find the probability that $x = 12$ using a normal approximation, we include all of the y-values between 11.5 and 12.5. If we were rounding fractional values to the nearest whole number, every number between 11.5 and 12.5 would get rounded to 12. When we do a continuity correction, we are including each fractional value with the closest whole number. (See Figure 7.6.5.) By including each y-value with the closest x-value, we are, in a sense, "filling in" the number line so that no y-values are left out. If we left gaps between the whole numbers, we would end up leaving out areas above those gaps, and our estimated probabilities would be off.

This is directly related to what we would do when constructing a histogram for a data set that has only integer values. The data values would be integers, but the class boundaries would be midway between consecutive integers.

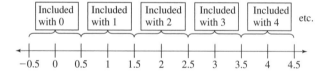

FIGURE 7.6.5 The y-values included with each x-value

8. If a tack is dropped on the floor, there are two possible outcomes: the tack lands on its side (point down), or the tack lands on its head (point up). The probability that a tack will land point up or point down can be determined by experimenting. Kalifa tossed a tack 100 times and recorded the results in the table below.

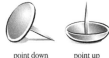

point down   point up

Outcome	Number of times it occurs
Tack lands point up	58
Tack lands point down	42

a. If you dropped Kalifa's tack once, what is the probability that it would land point up? What is the probability that it would land point down?

b. If you dropped Kalifa's tack 500 times, how many times would you expect it to land point up?

c. Is it equally likely that the tack will land point up or point down? Explain.

d. Is it possible to determine theoretical probabilities for this situation? Why or why not?

FIGURE 7.6.6 Tossing tacks

Classroom Connection

Look at Figure 7.6.6. It is taken from *Connected Mathematics: What Do You Expect?* (page 15). ◆

Focus on Understanding

1. Answer the questions in *Tossing Tacks* (Figure 7.6.6).
2. Is tossing a tack 500 times and counting $x =$ *the number of times it lands point up* a binomial experiment? Explain.
3. Find the mean and standard deviation of x. What, if anything, do these have to do with the answers to the questions in *Tossing tacks*?
4. Using the experimental probabilities from Kalifa's 100 tosses, determine whether the distribution of x is approximately normal.
5. Use a normal approximation to find the following probabilities. If possible, compare your results to those you would get from the TI-83 plus's binomial cdf.
   a. $P(x \leq 275)$
   b. $P(x \geq 310)$
   c. $P(280 < x < 300)$
   d. $P(x = 290)$
6. Rory, a middle school student, does not believe the answer to 5(d) can be correct. Since 290 is the expected value of x, he claims that $P(x = 290)$ must be higher. What should you tell him?

Chapter 7 Summary

This chapter was about random variables and the probability distributions associated with those variables. Just as the mean, variance, and standard deviation were used to describe the distribution of a data set in Chapter 3, a random variable also has a mean, variance, and standard deviation. There are two main types of probability distributions—discrete probability distributions (like binomial distributions) and continuous probability distributions (especially normal distributions). Under certain conditions, we can use a normal distribution to approximate probabilities for a binomial distribution, as we did when we found the probability of having at least 310 tacks out of 500 land point up in the last *Focus on Understanding* in this chapter.

Z-scores, introduced back in Chapter 3, play a critical role in the concept of finding the areas under the normal curve from this chapter. Z-scores will be an integral part of the remaining chapters as well as we move into finding confidence intervals and tests of hypotheses.

The key terms and ideas from this chapter are listed below:

chance experiment   184
random variable   184
probability distribution   186
expected value   189
mean of a random variable   190
deviation   193
squared deviation   193
variance of a random variable   195
standard deviation of a random variable   196
z-score   213
repeated identical trials   200

binomial experiment   202
binomial random variable   203
binomial pdf   203
mean of a binomial random variable   190
variance of a binomial random variable   203
standard deviation of a binomial random variable   204
binomial cdf   216
normal curve   208
standard normal curve   208
continuity correction   223

Assessment is an integral part of every curriculum from the elementary school all the way through college. The question always arises—*what is it that students should be able to do after completing this lesson/unit/chapter?* We have included here our intended learning goals for Chapter 7. Students who have a good grasp of the concepts developed in this chapter should be successful in responding to these items:

- Explain, describe, or give an example of what is meant by each of the terms in the vocabulary list.
- Simulate a chance experiment using colored chips, a calculator, or some other means.
- Construct a table for the probability distribution of a random variable, using experimental probabilities obtained from a simulation.
- Draw an area model to represent the outcomes of a chance experiment.
- Draw a tree diagram to represent the outcomes of a chance experiment.
- Construct a table for the probability distribution of a random variable, using theoretical probabilities.

- Determine the mean (expected value), variance, and standard deviation of a random variable.
- Identify a situation as a binomial experiment. Explain the features of the situation that make it a binomial experiment.
- Find probabilities in a binomial experiment.
- Determine the mean (expected value), variance, and standard deviation of a binomial random variable.
- Use z-scores to find probabilities for a normal random variable.
- Use areas from the standard normal curve to determine z-scores and values of a normal random variable.
- Determine whether a binomial distribution will be approximately normal.
- Use the continuity correction and a normal curve to determine probabilities in a binomial experiment.

Your course instructor may have additional or different assessable outcomes for your class. As teachers (or future teachers) you should think about the assessment outcomes and learning goals for each chapter as you work through them.

EXERCISES FOR CHAPTER 7

1. Using the *Data Analysis and Probability Standards for Grades 6–8* from the NCTM's *Principles and Standards for School Mathematics* found at www.nctm.org, identify the middle school objectives that are found in Chapter 7.

2. Using the state standards for mathematics for the content area of data analysis and probability for your state, identify the middle school objectives that are found in Chapter 7. The following website may be useful: www.doe.state.in.us. This website will allow you to access the web pages of the state departments of education for the 50 states. From the state web pages you should be able to find the state's mathematical standards.

3. Seventy percent of the customers at a local discount store pay with cash. Two customers are standing in the check-out line. It is possible that none, one, or both of these customers pay with cash.
    a. Use a simulation to find the experiment probability of none, one, or both customers paying with cash. Run your simulation 20 times. Construct a probability distribution for your experimental probabilities.
    b. Draw an area model to represent the outcomes of this experiment.
    c. Draw a tree diagram to represent the outcomes of this experiment.
    d. Construct the probability distribution for the theoretical probability of none, one, or both customers paying with cash.
    e. Which model—areas or tree diagrams—do you find easier to construct? Why?
    f. Are there some situations where one model might be easier for a student to use than the other? Explain.
    g. Is this situation more like Nikki's situation or Nicky's? Explain.

4. Screamin' Ice Cream keeps track of flavors that customers seem to like the most in order to keep preferred flavors in stock. They know from experience that 65% of their customers prefer vanilla ice cream. The tree diagram below shows all the possible outcomes of three customers ordering ice cream.

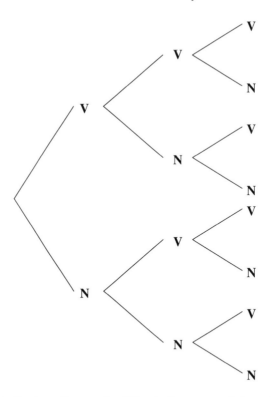

a. Complete the tree diagram by filling in the appropriate probabilities for each branch.
b. Use the tree diagram from part (a) to complete the probability distribution shown below:

Number of Customers Who Order Vanilla	Theoretical Probability
0	
1	
2	
3	

c. Explain how you could simulate this situation using colored chips.

5. Theron has been playing Texas Hold 'Em for more than two years and has kept track of his wins (and losses) to the nearest five dollars. He has learned from experience that the probability he will lose $10 is 10%, lose $5 is 15%, win $0 is 25%, win $10 is 15%, win $25 is 30%, and win $50 is 5%.
   a. Construct a probability distribution table using Theron's experimental probabilities.
   b. Based on this experimental data, on any given night what is Theron's expected value at the poker table?
   c. Find the variance and standard deviation of his probability distribution.

6. Mr. Lange knows from experience that 25% of parents do not show up for parent-teacher conferences. He currently has 12 students in his class (a small class). Consider the event of parents showing up for parent-teacher conferences

as a binomial experiment where the number of trials is 12 and the probability of a success, in this case having a set of parents not show up, to be 25%.
   a. Construct the probability distribution table for this binomial experiment.
   b. Find the mean of this probability distribution. What does this value tell you in this situation?
   c. Find the variance and standard deviation of this probability distribution.
7. Given the following chance experiments, determine which ones are binomial experiments and which are not. Be sure to explain why or why not for each!
   a. Kayla rolls a 12-sided die 20 times. She is interested in the event of rolling a number divisible by 3.
   b. Thirty percent of the customers at the local Dairy Knight purchase shakes, 45% purchase malts, and the rest purchase sodas. For 25 customers selected at random, a record is kept as to whether they ordered shakes, malts, or sodas.
   c. Dr. Susan estimates that 38% of her dental patients need some type of filling when they come in to visit her. She decides to take a sample of the next 10 patients who sit in her chair and see if they need fillings or not.
   d. The number of catalogs received per day at the Perkowski household can be considered a random variable ranging from 1 to 8. The experimental probabilities associated with this random variable are shown in the table below:

#catalogs	1	2	3	4	5	6	7	8
Prob.	0.05	0.10	0.15	0.23	0.28	0.09	0.06	0.04

   e. Cody rolls two, 8-sided dice 30 times. He is interested in the outcome of rolling a number greater than 10.
   f. A local community college has three statistics instructors: Mr. Blue, Mrs. Green, and Ms. White. Past enrollment patterns show that generally Mr. Blue has 30% of the students enrolled in statistics, Mrs. Green has 20%, and Ms. White has 50%. Past records also show that 10% of the students in Mr. Blue's class get A's, 50% of the students in Mrs. Green's class get A's, and 80% of the students in Ms. White's class get A's. Shayna, a college sophomore enrolled at this college, is interested in the event that she will get an A in statistics.
8. Use the information given in exercise 7.7 part (c) to answer the following:
   a. Construct the probability distribution table for the chance experiment.
   b. Find the number of patients that Dr. Susan can expect out of 10 to need fillings.
   c. Find the variance and standard deviation of the distribution.
9. Use the information given in exercise 7.7 part (d) to answer the following:
   a. Find the mean number of catalogs received per day.
   b. Find the variance and standard deviation for the number of catalogs received per day.
10. Use the information given in exercise 7.7 part (e) to answer the following:
    a. Find the expected number of rolls out of 30 where the sum is greater than 10.
    b. Find the variance and standard deviation for the number of rolls out of 30 where the sum is greater than 10.
11. Consider the number of minutes a person has to stand in the check-out line at a large discount store to be a normal random variable with a mean of 6.4 minutes and a standard deviation of 1.5 minutes. What is the probability that a customer will stand in line:
    a. Between 5 and 8 minutes?

b. Less than 3 minutes?
   c. More than 9 minutes?
   d. Less than 8 minutes?
12. Professor Henry knows from past experience that 30% of all homework papers handed in will be missing the name of the person who handed it in. In a class of 200, this is a problem! What is the probability that out of 200 homework papers:
    a. Between 50 and 70 papers will be missing the name?
    b. At least 75 papers will be missing the name?
    c. At most 45 papers will be missing the name?
    d. Exactly 65 papers will be missing the name?
13. Some years there seem to be more hurricanes than others. Suppose the number of hurricanes per year in the Atlantic basin can be treated as an approximately random variable (it is a little chunky) with a mean of 5.8 hurricanes per year and a standard deviation of 1.45 hurricanes per year. What is the probability that:
    a. Exactly 7 hurricanes will occur?
    b. At most 3 hurricanes will occur?
    c. At least 8 hurricanes will occur?
14. Suppose the score on a statistics exam can be treated as a random variable with a normal distribution. If the exam had a mean of 72.6 with a standard deviation of 10.6, what is a student's raw score if:
    a. They received a z-score of 1.25?
    b. They received a z-score of -2.04?
15. Assume that the length of time it takes a student to complete a new state assessment is a random variable with a normal distribution. The probability that a student will take more than 150 minutes on a new state assessment is 0.418. If the standard deviation for the length of time to complete this exam is known to be 10.5 minutes, find the mean length of time to complete the assessment.
16. In certain rural areas, students spend a tremendous amount of time riding the school bus to and from school. Suppose the length of time spent riding the bus to school is a normal random variable with a standard deviation of 15.3 minutes. If the probability is 0.9474 that a student will spend at most 65 minutes on the bus, find the mean length of time spent riding the bus to school.
17. The length of time spent per day on mathematics instruction varies greatly from school to school. Suppose the length of time spent on mathematics instruction per day in schools is a normal random variable with a mean of 27.3 minutes. If the probability is 0.0885 that at least 45 minutes are spent per day on mathematics instruction, find the standard deviation for the length of time spent per day on mathematics instruction.
18. The average teacher salary for a particular district is known to be $38,500 with a standard deviation of $880. If teacher salaries in this district are treated as a random variable with a normal distribution:
    a. Find the salary of a teacher with a salary z-score of 2.34.
    b. Find the salary of a teacher with a salary z-score of -1.65.
    c. Find the salary of a teacher with a salary z-score of 1.08.
    d. Find the salary of a teacher with a salary z-score of -1.32.
    e. What is the probability that a teacher in this district makes more than $41,000?
    f. What is the probability that a teacher in this district makes less than $35,000?
    g. What is the probability that a teacher in this district makes at least $37,000?
    h. One teacher comments that her salary is at the 95th percentile for the district. What is her salary?

19. A fast-food chain claims that 75% of all of its quick-win soda cups are winners. (There is a little tab you pull off the side of the cup which says you are a winner or not a winner.) Brent bought eight sodas in an effort to be a winner.
   a. What is the probability that at least 5 of the cups said he was a winner?
   b. What is the probability that between 3 and 6 of the cups (inclusive) said he was a winner?
   c. What is the probability that at most 3 cups said he was a winner?
   d. Is it reasonable to approximate this distribution using a normal curve? Why or why not?

20. At a small college it is known from previous years that 35% of all freshmen start in College Algebra. The new freshman class has 100 students in it.
   a. Is it reasonable to approximate this distribution using a normal curve? Why or why not?
   b. What is the probability that between 35 and 42 freshmen will start in College Algebra?
   c. What is the probability that at most 24 students will start in College Algebra?
   d. What is the probability that at least 48 students will start in College Algebra?
   e. This particular college tries to keep class size to 24 students for freshmen. Why might the probabilities in parts (c) and (d) be important to know?

CHAPTER 8

Distributions from Random Samples

8.1 RANDOM SAMPLING
8.2 THE DISTRIBUTION OF SAMPLE MEANS
8.3 THE DISTRIBUTION OF SAMPLE PROPORTIONS

Now that you are familiar with random variables and probability distributions, we are ready to explore the idea of random sampling. Remember that, in statistics, a **population** is a group of items (people, objects, etc.) that we are interested in. For practical reasons, it is sometimes either difficult or impossible to collect data from every member of the population. In such cases, a random sample is often used to approximate the population. The main issue we will address in this chapter is how accurate we can expect these approximations to be. How well does a random sample represent the population?

8.1 RANDOM SAMPLING

What Makes a Sample Random?

When used in our everyday language, the word "random" might be most easily described by what it is not. A *random* selection would not be systematic or logical. It would not have a clear order, pattern, or organization. It would not be predictable. Thus, in our everyday understanding of the word "random," most people would think of YXDKW as a random selection of letters, while ABCDE or ACEGI are not.

In statistics, however, the word "random" as in "random sample" has a very precise meaning that has to do with the process by which the sample is selected. In the statistical sense of the word, YXDKW, ABCDE, and ACEGI may all be random samples, depending on the process that was used to select them. In fact, a selection process that excludes the possibility of getting ABCDE or ACEGI would not be considered random at all.

Random Samples

What is the best way to choose a sample from a large population? In most situations, statisticians agree that it is preferable to use a procedure that gives each member of the population the same chance of being chosen. Sampling plans with this property are called *random sampling plans*. Samples chosen with a random sampling plan are called *random samples*.

FIGURE 8.1.1 Random samples

Read the paragraph in Figure 8.1.1, in which *Connected Mathematics* introduces the idea of random sampling (*Samples and Populations: Random Samples*, page 37). In it, the authors define a random sample as a sample chosen using "a procedure that gives each member of the population the same chance of being chosen." Most college textbooks use a subtly different definition, similar to the definition in the box below.

Random Samples

A **random sample of size n** is a sample of n items chosen from the population in such a way that every possible sample (of the same size) is equally likely to be chosen.

Are these two definitions equivalent? That is the subject of Exploration 8.1.

Classroom Exploration 8.1

Suppose a middle school class consists of 24 students seated in 6 rows of 4 students each. We plan to select a sample of 4 different students from the class. Here is a diagram of the class:

Row 1	Row 2	Row 3	Row 4	Row 5	Row 6
Amy	Eddie	Ira	Max	Quincy	Ursula
Bailey	Flo	Juan	Nick	Raul	Violet
Chrissy	George	Kareem	Ophelia	Sammy	Will
Dave	Holly	Lori	Paul	Tom	Xavier

1. How many different possible samples of 4 students can be chosen from the class? (Hint: Think back to Chapter 6. When choosing a sample of 4 students from the 24 students in the class, are repetitions allowed? Does order matter?) Suppose we compare two different sampling plans, *Plan A* and *Plan B*.

   Plan A: We write each student's name on a separate 3" by 5" index card. We shuffle the 24 cards and choose 4 of them. Our sample will be the 4 students whose names are selected.

> *Plan B*: We roll an ordinary six-sided number cube. If we roll a 1, our sample will be the four students in Row 1; if we roll a 2, we'll use the four students in Row 2; etc.
>
> 2. Using *Plan A*, are all of the possible samples you counted in #1 equally likely to be chosen? What is the probability of each sample?
> 3. Using *Plan B*, are all of the possible samples you counted in #1 equally likely to be chosen? List a sample of four students that is impossible to get using *Plan B*.
> 4. Using *Plan B*, what is the probability that Amy is a member of the chosen sample? Is this probability the same for Flo? Does each member of the population have the same probability of being chosen?
> 5. Using *Plan A*, how many of the possible samples contain Amy? What is the probability that Amy is in the selected sample? Would the answers to these questions be different for Flo? Does each member of the population have the same probability of being chosen?

As you have seen in the Exploration 8.1, the two different definitions are not equivalent. In general, if every possible sample (of the same size) is equally likely to be chosen, then every member of the population will be equally likely to be chosen. The converse, however, is not true. You can have a sampling plan where all members of the population are equally likely to be chosen, but all possible samples are not equally likely (in fact, some samples may never occur).

Perhaps the definition used in most college textbooks is a bit too complicated for middle school students. In any case, when we use the term **random sample**, we will be assuming the second definition, that all possible samples (of the same size) are equally likely. From this point on, just about everything we do in this book will be based on random samples. We will try to emphasize this, but if we sometimes slip and just say sample, you can probably infer that we mean a random sample.

Selecting Random Samples

Connected Mathematics continues the exploration of random samples by having students actually select some. Look at the excerpts from *Samples and Populations: Random Samples* (pages 38 and 39) in Figures 8.1.2 and 8.1.3.

Notice some important ideas that are introduced under "Think about this!" in Figure 8.1.2. First, they suggest that students use the data from random samples to make predictions about the population that the samples came from. Using a sample to make a prediction about the population is called **inference**. Inference is the subject of the field of statistics called **inferential statistics**, and our main focus for the rest of this book. Secondly, they get students thinking about different methods that could be used to select random samples (spinners, number cubes, random number tables, calculators). You will be doing the same thing shortly. Lastly, they get students thinking about the importance of sample size, something that we will address in detail as we go on in this chapter and the next.

Selecting a Random Sample

The table on the next page contains data for 100 eighth graders at Clinton Middle School. The data were collected on a Monday. They include the number of hours of sleep each student got the previous night and the number of movies, including television movies and videos, each student watched the previous week.

You could describe these data by calculating five-number summaries or means, and you could display the distribution of the data by making stem plots, histograms, or box plots. However, doing calculations and making graphs for the entire data set would require a lot of work.

> **Think about this!**
>
> Instead of working with the entire data set, you can select a random sample of students. You can look for patterns in the data for the sample and then use your findings to make predictions about the population.
> - What methods might you use to select a random sample of students?
> - How many students would you need in your sample in order to make accurate estimates of the typical number of hours of sleep and the typical number of movies watched for the entire population of 100 students?

FIGURE 8.1.2 Selecting a random sample

In *Connected Mathematics*, students go on to select random samples of different sizes and compare medians and five-number summaries for the different samples. As we proceed in this book, we are going to be more concerned with means and proportions than with medians and five-number summaries. In the following *Focus on Understanding*, we will concentrate on means. It is important to distinguish between the mean of the sample (or **sample mean**), symbolized by $\bar{x}$, and the mean of the population (**population mean**), symbolized by μ.

Focus on Understanding

1. Describe two different methods for selecting a random sample of five different students from the *Grade 8 Database* in Figure 8.1.3 (the population). Don't just say "Use spinners." or "Use a calculator." What would the spinners look like? How would you use them? How would you make sure that you get five *different* students?

2. Choose one of your methods from #1 and use it to select a random sample of size 5. Compute the mean number of sleep hours for your sample. Based on the mean of this single sample, what would be your best guess for the mean of the population?

3. Record your sample mean on the board. Continue selecting random samples and recording the results until your class has created a list of 60 to 100 sample means.

Grade 8 Database

Student number	Gender	Sleep (hours)	Movies	Student number	Gender	Sleep (hours)	Movies
01	boy	11.5	14	51	boy	5.0	4
02	boy	2.0	8	52	boy	6.5	5
03	girl	7.7	3	53	girl	8.5	2
04	boy	9.3	1	54	boy	9.1	15
05	boy	7.1	16	55	girl	7.5	2
06	boy	7.5	1	56	girl	8.5	1
07	boy	8.0	4	57	girl	8.0	2
08	girl	7.8	1	58	girl	7.0	7
09	girl	8.0	13	59	girl	8.4	10
10	girl	8.0	15	60	girl	9.5	1
11	boy	9.0	1	61	girl	7.3	5
12	boy	9.2	10	62	girl	7.3	4
13	boy	8.5	5	63	boy	8.5	3
14	girl	6.0	15	64	boy	9.0	3
15	boy	6.5	10	65	boy	9.0	4
16	boy	8.3	2	66	girl	7.3	5
17	girl	7.4	2	67	girl	5.7	0
18	boy	11.2	3	68	girl	5.5	0
19	girl	7.3	1	69	boy	10.5	7
20	boy	8.0	0	70	girl	7.5	1
21	girl	7.8	1	71	boy	7.8	0
22	girl	7.8	1	72	girl	7.3	1
23	boy	9.2	2	73	boy	9.3	2
24	girl	7.5	0	74	boy	9.0	1
25	boy	8.8	1	75	boy	8.7	1
26	girl	8.5	0	76	boy	8.5	3
27	girl	9.0	0	77	girl	9.0	1
28	girl	8.5	0	78	boy	8.0	1
29	boy	8.2	2	79	boy	8.0	4
30	girl	7.8	2	80	boy	6.5	0
31	girl	8.0	2	81	boy	8.0	0
32	girl	7.3	8	82	girl	9.0	8
33	boy	6.0	5	83	girl	8.0	0
34	girl	7.5	5	84	boy	7.0	0
35	boy	6.5	5	85	boy	9.0	6
36	boy	9.3	1	86	boy	7.3	0
37	girl	8.2	3	87	girl	9.0	3
38	boy	7.3	3	88	girl	7.5	5
39	girl	7.4	6	89	boy	8.0	0
40	girl	8.5	7	90	girl	7.5	6
41	boy	5.5	17	91	boy	8.0	4
42	boy	6.5	3	92	boy	9.0	4
43	boy	7.0	5	93	boy	7.0	0
44	girl	8.5	2	94	boy	8.0	3
45	girl	9.3	4	95	boy	8.3	3
46	girl	8.0	15	96	boy	8.3	14
47	boy	8.5	10	97	girl	7.8	5
48	girl	6.2	11	98	girl	8.5	1
49	girl	11.8	10	99	girl	8.3	3
50	girl	9.0	4	100	boy	7.5	2

FIGURE 8.1.3 Data from 100 eighth graders

4. Make a dot plot (also called a line plot) for the sample means generated by your class.
5. Repeat #2 through #4, but this time use samples of size 20. Make your dot plot directly above the dot plot for samples of size 5, using the same horizontal axis.
6. Compare the two dot plots of sample means. Comment on the shape, spread, and center of the two distributions.

7. Based on your dot plots, what would be your guess for the mean of the population? Which would seem better for predicting the mean of a population, a random sample of size 5 or a random sample of size 20? Explain, based on your dot plots.

Figure 8.1.4 shows some of the results that a typical class might get for the preceding *Focus on Understanding*. Since the random samples selected by your class will not be the same as the ones we selected, you should not expect your results to match ours exactly. You should, however, still see the same general ideas. We used 100 random samples of each size.

We will also tell you that the mean of the population (sleep hours in the Grade 8 Database) is 7.962 hours. While we are pretty sure none of you guessed exactly 7.962 by looking at your dot plots, we bet most people were pretty close. Both distributions seem to have their center slightly below 8. We might guess that the mean of each of these distributions is pretty close to the 7.962, the mean of the population.

What about the spread and shape of the two distributions? The means from the smaller samples are generally more spread out. The distribution of means from the larger samples is more mound-shaped (maybe close to normal) and clumped together closer to the center. Compare the ranges of the two distributions:

For the means of the smaller samples: *Range* = max − min = 9.06 − 5.76 = 3.30
For the means of the larger samples: *Range* = max − min = 8.77 − 7.18 = 1.59

Based on their ranges, we'd say that the means from samples of size 5 have about twice as much spread as the means from samples of size 20.

As we noted previously, you probably were able to get a pretty good guess for the mean of the population by looking at the dot plots of the sample means. Before you get too happy about this, however, remember that in most practical situations, we would not be looking at 100 random samples as we did above, but only one. This is fine, if we chose a sample whose mean is close to the center of the distribution. However, if we have chosen a sample whose mean is at the far left or far right end of the distribution, our guess for the population mean will be off by quite a bit. This is why larger samples are better for predicting the mean of the population; the means from larger samples are less spread out, staying closer to the mean of the population. Hence, we are more likely to get a sample mean close to the true population mean.

Let's explore this idea a bit further. What percent of the 100 samples of size 5 (in Figure 8.1.4) are within half an hour of the population mean? (See Figure 8.1.5.) Looking at the list of sample means in Figure 8.1.4, there are 45 of the 100 samples of size 5 with means between 7.462 and 8.462 (starting with 7.54 and ending with 8.46). Therefore, 45% of these samples of size 5 have means within a half-hour of the population mean. We can interpret this as the experimental probability that a sample of size 5 has a mean within 0.5 of the population mean.

Let's do the same thing for the samples of size 20. Again, looking at the list in Figure 8.1.4, there are only 9 sample means that are <u>not</u> between 7.462 and 8.462

240 Chapter 8 Distributions from Random Samples

Means for 100 Random Samples of Size 5 from Sleep Hours Data (sorted)

5.76	7.04	7.34	7.54	7.80	7.96	8.18	8.36	8.52	8.76
6.58	7.10	7.36	7.60	7.80	8.00	8.20	8.38	8.52	8.82
6.66	7.10	7.36	7.66	7.82	8.00	8.20	8.44	8.54	8.84
6.80	7.20	7.36	7.66	7.86	8.00	8.20	8.44	8.58	8.90
6.86	7.22	7.40	7.68	7.86	8.00	8.22	8.46	8.62	8.90
6.90	7.22	7.42	7.70	7.88	8.06	8.22	8.48	8.62	8.96
6.90	7.26	7.42	7.72	7.90	8.10	8.26	8.50	8.62	8.96
6.96	7.30	7.44	7.72	7.92	8.10	8.26	8.50	8.66	9.00
7.00	7.30	7.46	7.76	7.92	8.10	8.28	8.50	8.70	9.04
7.02	7.32	7.46	7.80	7.94	8.16	8.30	8.50	8.76	9.06

Means for 100 Random Samples of Size 20 from Sleep Hours Data (sorted)

7.180	7.600	7.675	7.770	7.860	7.930	8.020	8.110	8.170	8.300
7.265	7.605	7.700	7.775	7.865	7.940	8.020	8.110	8.185	8.310
7.365	7.615	7.705	7.785	7.870	7.945	8.030	8.110	8.185	8.355
7.395	7.615	7.720	7.790	7.870	7.970	8.035	8.125	8.190	8.395
7.445	7.625	7.730	7.800	7.870	7.975	8.050	8.130	8.215	8.400
7.445	7.650	7.745	7.815	7.895	7.980	8.070	8.145	8.275	8.445
7.490	7.660	7.755	7.835	7.905	7.985	8.080	8.150	8.275	8.455
7.515	7.665	7.760	7.840	7.915	7.985	8.085	8.160	8.280	8.490
7.535	7.670	7.760	7.850	7.915	8.010	8.085	8.165	8.285	8.655
7.585	7.675	7.765	7.850	7.930	8.015	8.095	8.170	8.300	8.770

Dot Plots for the Two Lists of Sample Means

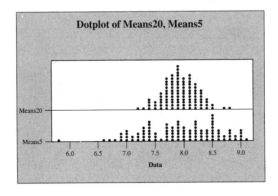

FIGURE 8.1.4 Some results for the means of random samples

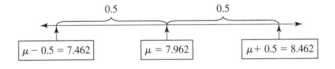

FIGURE 8.1.5 Values within half an hour of the population mean

(6 are too low, and 3 are too high). So, 91 of the 100 samples of size 20 have means within a half-hour of the population mean. In other words, 91% or 0.91 is the experimental probability that a sample of size 20 has a mean within 0.5 of the population mean.

Notice that samples of size 20 have a greater probability (0.91 as opposed to 0.45) of getting a sample mean within a fixed distance (0.5) of the population mean than samples of size 5. In general, we increase the probability of getting a good prediction about the population if we use larger samples.

Based on what we've done so far in this chapter, if we were using the mean of a random sample to predict the value of the population mean, would we be confident that the mean of the sample was exactly the same as the population mean? No. In fact, none of the sample means listed in Figure 8.1.4 matched the population mean exactly. That is why statisticians typically give a "margin of error" for their predictions.

Suppose that we allowed ourselves a half-hour margin of error when we make predictions about the mean number of sleep hours in the Grade 8 Database. For example, if we chose a random sample with a mean of 7.60, we would predict that the population mean was somewhere between 7.10 and 8.10. In general, we would predict that the population mean was between $\bar{x} - 0.5$ and $\bar{x} + 0.5$. If that was our plan, our prediction would be true as long as the value of $\bar{x}$ was within 0.5 of the population mean μ.

For example, if $\bar{x} = 7.44$, the distance from $\bar{x}$ to μ is $7.962 - 7.44 = 0.522$, so this value of $\bar{x}$ is not within 0.5 of μ. Our prediction would be that the population mean was between $7.44 - 0.5 = 6.94$ and $7.44 + 0.5 = 7.94$. Since $\mu = 7.962$ is not between 6.94 and 7.94, our prediction would have been wrong for this particular sample.

From the work we did earlier, 45% of the means from our samples of size 5 and 91% of the means from our samples of size 20 were within 0.5 of the population mean. Therefore, if our predictions included a half-hour margin of error, we would have made correct predictions 45% of the time for sample size 5 and 91% of the time for sample size 20.

In Chapter 9, we will continue to explore the relationship between the accuracy of predictions and the sample size. We will develop formulas relating the probability of correct predictions to both the sample size and the margin of error.

Focus on Understanding

Using the means for your class's random samples of size 5 and 20, answer the questions below and compare your results to those that we got above. Again, since they are based on different random samples, you should not expect to get exactly the same results, but they should be similar.

1. Find the range of the sample means from your samples of size 5. Compare it to the range of sample means for samples of size 20. Based on these ranges, which distribution has more spread (or variability)?

2. Find the experimental probability that the mean of a random sample of size 5 is within a half-hour of the population mean ($\mu = 7.962$). Do the same for random samples of size 20. Do the larger samples give a greater probability that $\bar{x}$ is close to μ?
3. If you allowed a half-hour margin of error when using a sample mean to predict the value of the population mean, what would you have predicted about μ from your very first sample of size 5? What about your first sample of size 20? Would these predictions have been true or false?
4. For samples of size 5, if you allowed a half-hour margin of error when using a sample mean to predict the value of the population mean, what percent of your predictions would have been true? Answer the same question for samples of size 20.

8.2 THE DISTRIBUTION OF SAMPLE MEANS

The mean of a random sample is often used to estimate or predict the mean of a population. When used in this way, we call $\bar{x}$ an *estimator* for μ. A particular value of $\bar{x}$ is called an *estimate* of μ. If, as in the previous section, we choose several samples from the same population, the value of $\bar{x}$ will vary from sample to sample. This makes $\bar{x}$ a random variable, and as a random variable, it has a probability distribution. In the previous section, when we took 100 sample means from samples of size 20, we were looking at 100 values of the random variable $\bar{x}$. By doing so, we began to develop some ideas about the distribution of $\bar{x}$. In this section, we will take a more thorough and systematic look at the probability distribution of $\bar{x}$.

Our first goal will be to show that, when based on random samples from a given population, $\bar{x}$ has three properties that make it particularly good for estimating μ:

- $\bar{x}$ is an unbiased estimator of μ (more about this shortly).
- The value of $\bar{x}$ tends to get closer to μ as the sample size n increases.
- When n is large enough, $\bar{x}$ has a familiar type of distribution.

Is $\bar{x}$ an Unbiased Estimator?

In general, an **estimator** is a statistic computed from a sample for the purpose of predicting the value of a population characteristic. An estimator is also a random variable; its value will vary from sample to sample and is determined by the outcome of a chance experiment (choosing the sample). As a random variable, an estimator has a probability distribution. There are three different ways in which the mean of this probability distribution can relate to the true value of the population characteristic. These three possibilities are illustrated in Figure 8.2.1.

In all three cases, the distribution of the estimator extends both above and below the true value of the population characteristic, indicating some estimates are above the true value and some are below. In Case 1, the average estimate lies below the true value of the population characteristic. On average, this statistic

Case 1: The statistic underestimates the true value on average.

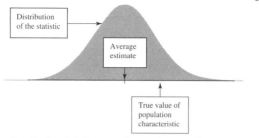

Case 2: The statistic overestimates the true values on average.

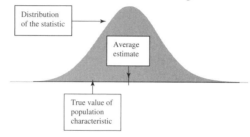

Case 3: The mean value of the statistic hits the true value exactly.

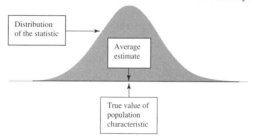

FIGURE 8.2.1 Three possibilities for the distribution of an estimator

underestimates the true value of the population characteristic. In Case 2, the average estimate is greater than the true value of the population characteristic; the statistic overestimates the true value on average.

The last case is the one we really want. In Case 3, even though some estimates are above the true value and others are below, the underestimates and overestimates balance out so that the average estimate hits the true value exactly. On average, this statistic neither underestimates nor overestimates the true value. A statistic with this property is called an **unbiased estimator**.

In the previous section, the average value of our sample means was close to the population mean, but the two were probably not exactly the same. Remember, however, that we were not looking at all possible values of $\bar{x}$ (all possible estimates), but only the values of $\bar{x}$ for the samples we happened to select. We will now be looking at all possible samples (of a given size).

EXAMPLE 8.2.1 Change for a Dollar?

When our son gets pennies and nickels in change, he usually doesn't spend them, but lets them pile up in a large container in his room (the population). Suppose that 80% of these coins are pennies and 20% are nickels. If x = *the value of a coin* in this population, the probability distribution for x is given at the right. Let's start by finding the mean of this population.

x	$P(x)$
1	0.8
5	0.2

$$\mu = \sum [x \cdot P(x)]$$
$$= 1(0.8) + 5(0.2)$$
$$= 1.8$$

Suppose for the moment that we didn't know the mean of the population and were planning on estimating μ by using $\bar{x}$ from a random sample. To keep things very simple, we'll start by considering random samples of size 2. We'll also assume that the coins are chosen independently (either the coins are chosen with replacement, or there are enough coins that removing one coin does not significantly affect the probabilities for the next coin). Figure 8.2.2 shows a tree diagram for the process of selecting a random sample of two coins. For each sample, we've computed the value of $\bar{x}$. ∎

There are four possible samples of size 2, corresponding to the four different paths through the tree in Figure 8.2.2. Figure 8.2.3 shows the four possible samples along with their means and probabilities. You should understand how the probabilities are computed from the tree diagram.

Based on the information in Figure 8.2.3, you should be able to determine the probability distribution of $\bar{x}$ and answer the questions in the *Focus on Understanding* that follows.

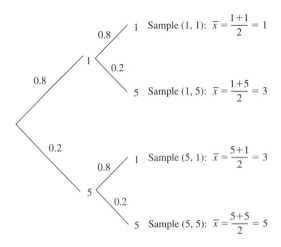

FIGURE 8.2.2 Selecting a random sample of size 2

Section 8.2 The Distribution of Sample Means

Sample	$\bar{x}$	Probability
(1, 1)	$\frac{1+1}{2} = 1$	$(0.8)(0.8) = 0.64$
(1, 5)	$\frac{1+5}{2} = 3$	$(0.8)(0.2) = 0.16$
(5, 1)	$\frac{5+1}{2} = 3$	$(0.2)(0.8) = 0.16$
(5, 5)	$\frac{5+5}{2} = 5$	$(0.2)(0.2) = 0.04$

FIGURE 8.2.3 The four possible samples of size 2

Focus on Understanding

$\bar{x}$	$P(\bar{x})$

1. Use the information in Figure 8.2.3 to complete the table at the right for the probability distribution of $\bar{x}$.
2. Recall that the samples came from a population with $\mu = 1.8$. In this case, how likely is it for the sample mean to be the same as the population mean? In other words, what is $P(\bar{x} = \mu)$?
3. For some samples, the sample mean underestimates the population mean. Find $P(\bar{x} < \mu)$.
4. For other samples, the sample mean overestimates the population mean. Find $P(\bar{x} > \mu)$.

In this case, the sample mean is more likely to underestimate the population mean than overestimate it. Because of this, we might expect that $\bar{x}$ would tend to underestimate μ on average. Let's look at the long-term average value of $\bar{x}$ and see. To find the long-term average value of $\bar{x}$, symbolized by $E[\bar{x}]$ or $\mu_{\bar{x}}$, we find the expected value of $\bar{x}$ as a random variable, just as we did in Section 7.2.

$$E[\bar{x}] = \mu_{\bar{x}} = \sum [\bar{x} \cdot P(\bar{x})]$$
$$= 1(0.64) + 3(0.32) + 5(0.04)$$
$$= 1.8 = \mu$$

Notice that the expected value of the sample mean is the same as the mean of the population. Even though some samples underestimate the population mean and others overestimate the population mean, in the long run the overestimates and underestimates balance out. The average value of the sample mean hits the population mean exactly. The sample mean is an unbiased estimator of the population mean. This is true in general, not just in this situation, and is important enough to put in a special box like the one below.

The Expected Value of $\bar{x}$

If $\bar{x}$ is computed from a random sample taken from a population with mean μ, then $\bar{x}$ is an unbiased estimator of μ:

$$E[\bar{x}] = \mu \quad \text{or} \quad \mu_{\bar{x}} = \mu$$

Note: The result above depends on the fact that the samples are <u>random</u>. If sampling is done in a biased way, the sample mean will be biased as well. The best way to avoid biased sampling is to use random samples.

Does $\bar{x}$ Get Closer to μ as *n* Increases?

To determine whether the values of $\bar{x}$ tend to get closer to μ as the sample size increases, we need to look at how the variance of the distribution of $\bar{x}$ compares to the variance of the population. We've discovered that $\mu_{\bar{x}} = \mu$, that is, the mean of the distribution of $\bar{x}$ is the same as the mean of the population. Could it also be true that the variance of the distribution of $\bar{x}$ is the same as the variance of the population? In other words, does $\sigma_{\bar{x}}^2 = \sigma^2$?

We'll start by computing the variance σ^2 for the population being sampled. For your convenience, we've shown the distribution for the population at the right. We'll use the computing formula for variance: $Var(x) = \sigma^2 = \sum[x^2 \cdot P(x)] - \mu^2$. Remember from before that $\mu = 1.8$.

x	P(x)
1	0.8
5	0.2

$$Var(x) = \sigma^2 = \sum[x^2 \cdot P(x)] - \mu^2$$
$$= [1^2(0.8) + 5^2(0.2)] - 1.8^2$$
$$= 2.56$$

So, the population we're sampling has a variance of 2.56. We'll need to compare this to the variance of the distribution of $\bar{x}$. Again, the distribution of $\bar{x}$ is given at the right. Using the same method as above, we'll compute $Var(\bar{x})$. Remember, since $\bar{x}$ is an unbiased estimator of μ, $\mu_{\bar{x}} = \mu = 1.8$.

$\bar{x}$	$P(\bar{x})$
1	0.64
3	0.32
5	0.04

$$Var(\bar{x}) = \sigma_{\bar{x}}^2 = \sum[\bar{x}^2 \cdot P(\bar{x})] - (\mu_{\bar{x}})^2$$
$$= [1^2(0.64) + 3^2(0.32) + 5^2(0.04)] - 1.8^2$$
$$= 1.28$$

Notice that this value is <u>not</u> the same as the variance of the population. Instead, it is exactly half as much; that is: $\sigma_{\bar{x}}^2 = \frac{\sigma^2}{2}$. Let's think about this. We took samples of size <u>2</u>, and $Var(\bar{x})$ was σ^2 divided by <u>2</u>. What do you suppose will happen if we took samples of size <u>3</u>? Maybe $\sigma_{\bar{x}}^2 = \frac{\sigma^2}{3}$? Let's check and see.

Sample	$\bar{x}$	Probability
(1, 1, 1)	$\frac{1+1+1}{3} = 1$	$(0.8)(0.8)(0.8) = 0.512$
(1, 1, 5)	$\frac{1+1+5}{3} = \frac{7}{3}$	$(0.8)(0.8)(0.2) = 0.128$
(1, 5, 1)	$\frac{1+5+1}{3} = \frac{7}{3}$	$(0.8)(0.2)(0.8) = 0.128$
(1, 5, 5)	$\frac{1+5+5}{3} = \frac{11}{3}$	$(0.8)(0.2)(0.2) = 0.032$
(5, 1, 1)	$\frac{5+1+1}{3} = \frac{7}{3}$	$(0.2)(0.8)(0.8) = 0.128$
(5, 1, 5)	$\frac{5+1+5}{3} = \frac{11}{3}$	$(0.2)(0.8)(0.2) = 0.032$
(5, 5, 1)	$\frac{5+5+1}{3} = \frac{11}{3}$	$(0.2)(0.2)(0.8) = 0.032$
(5, 5, 5)	$\frac{5+5+5}{3} = 5$	$(0.2)(0.2)(0.2) = 0.008$

FIGURE 8.2.4 The eight possible samples of size 3

Remember, in this case we are drawing samples from a population that is 80% pennies and 20% nickels. The eight possible samples of size 3 are listed in Figure 8.2.4, along with their sample means and probabilities. We are finding the probabilities exactly as we did before, except that we didn't draw the tree diagram. (It's a little big, but you could draw it if you wanted to.)

Again, treating $\bar{x}$ as a random variable, we get the probability distribution at the right. Once again, it is never true that $\bar{x} = \mu$. Some samples have means that are below $\mu = 1.8$, and some samples have means above 1.8. But, based on what we did earlier, we would expect that $\mu_{\bar{x}} = 1.8$. Let's check and see.

$\bar{x}$	$P(\bar{x})$
1	0.512
$\frac{7}{3}$	0.384
$\frac{11}{3}$	0.096
5	0.008

$$\mu_{\bar{x}} = \sum [\bar{x} \cdot P(\bar{x})]$$
$$= 1(0.512) + \frac{7}{3}(0.384) + \frac{11}{3}(0.096) + 5(0.008)$$
$$= 1.8 = \mu$$

So, just as we expected! Now, we'll look at $Var(\bar{x})$.

$$\sigma_{\bar{x}}^2 = \sum [\bar{x}^2 \cdot P(\bar{x})] - (\mu_{\bar{x}})^2$$
$$= \left[1^2(0.512) + \left(\frac{7}{3}\right)^2(0.384) + \left(\frac{11}{3}\right)^2(0.096) + 5^2(0.008)\right] - 1.8^2$$
$$= 0.8533$$
$$= \frac{2.56}{3}$$

Just as we suspected, when we use samples of size 3, $Var(\bar{x})$ is σ^2 divided by 3. This is true in general, and worthy to be displayed in a box.

The Variance and Standard Deviation of the Distribution of Sample Means

If $\bar{x}$ is computed from a random sample of size n taken from a population with variance σ^2, then:

$$Var[\bar{x}] = \frac{\sigma^2}{n} \quad \text{or} \quad \sigma_{\bar{x}}^2 = \frac{\sigma^2}{n}$$

It follows that:

$$\sigma_{\bar{x}} = \frac{\sigma}{\sqrt{n}}$$

Pause to think about the consequences of this result. As n increases, the denominators in the formulas above increase, decreasing the variance and standard deviation of the distribution of $\bar{x}$. Variance and standard deviation are measures of spread; the smaller they are, the closer the values are to the mean of the distribution. Thus, as n increases, the values of $\bar{x}$ will be less spread out, staying closer to their average value, μ. So, we have shown that the values of $\bar{x}$ tend to get closer to μ as the sample size increases. We've shown what middle school students' intuition tells them: larger samples tend to give more accurate estimates.

Focus on Understanding

In Section 8.1, we noted that the means from samples of size 5 seemed to be about twice as spread out as the means for samples of size 20. (See Figure 8.1.4.) Explain why this should be so. (Hint: Look at $\sigma_{\bar{x}}$ for $n = 5$ and $n = 20$ and do some algebra.)

Does $\bar{x}$ have a Familiar Distribution?

We need to discuss one more issue about the distribution of $\bar{x}$, and that has to do with the shape of the distribution. Unfortunately, we can't really see the result we want unless the sample size is large. Thinking of all the work we had to do just for samples of size 2 and 3, just imagine what we'd have to do for samples of size 100. Luckily, we don't have to do this by hand; that's what computers are for. Figure 8.2.5 shows a histogram of the distribution of $\bar{x}$ using samples of size 100.

What does its shape remind you of? Is the shape of this distribution related to the distribution of the population? A histogram for the population is shown in Figure 8.2.6. It certainly doesn't look like the histogram for $\bar{x}$ in Figure 8.2.5.

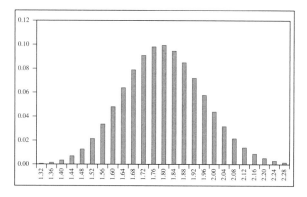

FIGURE 8.2.5 The distribution of $\bar{x}$ for samples of size 100

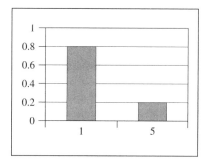

FIGURE 8.2.6 The distribution of coins in the population being sampled

What we see in Figure 8.2.5 can be explained by the **Central Limit Theorem**.

The Central Limit Theorem

Regardless of the distribution of the population being sampled, the distribution of sample means taken from random samples of size n is approximately normal when n is large.

So, how large a sample do we need to get a normal distribution for $\bar{x}$? The standard rule of thumb says $n \geq 30$ is large enough. It actually depends on what the distribution of the population looks like. In general, the closer the distribution of the population is to a normal distribution, the smaller n needs to be to get a normal distribution for $\bar{x}$. In fact, if the population has a normal distribution, the distribution of $\bar{x}$ will be normal for any value of n. On the other hand, if the distribution of the population is very skewed (and therefore far from normal), a sample size quite a bit larger than 30 may be necessary.

In this book, we will consider the distribution of $\bar{x}$ to be close enough to normal if either the population has a normal distribution or $n \geq 30$.

Summary of Facts about the Distribution of $\bar{x}$

If $\bar{x}$ is computed for random samples of size n taken from a population with mean μ and variance σ^2, the distribution of the possible values of $\bar{x}$ has the following characteristics:

1. $\mu_{\bar{x}} = \mu$

   The mean of the distribution of $\bar{x}$ is the same as the mean of the population being sampled.

   $\bar{x}$ is an unbiased estimator of μ.

2. $\sigma_{\bar{x}}^2 = \dfrac{\sigma^2}{n}$ and $\sigma_{\bar{x}} = \dfrac{\sigma}{\sqrt{n}}$

   Therefore, the values of $\bar{x}$ tend to get closer to μ as the sample size n increases.

3. The distribution of $\bar{x}$ is normal (or close enough to it) if either of the following is true:

   - The population has a normal distribution or
   - The sample size is large ($n \geq 30$).

Applying What We Have Learned

EXAMPLE 8.2.2 Nicky's Average Score

Remember Nicky, the 60% free throw shooter in a one-and-one situation? Nicky was introduced in *Connected Mathematics: What Do You Expect* (page 50) and in Section 7.1 of this book. In both places, students were asked to simulate Nicky's one-and-one situation several times and use the results to find experimental probabilities for Nicky's score. Suppose that Lauren simulates Nicky's one-and-one situation 50 times, recording the number of points scored in each. However, instead of using the data to find experimental probabilities, Lauren will use the average score from the data to estimate the mean number of points scored by a 60% free throw shooter in a one-and-one situation. ■

Focus on Understanding

First, recall that in Section 7.1, you found the probability distribution for Nicky's score. For your reference, this distribution is given in the table at the right. We also found, in Section 7.2, that the mean of this distribution was $\mu = 0.96$.

x	$P(x)$
0	0.40
1	0.24
2	0.36

1. Using the probability distribution above, find the variance and standard deviation for Nicky's score.

2. Lauren is going to be using a random sample of size 50 (from her simulation) to compute $\bar{x}$. What type of probability distribution does $\bar{x}$ have? Find the mean and standard deviation of this distribution.

As we go on, we will use the results of the previous *Focus on Understanding*. Suppose that we now want to find the probability that Lauren will get a sample mean that is within 0.1 of the actual mean; that is, we want to find $P(\mu - 0.1 \leq \bar{x} \leq \mu + 0.1)$:

$$P(\mu - 0.1 \leq \bar{x} \leq \mu + 0.1)$$
$$= P(0.96 - 0.1 \leq \bar{x} \leq 0.96 + 0.1)$$
$$= P(0.86 \leq \bar{x} \leq 1.06)$$

In the *Focus on Understanding*, you found that $\bar{x}$ has an approximately normal distribution with $\mu_{\bar{x}} = \mu = 0.96$ and $\sigma_{\bar{x}} = \frac{\sigma}{\sqrt{n}} = \frac{0.8709}{\sqrt{50}} = 0.1232$. Using this information, we can find the z-values corresponding to $\bar{x} = 0.86$ and $\bar{x} = 1.06$.

$$z_1 = \frac{\bar{x} - \mu_{\bar{x}}}{\sigma_{\bar{x}}} = \frac{0.86 - 0.96}{0.1232} = -0.81$$

$$z_2 = \frac{\bar{x} - \mu_{\bar{x}}}{\sigma_{\bar{x}}} = \frac{1.06 - 0.96}{0.1232} = 0.81$$

The probability we want is the area between these two z-values. (See Figure 8.2.7.) Following the usual procedure for finding areas under a normal curve:

$$P(0.86 \leq \bar{x} \leq 1.06) = P(-0.81 \leq z \leq 0.81)$$
$$= Area2 - Area1$$
$$= 0.7910 - 0.2090 \quad \text{(from Table 1)}$$
$$= 0.582$$

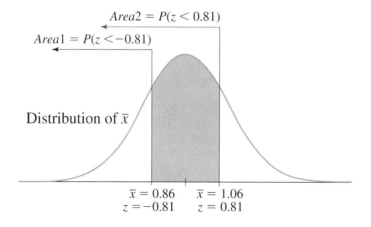

FIGURE 8.2.7 Finding $P(0.86 \leq \bar{x} \leq 1.06)$

So, if Lauren used a random sample 50 simulations to estimate the average number of points Nicky scores in one-and-one situations, she would have a 58.2% probability of getting within 0.1 of Nicky's true average ($\mu = 0.96$). I suppose that's pretty good, but that still leaves a probability of more than 40% that the estimate would be off by more than 0.1. Suppose we wanted to improve Lauren's chance of getting close to the true mean? Based on our work earlier in this section, we know that the values of $\bar{x}$ tend to get closer to μ as the sample size increases. In the following *Focus on Understanding*, you'll look at what happens if we use a larger sample.

Focus on Understanding

Suppose that four students each simulate Nicky's one-and-one situation 50 times, then combine their data together to get a more accurate estimate.

1. You are going to be finding the probability that the mean from their combined data will be within 0.1 of the true mean, just as we did earlier using Lauren's single sample of size 50, but before you do, decide whether this new probability will be smaller or larger than 0.582 or 58.2%, the probability we got earlier. Explain.

2. Okay, now go ahead and find the probability that the combined data will have a mean within 0.1 of the true mean μ. Is your answer consistent with what you decided in #1?

You've now seen that, by increasing the sample size, we can increase the probability that the sample mean $\bar{x}$ will be close to the population mean μ. In general, larger samples give more accurate estimates. Notice that we judge the accuracy of an estimate by looking at two different numbers:

- The maximum distance that we expect between our estimate and the true value (0.1 in the example above). This distance is called the **error bound** or **margin of error**.
- The probability that our estimate is within this distance of the true value (in the example above, 58.2% when $n = 50$, 89.5% for $n = 200$). This probability is called the **confidence level**. In general, as the sample size increases, we grow more confident (as measured by the confidence level) that our estimate will be close to the true value.

Both the error bound and the confidence level will be discussed more fully in Chapter 9. You will see that the sample size, the error bound, and the confidence level are related to each other; if we know any two of these, we can determine the third. So far, we have been given the error bound and the sample size and been asked to determine the confidence level. In such a situation, we follow a procedure similar to the first type of problem we discussed in Section 7.5:

The difference is that we are dealing with the distribution of $\bar{x}$, not x, so when we compute the z-score, we need to use the mean and standard deviation of the distribution of $\bar{x}$ $\left(\mu_{\bar{x}} = \mu \text{ and } \sigma_{\bar{x}} = \frac{\sigma}{\sqrt{n}}\right)$. Therefore, $z = \frac{\bar{x} - \mu_{\bar{x}}}{\sigma_{\bar{x}}} = \frac{\bar{x} - \mu}{\left(\frac{\sigma}{\sqrt{n}}\right)}$.

This type of problem could be diagrammed as follows:

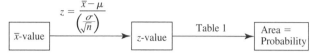

In Chapter 9, we will be reversing this procedure, starting with the confidence level (the probability) and working from right to left in the diagram above. Before we get there, however, we need to know some things about the distribution of a sample proportion (an experimental probability). That is the subject of the next section.

8.3 THE DISTRIBUTION OF SAMPLE PROPORTIONS

Just as we use the mean of a random sample to estimate a population mean, we can use a proportion from a random sample to estimate a population proportion. In fact, that is exactly what we do whenever we use an experimental probability to approximate the true (theoretical) probability, as in the *Classroom Connection* that follows.

Classroom Connection Read through the excerpt from *Connected Mathematics: How Likely Is It?* (page 32), as shown in Figure 8.3.1.

Classroom Exploration 8.3

1. Simulate drawing a random sample of size 10 (with replacement) from the bucket in *Drawing More Blocks* (Figure 8.3.1). Describe your simulation method, and give the experimental probability of drawing a yellow block, based on this single sample of size 10.
2. Suppose that you didn't know the contents of the bucket, but only the result of your sample of size 10. Based on your sample, what would be your best guess for the proportion of yellow blocks in the bucket?
3. Record your sample proportion on the board, along with those of the other students in your class.
4. Repeat #1 through #3, but with a sample of size 50.
5. Make dot plots for both of your lists of sample proportions, using the same horizontal axis. Comment on the center and spread of the two distributions.
6. Based on your dot plots, explain why samples of size 50 might be better for estimating the proportion of yellow blocks in the bucket than samples of size 10.

Drawing More Blocks deals with the relationship between theoretical probability and experimental probability. It illustrates the idea that experimental

254 Chapter 8 Distributions from Random Samples

4.2 Drawing More Blocks

Your teacher put eight blocks in a bucket. All the blocks are the same size. Three are yellow, four are red, and one is blue.

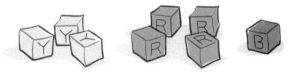

Problem 4.2

A. When you draw a block from the bucket, are the chances equally likely that it will be yellow, red, or blue? Explain your answer.

B. What is the total number of blocks? How many blocks of each color are there?

C. What is the *theoretical probability* of drawing a blue block? A yellow block? A red block? Explain how you found each answer.

Now, as a class or in groups, take turns drawing a block from the bucket. After each draw, return the block to the bucket. Keep a record of the blocks that are drawn. If you work in a group, take turns drawing blocks until you have 40 trials.

D. Based on your data, what is the *experimental probability* of drawing a blue block? A yellow block? A red block?

E. Compare the theoretical probabilities you found in part C to the experimental probabilities you found in part D. Are the probabilities for each color close? Are they the same? If not, why not?

■ **Problem 4.2 Follow-Up**

Suppose you and your classmates each took three turns drawing a block from the bucket, replacing the block each time, and then used the large amount of data you collected to find new experimental probabilities for drawing each color. You found the theoretical probability of drawing each color in part C. Do you think these new experimental probabilities would be closer to the theoretical probabilities than the experimental probabilities you found in part D were? Explain your reasoning.

FIGURE 8.3.1 Theoretical vs. experimental probability

probabilities can be used to estimate theoretical probabilities and that this estimate improves if we collect more data by including more repetitions of the experiment. These are the same ideas we addressed in the previous section about the distribution of sample means, except this time we're dealing with proportions, rather than means.

Before we go further, we need to introduce some symbols and terminology. Suppose we have a population, and some members of the population have a characteristic which makes them special, different from the rest of the population. For example, in a population of blocks, some are yellow. The proportion of yellow blocks in the population would be symbolized by the letter p. This proportion is exactly the same as the theoretical probability of choosing a yellow block from the population. In this case:

$$p = \frac{\text{number of yellow blocks in the population}}{\text{number of blocks in the population}} = \frac{3}{8} = 0.375$$

When we use a random sample from the population to find the experimental probability of choosing a yellow block, we are using the proportion of yellow blocks in the sample to estimate p, the population proportion of yellow blocks. Statisticians often use a "hat" symbol to indicate an estimate. In the previous section, we might have used the symbol $\hat{\mu}$ (*mu-hat*) for the sample mean, instead of $\bar{x}$, since the sample mean is used as an estimate for μ. In the same way, we will use the symbol $\hat{p}$ (*p-hat*) to stand for the proportion of special items in a sample, since the sample proportion is generally used to estimate the population proportion p.

Suppose, for example, we choose a random sample of 10 blocks (with replacement), and we got 3 yellow blocks, 5 red, and 2 blue. From this sample, our best estimate of the proportion of yellow blocks in the population would be the proportion of yellow blocks in the sample:

$$p \approx \hat{p} = \frac{\text{number of yellow blocks in the sample}}{\text{number of blocks in the sample}} = \frac{3}{10} = 0.3$$

Just as $\bar{x}$ is a random variable, so is $\hat{p}$; its value varies from sample to sample. In this section, we will look at the probability distribution of $\hat{p}$ under random sampling. Our first goal will be to show that $\hat{p}$ has properties that make it extremely useful for estimating p:

- $\hat{p}$ is an unbiased estimator of p.
- The value of $\hat{p}$ tends to get closer to p as the sample size n increases.
- When n is large enough, $\hat{p}$ has a familiar type of distribution. (You can probably guess what it will be.)

Is $\hat{p}$ an Unbiased Estimator?

Let's go back to the situation in *Change for a Dollar*. We'll return to *Drawing More Blocks* later.

EXAMPLE 8.3.1 Change for a Dollar Revisited

Recall that we are dealing with a population that is 80% pennies and 20% nickels. Before, we were interested in the mean value of the coins. This time, we want to focus on the probability of selecting a nickel. The theoretical probability of choosing a nickel is the same as the proportion of nickels in the population: $p = 0.20$ or 20%.

But suppose we didn't know the proportion of nickels in the population. We might try to estimate the probability of choosing a nickel by doing an experiment. We could choose coins repeatedly from the population and use the experimental probability as an estimate of the theoretical probability. Or, to put it another way, we could take a random sample and use the sample proportion as an estimate of the population proportion. For example, if we chose a random sample of 100 coins and got 22 nickels in the sample, we would estimate the population proportion of nickels as:

$$p \approx \hat{p} = \frac{22}{100} = 0.22$$

Sample	$\hat{p}$	Probability
1	$\frac{0}{1} = 0$ or 0%	0.8
5	$\frac{1}{1} = 1$ or 100%	0.2

FIGURE 8.3.2 The two possible samples of size 1

We're going to start with an even simpler case than in the last section—with a sample of size 1. The two possible samples of size 1 are listed in Figure 8.3.2 (1 if we choose a penny, 5 if we choose a nickel), along with its proportion of nickels and the probability of that sample. ∎

Focus on Understanding

The probability distribution of the random variable $\hat{p}$ is given at the right. Recall that the population proportion p is 20% or 0.20.

$\hat{p}$	$P(\hat{p})$
0	0.8
1	0.2

1. What is the probability that $\hat{p}$ is exactly the same as p? That is, what is $P(\hat{p} = p)$?
2. What is the probability that $\hat{p}$ underestimates p? That is, what is $P(\hat{p} < p)$?
3. What is the probability that $\hat{p}$ overestimates p? That is, what is $P(\hat{p} > p)$?
4. To show that $\hat{p}$ is an unbiased estimator of p, we must show that the average or expected value of the random variable $\hat{p}$ is exactly the same as p. Find $E(\hat{p})$. Is it the same as p?

In the last section we found that, although the sample mean $\bar{x}$ sometimes underestimates the population mean μ and sometimes overestimates it, the average value of $\bar{x}$ hits μ exactly; it is an unbiased estimator. You saw the same kind of result in the previous *Focus on Understanding*. Although the sample proportion $\hat{p}$ sometimes underestimates the population proportion p and sometimes overestimates it, the average value of $\hat{p}$ is exactly p. The sample proportion $\hat{p}$ is an unbiased estimator of the population proportion p. We'll verify this again for a larger sample size later, but it does hold true in general, and is important enough to display in the following box.

The Expected Value of $\hat{p}$

If $\hat{p}$ is to be computed from a random sample and used to estimate a population proportion p, then $\hat{p}$ is an unbiased estimator of p:

$$E[\hat{p}] = p \quad \text{or} \quad \mu_{\hat{p}} = p$$

Section 8.3   The Distribution of Sample Proportions   257

Note: Once again, the previous result depends on the fact that the samples are *random*. If sampling is done in a biased way, the sample proportion will be biased as well.

Does $\hat{p}$ Get Closer to p as n Increases?

To determine whether $\hat{p}$ gets closer to p as n increases, we need to look at the variance of the distribution of $\hat{p}$. The distribution of $\hat{p}$ for samples of size 1 is reproduced at the right. Using the computing formula for variance:

$\hat{p}$	$P(\hat{p})$
0	0.8
1	0.2

$$\sigma_{\hat{p}}^2 = \sum [\hat{p}^2 \cdot P(\hat{p})] - (\mu_{\hat{p}})^2$$
$$= [0^2(0.8) + 1^2(0.2)] - (0.2)^2$$
$$= 0.16$$

Notice that $pq = (0.2)(0.8) = 0.16$ also. This is not a coincidence; there's something going on here. When $n = 1$:

$$\hat{p} = \frac{\text{number of nickels in sample}}{\text{number of coins in sample}} = \frac{\text{number of nickels in sample}}{1}$$
$$= \text{number of nickels in sample}$$

The number of nickels in the sample is a binomial random variable. For a binomial random variable, the variance is given by $\sigma^2 = npq$:

For $n = 1$: $Var[\hat{p}] = \sigma_{\hat{p}}^2 = npq = 1 \cdot pq = pq$

We need to see what happens to $Var[\hat{p}]$ as n increases, so let's go to $n = 2$. That will be the subject of the following *Focus on Understanding*.

Focus on Understanding

1. Draw a tree diagram for the process of selecting a random sample of size 2 from a population that is 20% nickels and 80% pennies.
2. Make a table listing the four possible samples of size 2, the value of $\hat{p}$ (the proportion of nickels) for each sample, and the probability of each sample.
3. Give the probability distribution for $\hat{p}$.
4. Use the probability distribution from #3 to find $E[\hat{p}]$. Does $E[\hat{p}] = p$, as expected?
5. Find $Var[\hat{p}]$.

In the previous *Focus on Understanding* you should have found that $Var[\hat{p}] = 0.08$. Recall that when we used samples of size 1, we got $Var[\hat{p}] = pq = 0.16$. Notice that when we take samples of size 2, the variance is *half* as much. That is:

$$\text{For } n = 2: Var[\hat{p}] = \frac{pq}{2} = \frac{(0.2)(0.8)}{2} = 0.08$$

What do you suppose would happen if we took samples of size 3? We won't go through that one, but it does work out exactly as you would expect.

The Variance and Standard Deviation of the Distribution of $\hat{p}$

If $\hat{p}$ is to be computed from a random sample of size n and used to estimate a population proportion p, then:

$$Var[\hat{p}] = \frac{pq}{n} \quad \text{or} \quad \sigma_{\hat{p}}^2 = \frac{pq}{n}$$

It follows that:

$$\sigma_{\hat{p}} = \sqrt{\frac{pq}{n}}$$

So, just as we saw with sample means, increasing the sample size n will increase the denominators in the previous formulas, making the variance and standard deviation of $\hat{p}$ decrease. Since variance and standard deviation are measures of spread, the values of $\hat{p}$ will become less spread out as the sample size increases, getting closer and closer to p, the mean of the distribution. We have shown that $\hat{p}$ tends to get closer to p as the sample size increases. Once again, larger samples give more accurate estimates.

Does $\hat{p}$ Have a Familiar Distribution?

You probably know what's coming here, so we'll get right to it. Figure 8.3.3 shows a histogram of the distribution of $\hat{p}$ for samples of size 100. Does it look like a normal distribution? You bet! Just as with the distribution of $\bar{x}$, as the sample size increases, the distribution of $\hat{p}$ gets closer and closer to a normal distribution. However, the rule of thumb for how large a sample is necessary is different for proportions than for means ($n \geq 30$). For proportions, it is better to use the same criteria we used to decide whether a binomial distribution is approximately normal, that is: $n \cdot \min(pq) > 5$.

Summary of Facts about the Distribution of $\hat{p}$

If $\hat{p}$ is computed from a random sample of size n used to estimate a population proportion p, the distribution of $\hat{p}$ has these characteristics:

1. $E[\hat{p}] = \mu_{\hat{p}} = p$

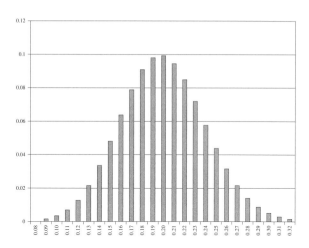

FIGURE 8.3.3 The distribution of sample proportions

The mean of the distribution of $\hat{p}$ is equal to the population proportion p. $\hat{p}$ is an unbiased estimator of p.

2. $Var[\hat{p}] = \sigma_{\hat{p}}^2 = \frac{pq}{n}$ and $\sigma_{\hat{p}} = \sqrt{\frac{pq}{n}}$

   Therefore, the values of $\hat{p}$ tend to get closer to p as the sample size increases.
3. The distribution of $\hat{p}$ is normal (or close enough) if: $n \cdot \min(pq) > 5$.

Focus on Understanding

Look back at *Drawing More Blocks*, the *Classroom Connection* at the beginning of this section.

1. What is the theoretical probability p for drawing a yellow block from the bucket?
2. In *Drawing More Blocks*, students will select a random sample of 40 blocks (with replacement) and compute $\hat{p}$, the proportion of yellow blocks in the sample. Is the distribution of $\hat{p}$ close to normal? (Show how you check this.) Find the mean and standard deviation of the distribution of $\hat{p}$.
3. We know that it is too much to expect for $\hat{p}$ to be exactly the same as p. Suppose we allowed ourselves a margin of error of 0.10. How low can $\hat{p}$ be and still be within 0.10 of p? How high can $\hat{p}$ be and still be within the margin of error? For a sample of size 40, find the probability that $\hat{p}$ is within 0.10 of the population proportion p.

4. In the follow-up question at the bottom of *Drawing More Blocks*, the students will use a larger sample to find the experimental probability $\hat{p}$. Suppose that the sample size is now 160. Will the probability that $\hat{p}$ is within 0.10 of p now be larger or smaller than the probability in #3 above? Find this probability.

Chapter 8 Summary

This chapter used the ideas from Chapter 7 to develop the idea of sampling distributions and the relationship of the sample statistics to the parameters of the original populations from which the samples were drawn. In the real world, samples are used to make predictions and decisions about the parameters of a population. For example, sample means can be used to make predictions about the corresponding population mean or sample proportions can be used to make predictions about the corresponding population proportion. One of the key ideas in this chapter was that as the sample size increases, the shape of the sampling distribution becomes more "normal." You also saw that, as the sample size increases, the standard error of the sampling distribution decreases.

The key terms and ideas from this chapter are listed below:

population   234
random sample of size n 235
inference   236
inferential statistics   236
sample mean   237
population mean   237
estimator   242
unbiased estimator   243

mean, variance, and standard
   deviation of $\bar{x}$ 253
Central Limit Theorem   249
population proportion   253
sample proportion   253
mean, variance, and standard
   deviation of $\hat{p}$ 258

Assessment is an integral part of every curriculum from the elementary school all the way through college. The question always arises—*what is it that students should be able to do after completing this lesson/unit/chapter?* We have included here our intended learning goals for Chapter 8. Students who have a good grasp of the concepts developed in this chapter should be successful in responding to these items:

- Explain, describe, or give an example of what is meant by each of the terms in the vocabulary list.
- List or count all the possible samples of a given size that can be taken from a given population.
- Determine the probability of a particular random sample being chosen or the probability that a random sample contains a particular object, item, or person.
- Create dot plots (line plots) using the means or proportions from random samples of a given size.
- Use sample means and sample proportions as estimates for population means and population proportions.

- Determine the probability distribution of $\bar{x}$, $\hat{p}$, or other sample statistics in a given situation.
- Use the probability distribution of a statistic to determine whether the statistic is an unbiased estimator.
- Find the mean, variance, and standard deviation of the distribution of $\bar{x}$ or $\hat{p}$ in a given situation and describe the shape of the distribution.
- Find the probability that the mean of a random sample will be within a given distance of the population mean.
- Find the probability that a proportion from a random sample will be within a given distance of the population proportion.

Your course instructor may have additional or different assessable outcomes for your class. As teachers (or future teachers) you should think about the assessment outcomes and learning goals for each chapter as you work through them.

EXERCISES FOR CHAPTER 8

1. Using the *Data Analysis and Probability Standards for Grades 6–8* from the NCTM's *Principles and Standards for School Mathematics* found at www.nctm.org, identify the middle school objectives that are found in Chapter 8.

2. Using the state standards for mathematics for the content area of data analysis and probability for your state, identify the middle school objectives that are found in Chapter 8. The following website may be useful: www.doe.state.in.us. This website will allow you to access the web pages of the state departments of education for the 50 states. From the state web pages you should be able to find the state's mathematical standards.

3. Suppose we have a list of 30 people, numbered from 1 to 30. We'd like to select a sample of 5 different people from the list. To select the sample, we plan to start by rolling an ordinary number cube. The first person selected will be the person whose number is rolled. Starting with that person, we'll select every sixth person from the list. For example, if we rolled a 4, we'd select person #4, 10, 16, 22, and 28.
    a. Using this plan, how many different samples are possible? List the people who would be in each of the possible samples.
    b. How many different samples contain person #17? What is the probability that a sample containing person #17 is selected?
    c. Does every person on the list have the same probability of being selected? Explain why or why not.
    d. Is a sample selected this way a random sample? Explain why or why not.
    e. Describe a method for choosing a random sample of 5 different people from the list. Explain why your sample would be random.
    f. How many different random samples of size 5 can be chosen from a list of 30 people?
    g. How many of the random samples contain person #17? What is the probability that person #17 is chosen?
    h. Does every person on the list have the same probability of being chosen? Explain why or why not.

4. There are 100 eighth graders enrolled at Mesa Middle School. Principal Owens would like to take a random sample of 30 eighth graders.
   a. How many random samples of size 30 are possible?
   b. Mrs. Bell's eighth grade class has 30 students. If Principal Owens chooses a random sample of 30 students, is it possible that all 30 are from Mrs. Bell's eighth grade class? What is the probability of this occurring?
5. A bag contains 5 prizes: A, B, C, D, and E. You are going to reach in and choose a sample of 2 prizes at random (without replacement).
   a. There are 10 possible samples of size 2. List them.
   b. Find the probability that the sample contains prize A.
   c. Does every prize have the same probability of being selected? Explain why or why not.
   d. Suppose prizes A, B, and C are each worth $1; D is worth $2, and E is worth $3. Find the mean value of the prizes in each of the samples in part (a) and make a table for the probability distribution of $\bar{x}$.
   e. Find μ, the mean value of the population of 5 prizes.
   f. Suppose that the sample mean is used to estimate the population mean. Find the probability that the sample mean hits the population mean exactly. That is, find $P(\bar{x} = \mu)$.
   g. Find $P(\bar{x} < \mu)$ and $P(\bar{x} > \mu)$. Is it more likely that the sample mean underestimates μ or overestimates μ?
   h. Find the mean of the distribution of $\bar{x}$.
   i. Is $\bar{x}$ an unbiased estimator of μ? Explain why or why not.
6. Using the same bagful of prizes from 5, you are going to be selecting a random sample of 3 prizes (without replacement).
   a. There are 10 possible samples of 3 prizes. List them.
   b. Find the median value $\tilde{x}$ for each of the samples in (a). Make a table for the probability distribution of the sample median $\tilde{x}$.
   c. Find M, the median of the population.
   d. Find the expected value of $\tilde{x}$.
   e. Is $\tilde{x}$ an unbiased estimator of M? Explain why or why not.
7. A spinner has five equal sectors numbered from 1 to 5. Samples of size 2 are taken using this spinner. The sampling distribution for the sample means is shown at the right.
   a. Compute the mean, variance, and standard deviation of this sampling distribution.
   b. Compute the mean, variance, and standard deviation of the original population {1, 2, 3, 4, 5}.
   c. Compare the mean of the sampling distribution with the mean of the original population. What do you notice?
   d. Describe how the variance of the sampling distribution compares to the variance of the original population. Don't just say it's smaller or larger, describe how much smaller or larger.
   e. If the sample size were changed from 2 to 4, what would you expect to happen to the mean of the new sampling distribution? Why?
   f. If the sample size were changed from 2 to 4, what would you expect to happen to the variance of the new sampling distribution? Why?

$\bar{x}$	$P(\bar{x})$
1.0	0.04
1.5	0.08
2.0	0.12
2.5	0.16
3.0	0.20
3.5	0.16
4.0	0.12
4.5	0.08
5.0	0.04

8. As a special promotion, a fast food chain gives a prize ticket to each customer. Customers scratch the tickets to reveal a hidden prize. Twenty percent of the tickets are worth $2 off their next purchase; twenty percent are worth $1 off; and the rest say "Thanks for playing. Please come again."

Sample	Probability	Sample Mean
(0, 0)		
(0, 1)		
(0, 2)		
(1, 0)		
(1, 1)		
(1, 2)		
(2, 0)		
(2, 1)		
(2, 2)		

  a. Make a table for the probability distribution for $x =$ *the value of one randomly selected ticket*.
  b. Find the mean of the distribution of x. What is the practical significance of the mean for the fast food chain?
  c. Find the variance and standard deviation for x.
  d. The nine possible samples of size 2 are listed in the table above. Complete the table with the probability and the mean for each sample. (Hint: A tree diagram might help for the probabilities.)
  e. Make a table for the probability distribution of $\bar{x}$.
  f. Find the mean of the distribution of $\bar{x}$. How does it compare to the mean for x?
  g. Find the variance of the distribution of $\bar{x}$. How does it compare to the variance of x?

9. A certain species of snake has a mean length of 18.2 inches with a standard deviation of 4.7 inches. A herpetologist doesn't know this, however, and plans to estimate the mean length of this snake population by using $\bar{x}$ from a random sample of 50 snakes.
  a. Find the mean and standard deviation of the distribution of $\bar{x}$.
  b. What type of probability distribution does $\bar{x}$ have? What information in this situation is important in determining this?
  c. Draw a picture of the distribution of $\bar{x}$ and label the mean of the distribution. What values of $\bar{x}$ are within 1 inch of the population mean?
  d. Find the probability that the mean of the herpetologist's sample will be within 1 inch of the population mean.
  e. Suppose the herpetologist decided to use a sample of size 100, instead of 50. Would this increase or decrease the probability that $\bar{x}$ is within 1 inch of μ?
  f. Redo (d) using a sample of size 100. Does it confirm your answer for (e)?

10. Suppose that the average salary of graduates of a certain university is $45,200 with a standard deviation of $7,400. The dean doesn't know this, and plans to estimate the average salary of graduates by using the mean of a random sample of 200 graduates.
  a. Find the mean and standard deviation of the distribution of $\bar{x}$.
  b. What type of probability distribution does $\bar{x}$ have? What information in this situation is important in determining this?
  c. Draw a picture of the distribution of $\bar{x}$ and label the mean of the distribution. What values of $\bar{x}$ are within $1,000 of the population mean?
  d. Find the probability that the mean of the dean's sample will be within $1,000 of the population mean.

e. Suppose the dean decided to use a sample of size 100, instead of 200. Would this increase or decrease the probability that $\bar{x}$ is within $1,000 of μ?
f. Redo (d) using a sample of size 100. Does it confirm your answer for (e)?

11. A teacher puts 5 blocks in a box, 2 red and 3 blue. A student is going to select a random sample of 2 blocks (without replacement) and use the sample to predict the proportion of red blocks in the box. (The student does not know how many blocks are in the box.)
    a. If Janice chooses 1 red block and 1 blue, what proportion of red blocks should she predict? What is the probability that a random sample has this proportion of red blocks?
    b. There are three different possible values for $\hat{p}$ in this situation. What are they?
    c. Make a table for the probability distribution of $\hat{p}$.
    d. Compare the probability distribution of $\hat{p}$ to the proportion of red blocks in the box (the population). What is the probability that $\hat{p}$ hits p exactly?
    e. Find $P(\hat{p} < p)$ and $P(\hat{p} > p)$. Is the sample proportion more likely to underestimate or overestimate the population proportion?
    f. Find the expected value of $\hat{p}$.
    g. Is $\hat{p}$ an unbiased estimator of p? Explain why or why not.

12. Using the same blocks as in 8.11, this time the teacher allows a student to choose a sample of size 3 (without replacement).
    a. Make a table for the probability distribution of $\hat{p}$.
    b. Find $P(\hat{p} < p)$ and $P(\hat{p} > p)$. Is the sample proportion more likely to underestimate or overestimate the population proportion?
    c. Is $\hat{p}$ an unbiased estimator of p? Explain why or why not.

13. Suppose that the probability of winning a certain carnival game is $p = 0.4$. Colby plans to play the game twice.
    a. Make a tree diagram showing the possible outcomes for 2 plays of the game. For each outcome, compute $\hat{p}$, the proportion of wins.
    b. Complete the table at the right to show the probability distribution of $\hat{p}$.
    c. In this case, what is the probability that the sample proportion $\hat{p}$ is the same as the population proportion p?
    d. Find $P(\hat{p} < p)$ and $P(\hat{p} > p)$. Is the sample proportion more likely to underestimate or overestimate the sample proportion?
    e. Use the probability distribution from (b) to find $E(\hat{p})$.
    f. Is $\hat{p}$ an unbiased estimator of p? Explain why or why not.
    g. Find the variance of the distribution of $\hat{p}$ from (b). Compare your result to what you would get from the formula $Var(\hat{p}) = \frac{pq}{n}$ given in Section 8.3.

$\hat{p}$	$P(\hat{p})$
0	
$\frac{1}{2}$	
1	

14. The probability of winning a different game is $p = 0.3$. Colby decides to play this game 3 times.
    a. Make a tree diagram showing the possible outcomes for 3 plays of the game. For each outcome, compute $\hat{p}$, the proportion of wins.
    b. Complete the table at the right to show the probability distribution of $\hat{p}$.
    c. Find $P(\hat{p} < p)$ and $P(\hat{p} > p)$. Is the sample proportion more likely to underestimate or overestimate the sample proportion?

$\hat{p}$	$P(\hat{p})$
0	
$\frac{1}{3}$	
$\frac{2}{3}$	
1	

d. Use the probability distribution from (b) to find $E(\hat{p})$.
e. Is $\hat{p}$ an unbiased estimator of p? Explain why or why not.
f. Find the variance of the distribution of $\hat{p}$ from (b). Compare your result to what you would get from the formula $Var(\hat{p}) = \frac{pq}{n}$ given in Section 8.3.

15. Nikki is an 80% free throw shooter, but a college scout doesn't know this. The scout plans to watch Nikki shoot free throws in practice and use her proportion of hits in this sample to estimate her free throw percentage.
   a. Suppose she takes 44 free throws in practice. Is this a large enough sample for the distribution of $\hat{p}$ to be approximately normal? Show how you determine this.
   b. Find the mean, variance, and standard deviation of the distribution of $\hat{p}$.
   c. Draw a picture of the distribution of $\hat{p}$ and label the mean of the distribution. What values of $\hat{p}$ are within 0.05 of p?
   d. Find the probability that the college scout's estimate will be within 0.05 of p.
   e. Suppose the scout decided to visit another practice, using a total of 88 free throws to compute $\hat{p}$. Should this increase or decrease the probability that the scout's estimate is within 0.05 of p?
   f. Redo part (d) using a sample size of 88 free throws. Does the result confirm your answer for (e)?

16. If a coin is tossed repeatedly, the experimental probability $\hat{p}$ should get close to the $p = 0.5$ as the number of tosses increases.
   a. Is a sample of 25 tosses large enough for the distribution of $\hat{p}$ to be approximately normal? Show how you determine this.
   b. Find the mean, variance, and standard deviation of the distribution of $\hat{p}$.
   c. Draw a picture of the distribution of $\hat{p}$ and label the mean of the distribution. What values of $\hat{p}$ are within 0.05 of p?
   d. Find the probability that $\hat{p}$ will be within 0.05 of p.
   e. Refigure the probability in (d), based on a sample of 100 tosses.
   f. Refigure this probability, based on a sample of 400 tosses.

CHAPTER 9

Estimating with Confidence

9.1 CONFIDENCE INTERVALS FOR PROPORTIONS
9.2 CONFIDENCE INTERVALS FOR MEANS
9.3 SAMPLE SIZE

In the previous chapter, we explored the distribution of sample means and sample proportions under random sampling. We found that, although it is unlikely for $\bar{x}$ to be the same as μ or for $\hat{p}$ to be the same as p, $\bar{x}$ and $\hat{p}$ still have properties that make them good estimators for μ and p. In this chapter, we will be dealing with situations where μ and p are unknown. We will show how to make reliable estimates for μ and p by using the values of $\bar{x}$ and $\hat{p}$ from random samples.

At this point in the book, we are firmly in the area of inferential statistics, drawing conclusions about populations by using data from samples. The ability to draw reliable conclusions about a population by using data from a sample is essential in our society. Businesses make decisions risking millions of dollars, based on conclusions they draw from samples. Medical researchers and drug manufacturers make countless decisions about the usefulness and safety of medicines, based on sample data. Politicians and other public policy makers also use data from samples to make decisions that affect the daily lives of millions of people. Such reliance on data from samples may seem misplaced in the light of what we saw in the previous chapter, that it is quite unusual for a sample mean to be the same as the mean of the population, or for a sample proportion to be the same as a population proportion.

What makes it all possible is estimation in a very precise way. Even though we know that a sample proportion is unlikely to be exactly the same as the proportion of the population it is drawn from, we still can use it to predict that the population mean falls within a particular range, an interval. We can also have a precise measure of the level of confidence that we should have in this prediction. The kind of prediction we are talking about is called a confidence interval.

9.1 CONFIDENCE INTERVALS FOR PROPORTIONS

What is a Confidence Interval?

Because it is unlikely for a random sample to produce a value of $\hat{p}$ that exactly matches the population proportion p, statisticians typically include a margin of error when using $\hat{p}$ to estimate p. For example, you might read about a poll showing that 54% of likely voters favor Proposition A, with a margin of error of $\pm 3\%$. This means that, in the sample, 54% of the likely voters favored Proposition A, but because of the way that proportions vary from sample to sample, it is more reasonable to believe that the true proportion for the entire population is somewhere between 51% and 57%. In this situation, $\hat{p} = 0.54$, and we believe that p lies in the interval between $\hat{p} - 0.03$ and $\hat{p} + 0.03$, or from 0.51 to 0.57. The interval $(\hat{p} - 0.03, \hat{p} + 0.03)$ or $(0.51, 0.57)$ is called a **confidence interval** for p. The **confidence level** for this interval is based on the process that produced it. Suppose that, by using a random sample to find $\hat{p}$, the interval $(\hat{p} - 0.03, \hat{p} + 0.03)$ has a 95% probability of containing the true value of p. In that case, we would call $(0.51, 0.57)$ a **95% confidence interval** for p.

Definition of Confidence Interval

A **confidence interval** for estimating a population proportion p is the interval $(\hat{p} - E, \hat{p} + E)$ where $\hat{p}$ is a sample proportion and E is the **error bound** or **margin of error**. This interval is sometimes abbreviated as $\hat{p} \pm E$. A confidence interval comes from a random process (choosing a random sample). The interval's **confidence level**, usually given as a percent, is the probability that the random process produces an interval that contains p.

In Exploration 8.3, you simulated drawing random samples of 40 blocks from a population that was 37.5% yellow blocks and used them to make a list of sample proportions. Figure 9.1.1 contains a sorted list of 100 such sample proportions. For each one, we have given the endpoints of a confidence interval, using an error bound of $E = 0.10$.

Classroom Exploration 9.1

In this exploration, you can use either the list given in Figure 9.1.1 or make a similar list by using the values of $\hat{p}$ from the samples of size 40 generated by your class in Exploration 8.3.

1. One sample proportion on the list is $\hat{p} = 0.400$. Describe a sample that would have produced this value of $\hat{p}$.
2. How many of the confidence intervals $(\hat{p} - E, \hat{p} + E)$ from the list contain $p = 0.375$? (Remember that the interval (a, b) does not include the endpoints. If the endpoints are included, the interval would be $[a, b]$.)

$\hat{p}$	$\hat{p}-E$	$\hat{p}+E$	$\hat{p}$	$\hat{p}-E$	$\hat{p}+E$	$\hat{p}$	$\hat{p}-E$	$\hat{p}+E$	$\hat{p}$	$\hat{p}-E$	$\hat{p}+E$
0.200	0.100	0.300	0.325	0.225	0.425	0.375	0.275	0.475	0.425	0.325	0.525
0.250	0.150	0.350	0.325	0.225	0.425	0.375	0.275	0.475	0.425	0.325	0.525
0.250	0.150	0.350	0.325	0.225	0.425	0.375	0.275	0.475	0.425	0.325	0.525
0.275	0.175	0.375	0.325	0.225	0.425	0.375	0.275	0.475	0.425	0.325	0.525
0.275	0.175	0.375	0.325	0.225	0.425	0.375	0.275	0.475	0.425	0.325	0.525
0.275	0.175	0.375	0.325	0.225	0.425	0.375	0.275	0.475	0.425	0.325	0.525
0.275	0.175	0.375	0.325	0.225	0.425	0.375	0.275	0.475	0.425	0.325	0.525
0.275	0.175	0.375	0.350	0.250	0.450	0.375	0.275	0.475	0.425	0.325	0.525
0.275	0.175	0.375	0.350	0.250	0.450	0.375	0.275	0.475	0.425	0.325	0.525
0.275	0.175	0.375	0.350	0.250	0.450	0.375	0.275	0.475	0.425	0.325	0.525
0.275	0.175	0.375	0.350	0.250	0.450	0.375	0.275	0.475	0.450	0.350	0.550
0.300	0.200	0.400	0.350	0.250	0.450	0.375	0.275	0.475	0.450	0.350	0.550
0.300	0.200	0.400	0.350	0.250	0.450	0.400	0.300	0.500	0.450	0.350	0.550
0.300	0.200	0.400	0.350	0.250	0.450	0.400	0.300	0.500	0.450	0.350	0.550
0.300	0.200	0.400	0.350	0.250	0.450	0.400	0.300	0.500	0.450	0.350	0.550
0.300	0.200	0.400	0.350	0.250	0.450	0.400	0.300	0.500	0.450	0.350	0.550
0.300	0.200	0.400	0.350	0.250	0.450	0.400	0.300	0.500	0.450	0.350	0.550
0.300	0.200	0.400	0.350	0.250	0.450	0.400	0.300	0.500	0.450	0.350	0.550
0.300	0.200	0.400	0.350	0.250	0.450	0.400	0.300	0.500	0.450	0.350	0.550
0.325	0.225	0.425	0.350	0.250	0.450	0.400	0.300	0.500	0.475	0.375	0.575
0.325	0.225	0.425	0.350	0.250	0.450	0.400	0.300	0.500	0.475	0.375	0.575
0.325	0.225	0.425	0.350	0.250	0.450	0.400	0.300	0.500	0.475	0.375	0.575
0.325	0.225	0.425	0.350	0.250	0.450	0.400	0.300	0.500	0.475	0.375	0.575
0.325	0.225	0.425	0.375	0.275	0.475	0.400	0.300	0.500	0.475	0.375	0.575
0.325	0.225	0.425	0.375	0.275	0.475	0.425	0.325	0.525	0.500	0.400	0.600

FIGURE 9.1.1 100 confidence intervals for p using E = 0.10

3. Based on #2, what is the experimental probability that the interval $(\hat{p} - 0.1, \hat{p} + 0.1)$ contains the true value of p?
4. In the same situation in the *Focus on Understanding* at the end of Section 8.3, you were asked to find the probability that the value of $\hat{p}$ was within 0.10 of the population proportion p. Your result should have been 0.8098. Compare this to your answer from #2. Are they close to each other? Should they be close? Explain.

Finding the Error Bound for a 95% Confidence Level

So far, we have been pretending as if the error bound was given and the confidence level was determined from it. In reality, we would typically start with the confidence level we want and use it to find the margin of error. We will start with the most commonly used confidence level: 95%.

When using the error bound to find the confidence level, we follow the same procedure as we did for finding probabilities in Section 7.5:

Section 9.1 Confidence Intervals for Proportions 269

The difference now is that we are dealing with the distribution of $\hat{p}$, not x, so when we compute the z-value, we use the mean and standard deviation of the distribution of $\hat{p}$, $\mu_{\hat{p}} = p$, and $\sigma_{\hat{p}} = \sqrt{\frac{pq}{n}}$. This type of problem could be diagrammed as follows:

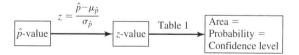

We will now be reversing this procedure, starting with the confidence level and working backwards, from right to left in the diagram. Just as when we solved the same type of problem in Section 7.5, since we will now be using the z-value to find $\hat{p}$, it would be helpful if the equation $z = \frac{\hat{p} - \mu_{\hat{p}}}{\sigma_{\hat{p}}}$ was solved for $\hat{p}$. Doing a little algebra (multiplying both sides by $\sigma_{\hat{p}}$ and adding $\mu_{\hat{p}}$ to both sides) gives $\hat{p} = \mu_{\hat{p}} + z\sigma_{\hat{p}}$. This equation is similar to the equation $x = \mu + z\sigma$ back in 7.5. So, our plan will follow the diagram below, working from right to left.

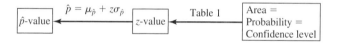

Suppose we are still drawing a sample of size 40 from the same old bucket, but this time we'd like to find the margin of error E for a 95% confidence interval. Figure 9.1.2 shows some key features of this situation in the distribution of $\hat{p}$.

Because we want $\hat{p}$ to have a 95% probability of being within the distance E of p, the central area in the figure must be 0.95. We'd like to use Table 1 to find z_1 and z_2, the z-values at the left and right ends of this central area. The problem is that

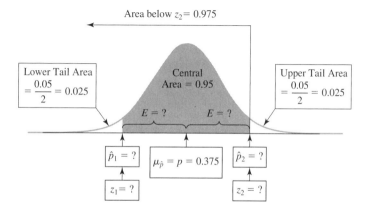

FIGURE 9.1.2 Using the distribution of $\hat{p}$ to find an error bound for a 95% confidence interval

Table 1 gives the area to the left, not the central area. Since the total area under the curve is 1 and the central area is 0.95, the remaining area is $1 - 0.95 = 0.05$. Because of the symmetry of the diagram, half of this area must be in each tail. Therefore, the area in each tail is $\frac{0.05}{2} = 0.025$. Since the upper tail area is also 0.025, the area below z_2 is $1 - 0.025 = 0.975$.

Finding an area of 0.975 in Table 1, we find that it is the area to the left of $z = 1.96$. This is z_2 in the diagram. We could also have found z_2 by using the TI-83 plus. Hit 2nd and DISTR. Select 3: invNorm(and enter invNorm(0.975). You should get the same result.

Either way you find it, by using $z_2 = 1.96$ and the equation $\hat{p} = \mu_{\hat{p}} + z\sigma_{\hat{p}}$, we can solve for $\hat{p}_2$.

$$\hat{p}_2 = \mu_{\hat{p}} + z_2 \sigma_{\hat{p}}$$
$$= 0.375 + 1.96(0.07655)$$
$$= 0.525$$

Finally, looking at Figure 9.1.2, the error bound E is the distance between p and $\hat{p}_2$.

$$E = |\hat{p}_2 - p| = |0.525 - 0.375| = 0.150$$

Focus on Understanding

1. In the example we just did, find the error bound E by using $\hat{p}_1$ instead of $\hat{p}_2$. Did you get the same result?
2. Figure 9.1.3 shows the same list of sample proportions as in Figure 9.1.1, but with $E = 0.15$. $E = 0.15$ is the error bound we calculated for a 95% confidence interval. What should you expect to find when you look at this list of 100 confidence intervals?
3. How many of the confidence intervals in Figure 9.1.3 contain $p = 0.375$? Based on this, what is the experimental probability that the interval $(\hat{p} - 0.15, \hat{p} + 0.15)$ contains the true value of p? Is this close to what you expected? Why is it not exactly 0.95?

Looking at the confidence intervals in Figure 9.1.3, you might notice that the value of p does not seem to be pinned down very closely. For example, if we chose a sample with $\hat{p} = 0.35$, we would believe (with 95% confidence) that p was somewhere between 0.2 and 0.5, or in other words, that somewhere between 20% and 50% of the blocks were yellow. That's quite a large range! We might conclude that, for a random sample of size 40, maybe 95% confidence is too much to expect. To pin down the value of p to a smaller interval of possibilities and maintain the same level of confidence in our results, we would need a larger sample size. That is the subject of the following *Focus on Understanding*.

Section 9.1   Confidence Intervals for Proportions   271

$\hat{p}$	$\hat{p} - E$	$\hat{p} + E$	$\hat{p}$	$\hat{p} - E$	$\hat{p} + E$	$\hat{p}$	$\hat{p} - E$	$\hat{p} + E$	$\hat{p}$	$\hat{p} - E$	$\hat{p} + E$
0.200	0.050	0.350	0.325	0.175	0.475	0.375	0.225	0.525	0.425	0.275	0.575
0.250	0.100	0.400	0.325	0.175	0.475	0.375	0.225	0.525	0.425	0.275	0.575
0.250	0.100	0.400	0.325	0.175	0.475	0.375	0.225	0.525	0.425	0.275	0.575
0.275	0.125	0.425	0.325	0.175	0.475	0.375	0.225	0.525	0.425	0.275	0.575
0.275	0.125	0.425	0.325	0.175	0.475	0.375	0.225	0.525	0.425	0.275	0.575
0.275	0.125	0.425	0.325	0.175	0.475	0.375	0.225	0.525	0.425	0.275	0.575
0.275	0.125	0.425	0.325	0.175	0.475	0.375	0.225	0.525	0.425	0.275	0.575
0.275	0.125	0.425	0.350	0.200	0.500	0.375	0.225	0.525	0.425	0.275	0.575
0.275	0.125	0.425	0.350	0.200	0.500	0.375	0.225	0.525	0.425	0.275	0.575
0.275	0.125	0.425	0.350	0.200	0.500	0.375	0.225	0.525	0.425	0.275	0.575
0.275	0.125	0.425	0.350	0.200	0.500	0.375	0.225	0.525	0.450	0.300	0.600
0.300	0.150	0.450	0.350	0.200	0.500	0.375	0.225	0.525	0.450	0.300	0.600
0.300	0.150	0.450	0.350	0.200	0.500	0.400	0.250	0.550	0.450	0.300	0.600
0.300	0.150	0.450	0.350	0.200	0.500	0.400	0.250	0.550	0.450	0.300	0.600
0.300	0.150	0.450	0.350	0.200	0.500	0.400	0.250	0.550	0.450	0.300	0.600
0.300	0.150	0.450	0.350	0.200	0.500	0.400	0.250	0.550	0.450	0.300	0.600
0.300	0.150	0.450	0.350	0.200	0.500	0.400	0.250	0.550	0.450	0.300	0.600
0.300	0.150	0.450	0.350	0.200	0.500	0.400	0.250	0.550	0.450	0.300	0.600
0.325	0.175	0.475	0.350	0.200	0.500	0.400	0.250	0.550	0.475	0.325	0.625
0.325	0.175	0.475	0.350	0.200	0.500	0.400	0.250	0.550	0.475	0.325	0.625
0.325	0.175	0.475	0.350	0.200	0.500	0.400	0.250	0.550	0.475	0.325	0.625
0.325	0.175	0.475	0.350	0.200	0.500	0.400	0.250	0.550	0.475	0.325	0.625
0.325	0.175	0.475	0.375	0.225	0.525	0.400	0.250	0.550	0.475	0.325	0.625
0.325	0.175	0.475	0.375	0.225	0.525	0.425	0.275	0.575	0.500	0.350	0.650

FIGURE 9.1.3 100 confidence intervals for p using $E = 0.15$

Focus on Understanding

Suppose that we are still drawing blocks from the same old bucket, but this time our sample size is 160. You are going to be finding the error bound E for estimating p with a 95% confidence interval, but before you do, answer the questions below.

1. When the sample size was 40, we found the error bound for a 95% confidence interval to be $E = 0.15$. Now that the sample size is 160, would you expect E to be larger or smaller than 0.15?

2. At the end of Section 8.3, you found that for samples of size 160, the probability that $\hat{p}$ would be within 0.10 of p was 0.9910. What does this say about the confidence level for estimating p using a margin of error of $E = 0.10$?

3. Based on #2, would you expect the error margin E for a 95% confidence interval to be larger or smaller than 0.10?

4. Using the same procedure as in the example above, find the error bound E for a 95% confidence interval when $n = 160$ instead of 40. (Hint: What does changing the sample size affect?) Does your result confirm your answers to #1 and #3 above?

Finding a Formula for the Error Bound of a 95% Confidence Interval

We'd now like to find a general formula for the error bound of a 95% confidence interval. Since the confidence level is still 95%, the areas we found earlier are still the same, and z_2 is still 1.96. We found $\hat{p}_2$ before by substituting values for $\mu_{\hat{p}}$, z_2, and $\sigma_{\hat{p}}$ into the equation $\hat{p}_2 = \mu_{\hat{p}} + z_2 \sigma_{\hat{p}}$. We are going to do exactly the same thing here, but using $\mu_{\hat{p}} = p$, $z_2 = 1.96$, and $\sigma_{\hat{p}} = \sqrt{\frac{pq}{n}}$. Therefore:

$$\hat{p}_2 = \mu_{\hat{p}} + z_2 \sigma_{\hat{p}}$$

$$= p + 1.96\sqrt{\frac{pq}{n}}$$

Lastly, we found the error bound E as the distance between p and $\hat{p}_2$. Computing this distance by substituting what we just found for $\hat{p}_2$:

$$E = |\hat{p}_2 - p|$$

$$= \left| p + 1.96\sqrt{\frac{pq}{n}} - p \right|$$

$$= \left| 1.96\sqrt{\frac{pq}{n}} \right|$$

The absolute value signs are not needed in this case. We have the general formula we wanted:

Error Bound for the 95% Confidence Interval ($\hat{p} - E, \hat{p} + E$):

$$E = 1.96\sqrt{\frac{pq}{n}}$$

Focus on Understanding

1. Redo #4 from the previous *Focus on Understanding* by using the formula we just derived. Is your result the same?
2. You might have noticed that the error bound that you got for a sample of size 160 was exactly half as much as the error bound for a sample of size 40. Show that this will always happen whenever the sample size is multiplied by 4.
3. What happens to the error bound if the sample size is multiplied by 9? 16? Generalize this result.

Section 9.1   Confidence Intervals for Proportions   273

What if p is Unknown?

We must admit that there is a flaw in the formula we've just found. Imagine a realistic situation where we wanted to find an error bound for estimating a population proportion. Would we know the value of p? No; if we knew the value of p, why would we be trying to estimate it? Yet, the formula for the error bound calls for us to use the value of p and q! In practice, since p and q are unknown, we'd use the best estimates, the sample proportions $\hat{p}$ and $\hat{q}$. Likewise, since p and q are usually unknown, we would use $\hat{p}$ and $\hat{q}$ to determine whether the distribution of $\hat{p}$ is close to normal. We would check that $n \cdot \min(\hat{p}, \hat{q}) > 5$.

A More Realistic Error Bound for the 95% Confidence Interval ($\hat{p}$ − E, $\hat{p}$ + E):

$$E = 1.96 \sqrt{\frac{\hat{p}\hat{q}}{n}}$$

In case you are wondering what effect using $\hat{p}$ and $\hat{q}$ in place of p and q has on the accuracy of the confidence intervals, the answer is "Not much." Figure 9.1.4 uses

$\hat{p}$	$\hat{p} - E$	$\hat{p} + E$	$\hat{p}$	$\hat{p} - E$	$\hat{p} + E$	$\hat{p}$	$\hat{p} - E$	$\hat{p} + E$	$\hat{p}$	$\hat{p} - E$	$\hat{p} + E$
0.200	0.076	0.324	0.325	0.180	0.470	0.375	0.225	0.525	0.425	0.272	0.578
0.250	0.116	0.384	0.325	0.180	0.470	0.375	0.225	0.525	0.425	0.272	0.578
0.250	0.116	0.384	0.325	0.180	0.470	0.375	0.225	0.525	0.425	0.272	0.578
0.275	0.137	0.413	0.325	0.180	0.470	0.375	0.225	0.525	0.425	0.272	0.578
0.275	0.137	0.413	0.325	0.180	0.470	0.375	0.225	0.525	0.425	0.272	0.578
0.275	0.137	0.413	0.325	0.180	0.470	0.375	0.225	0.525	0.425	0.272	0.578
0.275	0.137	0.413	0.325	0.180	0.470	0.375	0.225	0.525	0.425	0.272	0.578
0.275	0.137	0.413	0.350	0.202	0.498	0.375	0.225	0.525	0.425	0.272	0.578
0.275	0.137	0.413	0.350	0.202	0.498	0.375	0.225	0.525	0.425	0.272	0.578
0.275	0.137	0.413	0.350	0.202	0.498	0.375	0.225	0.525	0.425	0.272	0.578
0.275	0.137	0.413	0.350	0.202	0.498	0.375	0.225	0.525	0.450	0.296	0.604
0.300	0.158	0.442	0.350	0.202	0.498	0.375	0.225	0.525	0.450	0.296	0.604
0.300	0.158	0.442	0.350	0.202	0.498	0.400	0.248	0.552	0.450	0.296	0.604
0.300	0.158	0.442	0.350	0.202	0.498	0.400	0.248	0.552	0.450	0.296	0.604
0.300	0.158	0.442	0.350	0.202	0.498	0.400	0.248	0.552	0.450	0.296	0.604
0.300	0.158	0.442	0.350	0.202	0.498	0.400	0.248	0.552	0.450	0.296	0.604
0.300	0.158	0.442	0.350	0.202	0.498	0.400	0.248	0.552	0.450	0.296	0.604
0.300	0.158	0.442	0.350	0.202	0.498	0.400	0.248	0.552	0.450	0.296	0.604
0.300	0.158	0.442	0.350	0.202	0.498	0.400	0.248	0.552	0.450	0.296	0.604
0.325	0.180	0.470	0.350	0.202	0.498	0.400	0.248	0.552	0.475	0.320	0.630
0.325	0.180	0.470	0.350	0.202	0.498	0.400	0.248	0.552	0.475	0.320	0.630
0.325	0.180	0.470	0.350	0.202	0.498	0.400	0.248	0.552	0.475	0.320	0.630
0.325	0.180	0.470	0.350	0.202	0.498	0.400	0.248	0.552	0.475	0.320	0.630
0.325	0.180	0.470	0.375	0.225	0.525	0.400	0.248	0.552	0.475	0.320	0.630
0.325	0.180	0.470	0.375	0.225	0.525	0.425	0.272	0.578	0.500	0.345	0.655

FIGURE 9.1.4 100 confidence intervals for p using $E = 1.96\sqrt{\dfrac{\hat{p}\hat{q}}{n}}$

the same list of 100 sample proportions as Figure 9.1.3, but uses the more realistic formula for E.

Comparing the two lists shows that the "realistic" confidence intervals are sometimes narrower and sometimes wider than those we computed earlier. For example, when $\hat{p} = 0.200$, the "realistic" error bound is:

$$E = 1.96\sqrt{\frac{\hat{p}\hat{q}}{n}} = 1.96\sqrt{\frac{(0.2)(0.8)}{40}} = 0.124$$

This is less than the error bound $E = .150$ that we got earlier by using p and q, resulting in a narrower confidence interval. On the other hand, when $p = 0.500$:

$$E = 1.96\sqrt{\frac{\hat{p}\hat{q}}{n}} = 1.96\sqrt{\frac{(0.5)(0.5)}{40}} = 0.155 > 0.150$$

So $\hat{p} = 0.500$ results in a wider confidence interval.

However, notice that, just as you saw in Figure 9.1.3, only one of the confidence intervals, the one for $\hat{p} = 0.200$, fails to contain p. In this case, using $\hat{p}$ and $\hat{q}$ in place of p and q had no effect on whether or not the confidence interval contained p. Using estimates in place of p and q in the error bound formula has even less effect when the sample size is larger. In general, the instances with wider intervals and the instances with narrower intervals tend to balance out, and the probability for the interval to contain the true value of p remains close to 0.95.

Classroom Connection

You might recognize Figure 9.1.5. We used *Tossing Tacks* before, in Section 7.6. It comes from *Connected Mathematics: What Do You Expect?* (page 15). ◆

8. If a tack is dropped on the floor, there are two possible outcomes: the tack lands on its side (point down), or the tack lands on its head (point up). The probability that a tack will land point up or point down can be determined by experimenting. Kalifa tossed a tack 100 times and recorded the results in the table below.

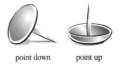

point down   point up

Outcome	Number of times it occurs
Tack lands point up	58
Tack lands point down	42

    a. If you dropped Kalifa's tack once, what is the probability that it would land point up? What is the probability that it would land point down?

    b. If you dropped Kalifa's tack 500 times, how many times would you expect it to land point up?

    c. Is it equally likely that the tack will land point up or point down? Explain.

    d. Is it possible to determine theoretical probabilities for this situation? Why or why not?

FIGURE 9.1.5 Tossing tacks

Section 9.1   Confidence Intervals for Proportions   275

Focus on Understanding

1. Look at part (a) in *Tossing Tacks*. Are the probabilities in part (a) theoretical or experimental?
2. Is a sample of size 100 large enough for the distribution of $\hat{p}$ to be close to normal? (Show how you determine this.)
3. Look at part (d). Since you know more than a middle school student, you can use Kalifa's data to estimate a theoretical probability with a 95% confidence interval. Find a 95% confidence interval for $p =$ *the probability that a tack lands point up*.
4. Find a 95% confidence interval for $p =$ *probability that a tack lands point down*.
5. How do the error bounds for your two intervals compare? Explain why this should be true in any situation where there are only two possible outcomes.

What if the Confidence Level is Not 95%?

Although 95% is the most commonly used confidence level, it is certainly not the only one. To generalize the work we have done above to other confidence levels, we need to introduce some new terminology and symbols.

When we use the confidence level to find the margin of error, we are concerned with central areas and tail areas. The **total tail area** is symbolized by α, the Greek letter *alpha*. Because of the symmetry of the normal curve, half of this area $\left(\frac{\alpha}{2}\right)$ is in each tail. The remaining area $(1 - \alpha)$ is the central area. Figure 9.1.6 shows these areas marked within the distribution of $\hat{p}$.

The z-value at the right end of the central area, which we called z_2 earlier, is called a **critical value**. It is a z-value with an **upper tail area** of $\frac{\alpha}{2}$, symbolized by $z_{\alpha/2}$.

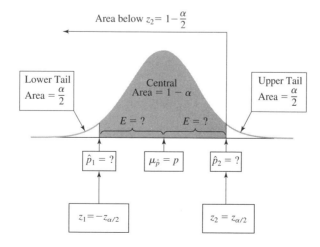

FIGURE 9.1.6 Using the distribution of $\hat{p}$ to find a general error bound

276 Chapter 9 Estimating with Confidence

Since $\hat{p}_1$ and $\hat{p}_2$ are both the same distance from the mean, z_1 is exactly the same as z_2, but negative ($z_1 = -z_{\alpha/2}$). To find a general formula for the error bound E, we could go through exactly the same process as we did earlier, using $z_2 = z_{\alpha/2}$ instead of $z_2 = 1.96$. The result will be exactly the same, except that we will have $z_{\alpha/2}$ in place of 1.96. So, instead of $E = 1.96\sqrt{\frac{\hat{p}\hat{q}}{n}}$, we would get $E = z_{\alpha/2}\sqrt{\frac{\hat{p}\hat{q}}{n}}$.

The confidence level for the interval comes from the central area $(1 - \alpha)$. Since confidence levels are generally given as percents, we need to multiply this by 100, making the confidence level $100(1 - \alpha)\%$. Putting all of this together, we have a general formula for an interval with confidence level $100(1 - \alpha)\%$.

A $100(1 - \alpha)\%$ Confidence Interval for Estimating p

Confidence Interval: $(\hat{p} - E, \hat{p} + E)$ or $\hat{p} \pm E$ Error Bound: $E = z_{\alpha/2}\sqrt{\frac{\hat{p}\hat{q}}{n}}$

- The **critical value** $z_{\alpha/2}$ is a z-value (standard normal) with an upper tail area of $\frac{\alpha}{2}$.
- The proportions $\hat{p}$ and $\hat{q}$ must come from a *random* sample.
- The sample size n must be large enough so that the distribution of $\hat{p}$ is close to normal: $n \cdot \min(\hat{p}, \hat{q}) > 5$

Suppose we want to use Kalifa's data from *Tossing Tacks* to find a 90% confidence interval for the probability that a tack lands point up. Because the confidence level is 90%, the central area would be 0.90, making the total tail area $\alpha = 0.10$. The upper tail area is $\frac{\alpha}{2} = 0.05$, so the area below $z_{\alpha/2}$ is $1 - 0.05 = 0.95$. (See Figure 9.1.7.)

Looking at the areas in Table 1, we find two that are close to 0.95:

- The area below $z = 1.64$ is 0.9495.
- The area below $z = 1.65$ is 0.9505.

Since both of these are equally close to 0.95, our best guess at $z_{\alpha/2}$ is midway between the two: $z_{\alpha/2} = 1.645$.

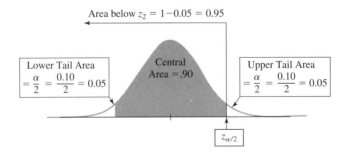

FIGURE 9.1.7 Finding $z_{\alpha/2}$

Section 9.1 Confidence Intervals for Proportions

You could also get $z_{\alpha/2}$ by using the TI-83 plus. Since the area to the left is 0.95, go to DISTR and enter invNorm(0.95). You should get the same result (with more decimal places if you want them).

Now we're ready to find the error bound:

$$E = z_{\alpha/2}\sqrt{\frac{\hat{p}\hat{q}}{n}} = 1.645\sqrt{\frac{(0.58)(0.42)}{100}} = 0.081$$

The 90% confidence interval is $0.58 \pm 0.081 = (0.499, 0.661)$. In other words, we would believe (with 90% confidence) that the true probability that the tack lands point up is somewhere between 0.499 and 0.661. In this case, "90% confidence" means that 90% of all random samples would result in confidence intervals that include the true probability of "point up," resulting in correct predictions for the location of p. It is certainly possible that Kalifa's sample is one of the 10% that lead to incorrect predictions, but our confidence comes from the fact that it is more likely that Kalifa's sample is one of the 90%.

Based on this confidence interval, would we be 90% confident in saying that the tack lands point up most of the time? Not quite. Notice that the 90% confidence interval includes values of p that are less than or equal to 0.5. So, as far as we can tell from this interval, the true probability of "point up" may be either above or below one-half.

Notice that this confidence interval is narrower than the 95% confidence interval you found earlier for the same situation. In general, a larger central area will result in a wider confidence interval. In other words, to have a higher level of confidence that the interval will contain p, we need a larger margin of error. Narrower intervals have lower confidence levels.

Focus on Understanding

1. If you used Kalifa's data to find an 80% confidence interval for the probability of "point up," would you expect it to be wider or narrower than the 90% confidence interval?
2. Find the 80% confidence interval, using either invNorm on the TI-83 plus or the closest z-value you can find in Table 1. Does it confirm your answer to #1?

A Shortcut: Using Table 2

Although we can find critical values by using Table 1, it is really not set up for this. Table 1 is really set up to find areas from z-values, not the other way around. It's organized based on z-values. Table 2 (inside the back cover of this book) is organized according to areas. The top of the table shows three different areas for the same point on the horizontal axis: the area to the left of that point, the upper tail area, and the central area. The first line below the areas shows critical values from standard

Table 2: Critical Values for Standard Normal and T-Distributions

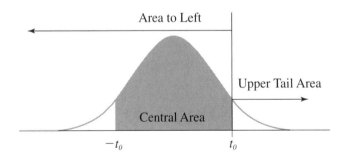

Area to Left of t_0	0.90	0.95	0.975	0.99	0.995	0.999	0.9995
Upper Tail Area	0.10	0.05	0.025	0.01	0.005	0.001	0.0005
Central Area	0.80	0.90	0.950	0.98	0.990	0.998	0.9990
Standard Normal	1.282	1.645	1.960	2.326	2.576	3.090	3.291

FIGURE 9.1.8 Using Table 2

normal curve (z-values). So, for example, if we wanted a 90% confidence interval, we would know that the central area should be 0.90. (See Figure 9.1.8.) Locating this central area, we find the critical value $z_{\alpha/2} = 1.645$, just as we found earlier.

Not only does Table 2 make finding critical values quicker and easier than Table 1, it's more accurate also. Notice that it gives the z-values to three decimal places, rather than two as in Table 1. However, it does not have nearly as many different areas as Table 1; it lists only the most common areas needed for confidence intervals. If we needed critical values for other areas, we would use Table 1 or a calculator like the TI-83 plus. We will do more with Table 2 in the next section.

Classroom Connection

Figure 9.1.9 is from *Mathematics in Context: Great Expectations* (page 31). You will use it in the following *Focus on Understanding*. ◆

Focus on Understanding

1. Do you think that the students' 1,000 mayflies are a random sample? Explain your position.

Section 9.1   Confidence Intervals for Proportions   279

2. Suppose you want to estimate the probability p that a mayfly lives at least 8 hours. Is the students' sample of 1,000 mayflies large enough to use the method in this section? (Show how you check.)

3. Whether you think it is or not, pretend that the 1,000 mayflies is a random sample for the rest of this problem. Find a 99% confidence interval for the probability p.

4. Based on your confidence interval, would you be 99% confident in concluding that the majority of mayflies live at least 8 hours?

E. Expectations

Expected Life of a Mayfly

For a science project, a group of students hatched 1,000 mayflies and carefully observed them. After six hours, all of the mayflies were alive, but then 150 died in the next hour. The students recorded their data in the table below.

Hours	6	7	8	9	10	11	12
Number Still Alive	1,000	850	600	250	100	20	0

12. Use the students' data to determine the expected life span of a mayfly. Explain how you found your answer.

FIGURE 9.1.9 Lifespan of a mayfly

9.2 CONFIDENCE INTERVALS FOR MEANS

In this section, we will find confidence intervals for estimating population means, just as we did for proportions in the previous section. Since we've already laid the groundwork, this should go a bit faster. We think that you'll find that the situation is similar to what you saw in 9.1 and the pieces will fall into place. Speaking of things that are similar and pieces falling into place, try the analogy puzzle in Exploration 9.2.

Classroom Exploration 9.2

Column 1	Match each item in Column 1 with the corresponding item in Column 2. Explain why the ones you choose go together. (Hint: This has something to do with the picture below.)	Column 2
p		$\frac{\sigma}{\sqrt{n}}$
$\hat{p}$		$n \geq 30$ or a normal population
$\sqrt{\frac{pq}{n}}$		$\bar{x}$
$n \cdot \min(p, q) > 5$		μ

Did you get it? It has to do with the distributions of $\hat{p}$ and $\bar{x}$ under random sampling. Figure 9.2.1 shows a side-by-side comparison.

As we work with the distribution of $\bar{x}$ to find confidence intervals for μ, you should find that the results are similar to those you saw in Section 9.1, except that:

- μ replaces p; that's what we're estimating.
- $\bar{x}$ replaces $\hat{p}$; that's the estimator.

Section 9.2 Confidence Intervals for Means

	Distribution of $\hat{p}$	Distribution of $\bar{x}$
Mean of the distribution	$\mu_{\hat{p}} = p$	$\mu_{\bar{x}} = \mu$
Standard deviation of distribution (also called the **standard error**)	$\sigma_{\hat{p}} = \sqrt{\dfrac{pq}{n}} \approx \sqrt{\dfrac{\hat{p}\hat{q}}{n}}$	$\sigma_{\bar{x}} = \dfrac{\sigma}{\sqrt{n}} \approx \dfrac{s}{\sqrt{n}}$
The distribution is close to normal if:	$n \cdot \min(p, q) > 5$	• The sample size is large ($n \geq 30$) or • The population has a normal distribution.

FIGURE 9.2.1 Comparing the distributions of $\hat{p}$ and $\bar{x}$

- $\dfrac{\sigma}{\sqrt{n}}$ replaces $\sqrt{\dfrac{pq}{n}}$; this is the standard deviation of the estimator, sometimes called the **standard error**.
- Lastly, there are different conditions to ensure that the distribution of the estimator is close to normal. When dealing with proportions, we wanted $n \cdot \min(p, q) > 5$. For means, we need either a large sample size ($n \geq 30$) or a normal distribution for the population we're sampling.

Finding a 95% Confidence Interval for μ

Since you've already seen a method for deriving a 95% confidence interval, we'll take a quicker approach this time. Figure 9.2.2 shows a normal distribution for $\bar{x}$, assuming that we have either a large sample size or a normal population. Notice that we have shaded the middle 95% of the distribution. From the work we did in 9.1, we know that the z-values at each end of this region are $z_1 = -1.96$ and $z_2 = 1.96$. For 95% of all random samples, the sample mean will lie in this central region, between $\bar{x}_1$ and $\bar{x}_2$. Therefore, the z-scores for 95% of the samples will be between -1.96 and 1.96.

For 95% of all random samples: $-1.96 < z = \dfrac{\bar{x} - \mu_{\bar{x}}}{\sigma_{\bar{x}}} < 1.96$

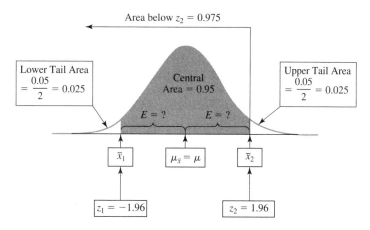

FIGURE 9.2.2 Using the distribution of $\bar{x}$ to find a 95% confidence interval

Since $\mu_{\bar{x}} = \mu$, we can substitute to get: $-1.96 < \frac{\bar{x}-\mu}{\sigma_{\bar{x}}} < 1.96$

To get a confidence interval for μ, we want to solve this compound inequality for μ. This is two inequalities in one: $-1.96 < \frac{\bar{x}-\mu}{\sigma_{\bar{x}}}$ and $\frac{\bar{x}-\mu}{\sigma_{\bar{x}}} < 1.96$. We'll solve the first one and leave the second one for you:

$$-1.96 < \frac{\bar{x} - \mu}{\sigma_{\bar{x}}}$$

$$-1.96\sigma_{\bar{x}} < \bar{x} - \mu \quad \text{(Multiplying both sides by } \sigma_{\bar{x}}\text{)}$$

$$-1.96\sigma_{\bar{x}} + \mu < \bar{x} \quad \text{(Adding } \mu \text{ to both sides)}$$

$$\mu < \bar{x} + 1.96\sigma_{\bar{x}} \quad \text{(Adding } 1.96\sigma_{\bar{x}} \text{ to both sides)}$$

Since $\sigma_{\bar{x}} = \frac{\sigma}{\sqrt{n}}$, we can substitute to get: $\mu < \bar{x} + 1.96\left(\frac{\sigma}{\sqrt{n}}\right)$.

Focus on Understanding

Solve the second part of the compound inequality for μ: $\frac{\bar{x}-\mu}{\sigma_{\bar{x}}} < 1.96$

Putting the two pieces of the compound inequality back together, we get:

$$\bar{x} - 1.96\left(\frac{\sigma}{\sqrt{n}}\right) < \mu < \bar{x} + 1.96\left(\frac{\sigma}{\sqrt{n}}\right)$$

This means that, for 95% of all random samples, the population mean μ will be in the interval between $\bar{x} - 1.96\left(\frac{\sigma}{\sqrt{n}}\right)$ and $\bar{x} + 1.96\left(\frac{\sigma}{\sqrt{n}}\right)$. This gives us a 95% confidence interval for estimating μ: $(\bar{x} - E, \bar{x} + E)$ or $\bar{x} \pm E$, where $E = 1.96\left(\frac{\sigma}{\sqrt{n}}\right)$.

Notice that this confidence interval has exactly the same form as the 95% confidence interval we first got for p, except that, since we are now estimating μ, $\bar{x}$ has replaced $\hat{p}$ as the estimator and $\frac{\sigma}{\sqrt{n}}$ has replaced $\sqrt{\frac{pq}{n}}$ as the standard error.

Focus on Understanding

Suppose that your state's Department of Education is developing a new test to measure the reading level of sixth graders. To pilot this test, they select a random sample of 50 sixth graders from within the state and have them take the test. The average test score for students in the sample was 70.5. Suppose we also know that, for all sixth graders in the state, the standard deviation will be 12.4.

1. Find a 95% confidence interval for estimating μ, the average score on this test for all sixth graders in the state.

2. Should we be certain that the confidence interval in #1 actually contains the true value of μ, the average test score for all sixth graders in the state? Interpret what is meant by "95% confidence" in this case.
3. What part of this problem is totally phony and unrealistic?

In the preceding *Focus on Understanding*, we do not know that the confidence interval we found actually contains the population mean. Our level of confidence (95%) comes from the fact that, by using the previous confidence interval formula, 95% of all random samples would produce intervals that include μ.

Generalizing the Confidence Interval for μ

The phony part of the problem is "we also know that, for all sixth graders in the state, the standard deviation will be 12.4." How in the world do we know that? It seems completely unrealistic that we would know the standard deviation for the population and still be trying to estimate the mean. Either we would know both μ and σ or we wouldn't know either one.

Just as we replaced $\sqrt{\frac{pq}{n}}$ with $\sqrt{\frac{\hat{p}\hat{q}}{n}}$ to get a more realistic confidence interval for p, we would like to replace $\frac{\sigma}{\sqrt{n}}$ with $\frac{s}{\sqrt{n}}$, using the standard deviation from the sample, rather than the standard deviation from the population. When the sample size is large, this is not a problem, since s is a good approximation of σ when n is large. For small samples, however, the situation is more complicated. For now, we'll stick with large samples; we'll deal with the small sample situation later in this section.

In the previous section, after getting a more realistic confidence interval, we generalized our results to confidence levels other than 95%. We can do exactly the same thing here. Figure 9.2.3 again shows the distribution of $\bar{x}$ under random sampling, but the total tail area is now α. Just as in the previous section, the area in

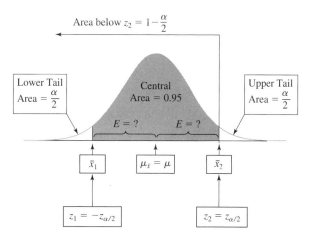

FIGURE 9.2.3 Using the distribution of $\bar{x}$ to find a general confidence interval

each tail is $\frac{\alpha}{2}$, and the central area is $1 - \alpha$. This means that the probability that $\bar{x}$ lies in the central region is $1 - \alpha$. In other words, for $100(1 - \alpha)\%$ of all random samples, the z-score for $\bar{x}$ will be between $-z_{\alpha/2}$ and $z_{\alpha/2}$. So, for $100(1 - \alpha)\%$ of all samples:

$$-z_{\alpha/2} < \frac{\bar{x} - \mu}{\sigma_{\bar{x}}} < z_{\alpha/2}$$

We can solve for μ exactly as we did before. We end up with:

$$\bar{x} - z_{\alpha/2}\left(\frac{\sigma}{\sqrt{n}}\right) < \mu < \bar{x} + z_{\alpha/2}\left(\frac{\sigma}{\sqrt{n}}\right)$$

Putting this all together, we have a realistic large sample confidence interval for estimating μ.

A Realistic, Large Sample $100(1 - \alpha)\%$ Confidence Interval for Estimating μ

Confidence Interval: $(\bar{x} - E, \bar{x} + E)$ or $\bar{x} \pm E$ Error Bound: $E = z_{\alpha/2}\left(\frac{s}{\sqrt{n}}\right)$

- The **critical value** $z_{\alpha/2}$ is a z-value (standard normal) with an upper tail area of $\frac{\alpha}{2}$.
- The values of $\bar{x}$ and s must come from a *random* sample.
- The sample size n must be large enough so that the distribution of $\bar{x}$ is close to normal and s is a good approximation of σ.
- By using the confidence interval formula above, $100(1 - \alpha)\%$ of all random samples would produce intervals that contain μ.

Classroom Connection

Figure 9.2.4 shows the mayfly data from *Math in Context: Great Expectations* (page 31). Students hatched 1,000 mayflies and observed how long they lived. ◆

Hours	6	7	8	9	10	11	12
Number Still Alive	1,000	850	600	250	100	20	0

12. Use the students' data to determine the expected life span of a mayfly. Explain how you found your answer.

FIGURE 9.2.4 Mayfly lifespan data

Focus on Understanding

1. Would you expect the lifespan of a mayfly (in hours) to be discrete or continuous data?

2. Fill in the table at the right, indicating the number of mayflies that fall in each group. The category 6–<7 includes mayflies whose lifespans were between 6 hours and 7 hours, but not including 7 hours. A mayfly that lived 6.99 hours would still be in the first group, but one that lived exactly 7 hours would be in the second group.

Lifespan in Hours	Number of Mayflies
6–<7	
7–<8	
8–<9	
9–<10	
10–<11	
11–<12	

3. Estimate the average lifespan for this sample of mayflies by using the center of each interval class (6.5 hours for the first group, etc.). Estimate the standard deviation the same way.

4. Suppose that the students working on this project actually measured the life of each mayfly to the nearest tenth of an hour and just grouped the data in their table. By using their raw data, they compute the average lifespan for their sample to be 8.25 hours with a sample standard deviation of 1.23. Use these values to estimate the lifespan of all mayflies with a 95% confidence interval.

5. What does the term "95% confidence" mean in this situation?

6. Would a 99% confidence interval be wider or narrower than the interval you found in #4?

7. Find a 99% confidence interval. Does it confirm your answer for #6?

Small Sample Sizes and the T-Distribution

We had quite a large sample size in the previous *Focus on Understanding* ($n = 1,000$). A large sample size has a couple of advantages. First, as we discussed in Section 8.2, when n is large, the Central Limit Theorem guarantees us that the distribution of $\bar{x}$ will be close to normal, regardless of what kind of distribution the population has. We don't need for the population to have a normal distribution when the sample size is large. Second, s tends to be very close to σ when n is large. For large samples, replacing σ with s makes very little difference because s and σ are almost identical.

The situation is different when the sample size is small. First, we cannot count on the Central Limit Theorem to ensure a normal distribution for $\bar{x}$. In order for $\bar{x}$ to have normal distribution, we need the population we're sampling to have a distribution that is close to normal. In practice, there is no way to know for certain that the population's distribution is normal; we just need to assume that it is. There are ways we can check to see whether this assumption seems reasonable. For example, we can look at a stem-and-leaf plot (or a histogram or dot plot) of the data to see whether it seems to be mound-shaped like a normal distribution. For very small samples, a plot of the data may not tell us very much. We often just need to assume that the population has a normal distribution, whether we have any justification for this assumption or not.

Second, when we derived the confidence interval for μ, we used the fact that the z-scores for sample means follow a standard normal distribution. When n is large, $z = \frac{\bar{x}-\mu}{(\sigma/\sqrt{n})}$ will have a standard normal distribution. However, replacing σ with s adds an additional variable to the situation; the value of s varies from sample to sample. Using s instead of σ gives $\frac{\bar{x}-\mu}{(s/\sqrt{n})}$ more variability than $\frac{\bar{x}-\mu}{(\sigma/\sqrt{n})}$, especially when the sample size is very small. Because of this additional variability, $\frac{\bar{x}-\mu}{(s/\sqrt{n})}$ does not have a standard normal distribution, even when the population's distribution is normal. Statisticians have studied the distribution of $\frac{\bar{x}-\mu}{(s/\sqrt{n})}$; they call it a **T-distribution**. Because of this, it is common to use the letter t to stand for $\frac{\bar{x}-\mu}{(s/\sqrt{n})}$, and to use z for $\frac{\bar{x}-\mu}{(\sigma/\sqrt{n})}$.

Figure 9.2.5 shows a T-distribution and a standard normal distribution together. Both are mound-shaped distributions, symmetrical about 0. Notice that, because of the additional variability that comes from using s in place of σ, the t distribution has more tail area than the standard normal distribution; it is more **tail-heavy**. In order to get an upper tail area of .025, for example, we need to go farther to the right in the T-distribution than we would in the standard normal.

There are actually many different T-distributions. Each one has a number, called the number of **degrees of freedom**. The higher the number of degrees of freedom, the closer the T-distribution gets to a standard normal curve. We could say that the standard normal distribution is a T-distribution with an infinite number of degrees of freedom. Table 2 shows critical values for T-distributions, as well as standard normal. Notice that, as the number of degrees of freedom increases, the critical values get closer and closer to the values for standard normal. For example, looking at the column for a 90% confidence interval (upper tail area = 0.05), $t_{.05} = 6.314$ for a T-distribution with 1 degree of freedom, gradually decreasing to $t_{.05} = 1.660$ for 100 degrees of freedom (as high as the table shows). Notice how close this is to $z = 1.645$ for standard normal.

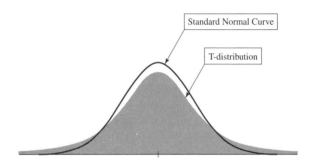

FIGURE 9.2.5 Comparing a T-distribution with the standard normal curve

When $\bar{x}$ and s are computed from random samples of size n, the values of t from $t = \frac{\bar{x}-\mu}{(s/\sqrt{n})}$ follow a T-distribution with $n - 1$ degrees of freedom, instead of the standard normal distribution we would have for $z = \frac{\bar{x}-\mu}{(\sigma/\sqrt{n})}$. As the sample size increases, this distribution will get closer and closer to standard normal. For small samples, the error bound for estimating μ will be exactly the same as for large samples, except that the critical value $z_{\alpha/2}$ will be replaced by $t_{\alpha/2}$, a critical value from a T-distribution with $n - 1$ degrees of freedom.

A Small-Sample $100(1 - \alpha)\%$ Confidence Interval for Estimating μ

Confidence Interval: $(\bar{x} - E, \bar{x} + E)$ or $\bar{x} \pm E$ Error Bound: $E = t_{\alpha/2}\left(\frac{s}{\sqrt{n}}\right)$

- The **critical value** $t_{\alpha/2}$ has an upper tail area of $\frac{\alpha}{2}$ in a T-distribution with $n - 1$ degrees of freedom. Note that the value of $t_{\alpha/2}$ depends on both n and α.
- The values of $\bar{x}$ and s must come from a <u>random</u> sample.
- If $n < 30$, we must assume that the population has a normal distribution. If the population does not have a normal distribution, this confidence interval will not be valid.
- By using the confidence interval formula above, $100(1 - \alpha)\%$ of all random samples would produce intervals that contain μ.

EXAMPLE Remember in Section 8.1, when we were selecting a random sample of size 5 from the Grade 8 Database in *Connected Mathematics*? Suppose the only sample we selected was:

    5.5   6.5   9.0   7.3   8.5

We're going to use this sample to find a 90% confidence interval for the mean number of sleep hours for the entire Grade 8 Database. Since the sample size is small, we must assume that the distribution of sleep hours is close to normal. We might check this assumption by looking at a dot plot of sleep hours for the entire class. Figure 9.2.6 shows this dot plot.

The dot plot appears somewhat mound-shaped, leading us to believe that a normally distributed population is not an unreasonable assumption (although outliers might be a concern). However, in a typical real-life situation, we would only have the sample to judge by, and we can tell very little about the shape of the distribution from so small a sample. In that case, we would just have to acknowledge that we are making an assumption about the population.

We need to find the mean and standard deviation of the sample. Using either the formulas from Section 3.2 or the features of the TI-83 plus, we find that $\bar{x} = 7.36$ and $s = 1.43$.

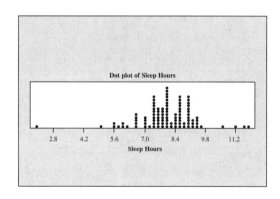

FIGURE 9.2.6 Dot plot of sleep hours for the Grade 8 database

We also need to find $t_{\alpha/2}$. Since we want a 90% confidence interval, the t-value we want has a central area of 0.90 and an upper tail area of $\frac{0.10}{2} = 0.05$. The number of degrees of freedom is $n - 1 = 5 - 1 = 4$. Looking at Table 2, we find the column with the correct areas and the row for 4 degrees of freedom. (See Figure 9.2.7.) The critical value we want is $t_{\alpha/2} = t_{.05} = 2.132$.

Finally, we are ready to find the error bound E and the 90% confidence interval:

$$E = t_{\alpha/2}\left(\frac{s}{\sqrt{n}}\right) = 2.132\left(\frac{1.43}{\sqrt{5}}\right) = 1.36$$

$$\bar{x} \pm E = 7.36 \pm 1.36 = (6.00, 8.72)$$

Based on our sample, we would be 90% confident in concluding that the average number of sleep hours in the Grade 8 Database is between 6.0 and 8.72 hours. This happens to be true; the average for the entire database is 7.962. In practice, however,

Area to Left of t_0	0.90	0.95	0.975	0.99	0.995	0.999	0.9995
Upper Tail Area	0.10	0.05	0.025	0.01	0.005	0.001	0.0005
Central Area	0.80	0.90	0.950	0.98	0.990	0.998	0.9990
Standard Normal	1.282	1.645	1.960	2.326	2.576	3.090	3.291
Degrees of Freedom for T							
1	3.078	6.314	12.706	31.821	63.657	318.309	636.619
2	1.886	2.920	4.303	6.965	9.925	22.327	31.599
3	1.638	2.353	3.182	4.541	5.841	10.215	12.924
4	1.533	2.132	2.776	3.747	4.604	7.173	8.610
5	1.476	2.015	2.571	3.365	4.032	5.893	6.869

FIGURE 9.2.7 Finding a critical value from a T-distribution

we would have no way of knowing whether or not this particular sample was one of the 90% that give correct predictions or one of the 10% that don't.

Could we use this same confidence interval to conclude (with 90% confidence) that, for all eighth graders in the United States, the average number of sleep hours is between 6.0 and 8.7 hours? If the Grade 8 Database itself was a random sample of U.S. eighth-graders, the answer would be yes. However, since the excerpt from *Connected Mathematics* back in 8.1 indicated that all of these students came from the same school, we would have to question whether what we see here is representative of the entire country. ∎

You may have noticed, once again, that we did not pin down the average number of sleep hours very closely. What do you expect for such a small sample? To achieve a more accurate estimate of a population mean, we need a larger sample like the one in the following *Classroom Connection*.

Classroom Connection

Figure 9.2.8 comes from *Connected Mathematics: Samples and Populations* (page 56). Use it for the following *Focus on Understanding*. ◆

3. Keisha opened a bag containing 60 chocolate chip cookies. She selected a sample of 20 cookies and counted the chips in each cookie.

Cookie	Chips
1	6
2	8
3	8
4	11
5	7
6	6
7	6
8	7
9	11
10	7

Cookie	Chips
11	8
12	7
13	9
14	9
15	8
16	6
17	8
18	10
19	10
20	8

Use the data from Keisha's sample to estimate the number of chips in the bag. Explain your answer.

FIGURE 9.2.8 Keisha's chocolate chip data

Focus on Understanding

1. Find the mean and standard deviation of Keisha's sample.
2. You are going to be estimating the mean number of chips per cookie with a 90% confidence interval. Before you look up the critical value $t_{\alpha/2}$, decide whether it will be less than, greater than, or the same as the critical value in the sleep-hours example we just finished. Explain how you decided.
3. Find the critical value. Does it confirm your answer to #2?

4. Find the confidence interval.
5. In using this confidence interval, what assumptions are we making about the sample and/or the population?
6. Use your confidence interval for the average number of chips per cookie to estimate the total number of chips in the remaining 40 cookies in the bag.

Deciding Which Distribution to Use

When should you use a T-distribution and when should you use a standard normal distribution? Here are some guidelines to help you decide.

- Since the T-distribution results from using s in place of σ, you should never use the T-distribution if σ is known. Granted, it is pretty unusual to be estimating μ when you know σ, but if it does happen, use a standard normal distribution to find the critical value.

- When σ is unknown, some books have students use the T-distribution only if $n < 30$ and use the standard normal distribution whenever $n \geq 30$. The reason for this is that tables for the T-distribution often only go up to 30 degrees of freedom. Although the critical values for the T-distribution are pretty close to those for standard normal for $n \geq 30$, it is more accurate to use the T-distribution whenever σ is unknown. Just use the closest number of degrees of freedom in Table 2. If the number of degrees of freedom is over 100, use standard normal. When using the TI-83 plus, use the T-distribution whenever σ is unknown, regardless of the sample size.

- When the sample size is small, some students are inclined to use a T-distribution when estimating p as well as μ. Since σ does not appear in the confidence interval formula for p, and we do not compute s when estimating p, you should not use a T-distribution when estimating p. If $n \cdot \min(\hat{p}, \hat{q}) > 5$, use a standard normal distribution. Otherwise, the method in this book does not apply.

Using the TI-83 plus

Not only will the TI-83 plus compute the mean and standard deviation, it will also compute confidence intervals. Enter the sleep-hours data from our sample of size 5 in L1. For your reference, the five values were 5.5, 6.5, 9.0, 7.3, and 8.5. Hit STAT and the right arrow key to move to TESTS. You would use 7: Z Interval if σ is known. Since σ is unknown, we want 8: T Interval. After making this selection, you should see:

```
TInterval
  Inpt:Data Stats
  List:L1
  Freq:1
  C-Level:.9
  Calculate
```

Since you have the raw data entered into a list, we want to select Data. Once Data is chosen, you should be able to enter L1 under List. Since the frequency of each data value is 1, enter 1 for Freq. The C-Level is the confidence level; to use 90% as we did before, enter 0.90 for C-Level. Move the cursor to Calculate and hit ENTER. You should see:

```
TInterval
 (5.9956,8.7244)
 x̄=7.36
 Sx=1.431083506
 n=5
```

This is consistent with what we found earlier.

Suppose you are given statistics, instead of raw data, as in the first confidence interval we computed in this section. For a random sample of 50 sixth graders, the mean reading test score was 70.5. For all sixth graders in the state, the standard deviation is 12.4. To this problem on the TI-83 plus, hit STAT and the right arrow key to select TESTS. Because you know the population standard deviation, you should use 7: Z Interval, rather than 8: T Interval. Since we have statistics rather than raw data, we would select Stats rather than Data. We enter 12.4 for σ, 70.5 for $\bar{x}$, and 50 for n. In that problem, we wanted a 95% confidence interval, so enter 0.95 for C-Level. Go to Calculate and hit ENTER. You should see:

```
ZInterval
 (67.063,73.937)
 x̄=70.5
 n=50
```

This should be consistent with your previous result.

Focus on Understanding

1. Use three different ways to find a 99% confidence interval for the average lifespan of a mayfly:

   a. Enter the midpoint of each interval class in L1 and the frequencies in L2. Use T Interval with Data. Freq should be L2 this time, rather than 1.
   b. Use T Interval again, but with Stats this time. Enter 8.25 hours for the sample mean and 1.23 for the sample standard deviation.
   c. Use Z Interval with Stats. Enter 8.25 hours for the sample mean and 1.23 for the population standard deviation.
   d. Compare the results of the three methods. Which do you think would be least accurate? Which two are almost the same? Explain why these two should be practically the same in this case.

Confidence Intervals—the Big Picture

You have now seen how to find confidence intervals to estimate both population proportions and population means. Although there are certainly differences between these, and variations depending on sample size and whether σ is known or unknown, you should see that the basic form for these confidence intervals is the same: (*estimator*) $\pm$ (*error bound*), where the error bound is $E = $ (*critical value*) $\cdot$ (*standard error*). If you go on to study statistics further, you will find that confidence intervals are used to estimate many other population characteristics besides means and proportions. Many of these confidence intervals have exactly the same form. By knowing the distribution of the estimator, you can determine the critical value and the standard error in these new situations. Once you know that, confidence intervals are a snap.

Interpreting Confidence Intervals Correctly

People often misinterpret what a confidence interval tells us. For example, suppose that we collect data on the speeds (in miles per hour) of a random sample of vehicles passing a certain point on an interstate highway. We use the data to find a 95% confidence interval for the mean, as we have done in this section, and the result is the interval (64.3, 66.7). The following list gives some common <u>incorrect</u> interpretations of this 95% confidence interval.

Some Common Wrong Interpretations

1. Ninety-five percent of the *vehicles in the sample* travel at speeds between 64.3 and 66.7 miles per hour.
2. Ninety-five percent of the *vehicles in the population* travel at speeds between 64.3 and 66.7 miles per hour.
3. The mean speed of vehicles in a sample will vary from sample to sample, but at this particular point on the interstate highway, 95% *of sample means* will be between 64.3 and 66.7 miles per hour.
4. The mean speed of vehicles in the population varies over time, but the population mean will be between 64.3 and 66.7 miles per hour 95% *of the time*.

None of these is correct. The 95% confidence interval does not contain 95% of the sample, as the first interpretation suggests, or 95% of the population, as in the second. While the third interpretation correctly observes that sample means do vary from sample to sample, it misinterprets what the 95% confidence interval tells us about this variation. The fourth interpretation is also incorrect. While it may be quite true that the average speed of vehicles varies with the time of day, if we truly selected a random sample of vehicles passing this location, the vehicles should have been selected at random times throughout the day. The population mean we are dealing with is the overall mean, taking all times of day into account. The population mean we are talking about is a fixed number; it does not vary.

The Correct Interpretation

This 95% confidence interval is a prediction about the location of the population mean. Based on our sample, we are predicting that the mean speed of all vehicles passing this point (the population mean) is between 64.3 and 66.7 miles per hour. The 95% confidence level comes from the method that produced it. Using this method, 95% of all random samples will produce intervals that include the population mean. We don't know whether this one particular sample is one of the 95% that do or one of the 5% that don't, but we can be reasonably sure of our prediction based on the overall percentage of correct predictions for the method we used to obtain it.

9.3 SAMPLE SIZE

> ### Classroom Exploration 9.3
>
> Discussion Question: Is a sample of size 1,000 always preferable to a sample of size 40? Based on what you've seen in Chapters 8 and 9, why might we prefer the larger sample? From a practical point of view, why might we prefer the smaller sample? Describe a situation where a sample of size 40 would be preferable to a sample of size 1,000.

A very practical question comes up before we ever choose a sample: How large a sample do we need? Getting a good answer to this question can be very important. If the sample we choose is too small, we cannot make useful predictions about the population. On the other hand, collecting data takes time and effort; in some cases, it takes money as well. If you have an unlimited supply of time, effort, and money, see me after class. If not, it is important to choose a sample size that is not needlessly large. In this section, we will explore the issue of sample size. Specifically, we will deal with the question of how large a sample size we need in order to get a specified degree of accuracy for our predictions.

When we use a sample to make predictions about the population, we measure the accuracy of our predictions with two numbers: the error bound and the confidence level. The smaller the error bound and the larger the confidence level, the more accurate the prediction. In many cases, we have an idea of how accurate we need our predictions to be before we ever collect a sample. We can use the error bound and the confidence level to find the required sample size.

Sample Size for Estimating p

In Section 9.1, we looked at *Tossing Tacks* (*Connected Mathematics: What Do You Expect* (page 15)). Kalifa had tossed a tack 100 times, and it had landed point up 58 times. We used this data to find a 90% confidence interval for the true probability of "point up." However, we found that there was a basic question about this situation that we could not answer with 90% confidence: Is the tack going to land point up most of the time? Because the 90% confidence interval contained possible values for p that were both smaller and larger than one-half, we could not conclude with confidence that p was more than 50%. Suppose we decide to take a larger sample to get a more precise estimate of p. Let's say we'd like to be 99% confident that our estimate is within 0.05 of the true value of p. We start with the error bound formula for estimating p and solve for n:

$$E = z_{\alpha/2}\sqrt{\frac{\hat{p}\hat{q}}{n}}$$

$$E^2 = (z_{\alpha/2})^2\frac{\hat{p}\hat{q}}{n} \quad \text{(Squaring both sides)}$$

$$E^2 n = (z_{\alpha/2})^2\hat{p}\hat{q} \quad \text{(Multiplying both sides by } n\text{)}$$

$$n = \frac{(z_{\alpha/2})^2\hat{p}\hat{q}}{E^2} \quad \text{(Dividing both sides by } E^2\text{)}$$

We now have a formula for the sample size; we just need to decide what numbers to plug into it. We wanted to estimate p to within 0.05 of its true value. This is the margin of error: $E = 0.05$. For $\hat{p}$ and $\hat{q}$ (estimates for p and q), we can use Kalifa's data:

$$\hat{p} = \frac{58}{100} = 0.58 \quad \text{and} \quad \hat{q} = \frac{42}{100} = 0.42$$

Lastly, $z_{\alpha/2}$ is determined by the confidence level. Since we want to be 99% confident, the confidence interval would be based on the middle 99% of a normal distribution. The upper tail area is $\frac{\alpha}{2} = \frac{0.01}{2} = 0.005$. Looking up the critical value in Table 2, we find $z_{\alpha/2} = z_{.005} = 2.576$. Plugging everything into the formula, we get:

$$n = \frac{(z_{\alpha/2})^2\hat{p}\hat{q}}{E^2} = \frac{(2.576)^2(0.58)(0.42)}{(0.05)^2} = 646.59$$

Therefore, in order to get the level of accuracy we want, we need to toss the tack 646.59 times. Wait a minute! How do you toss a tack 0.59 times? You can't; n has to be a whole number. When finding the sample size, it is standard procedure to round *up* to the nearest whole number. The logic behind this is that rounding down would lead to a prediction that is slightly less accurate than we wanted, whereas rounding up the sample size improves the accuracy slightly. It is better to err on the side of greater accuracy. So, we would plan on tossing the tack 647 times. Hey, we're only tossing tacks! It's not like every extra toss costs us $100. We could even toss it 650, or even 1,000 times, just to make the numbers work out pretty. The *point* is (no pun intended) that, based on the data from Kalifa's sample, we need at least 647 tosses to get the desired level of accuracy.

We used the data from Kalifa's sample to find the values of $\hat{p}$ and $\hat{q}$ to use in the formula. What if we had no prior sample to use? After all, we're doing this to decide how large a sample to use in the first place. It would not be unusual to have no prior information at all about p and q. That is the subject of the next *Focus on Understanding*.

Focus on Understanding

1. The graph at the right shows the parabola $y = x(1 - x)$. Notice that the largest value of y is 0.25, which occurs when $x = 0.5$. Explain why this graph shows that the largest possible value for $\hat{p}\hat{q}$ is 0.25, and that this will occur when $\hat{p}$ and $\hat{q}$ are both 0.5.

2. Based on the idea that it is better to end up with more accuracy than we really need, rather than less, explain why $\hat{p} = 0.5$ and $\hat{q} = 0.5$ would be a good choice in the sample size formula $n = \frac{(z_{\alpha/2})^2 \hat{p}\hat{q}}{E^2}$.

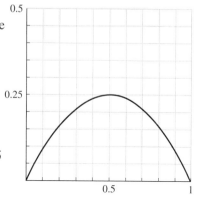

The following box sums up the procedure for determining the sample size needed to estimate a population proportion with a specified error bound and confidence level.

Finding the Sample Size Needed to Estimate *p*

The sample size needed to estimate p to within a margin of error E with $100(1 - \alpha)\%$ confidence is:

$$n = \frac{(z_{\alpha/2})^2 \hat{p}\hat{q}}{E^2}$$

- Use the best available estimates for $\hat{p}$ and $\hat{q}$. If no estimates are available, use $\hat{p} = 0.5$ and $\hat{q} = 0.5$.
- Round *n up* to the next whole number.

Classroom Connection

Figure 9.3.1 is from *Mathematics in Context: Great Expectations* (page 51). Use it in the *Focus on Understanding* that follows. ◆

Focus on Understanding

1. Answer #1 in Figure 9.3.1.
2. Answer #2 in Figure 9.3.1.

> In the small town of Adorp (population 2,000), a journalist from the local newspaper went to the mall and surveyed 100 people about building a new mall on the other side of town. Twenty-five people said they would like a new mall. The next day the journalist wrote an article about this issue with the headline:
>
> "A Large Majority of the People in Adorp Do NOT Want a New Mall."
>
> 1. Is this headline a fair statement? Explain your answer.
>
> A television reporter wanted to conduct her own survey. She called a sample of 20 people, whose names she randomly selected from the Adorp telephone directory. Sixty percent said they were in favor of the new shopping mall.
>
> 2. Is it reasonable to say that 60% of all the people in Adorp are in favor of the new mall? Explain why or why not.
>
> **FIGURE 9.3.1** Drawing conclusions from samples
>
> 3. Let p = *the proportion of people in Adorp who favor the new mall*. Based on the results of the television reporter's sample of size 20, how large a sample would be needed to estimate p to within 0.10 with 95% confidence?
> 4. Answer the same question, but this time suppose that you have no prior sample to use for estimating $\hat{p}$ and $\hat{q}$.

Sample Size for Estimating μ

When we are estimating a population mean, instead of a proportion, we can derive a formula for the sample size in exactly the same way. We would start with the error bound formula $E = z_{\alpha/2} \left(\dfrac{\sigma}{\sqrt{n}} \right)$ and solve for n. We'll leave that for an exercise and just tell you the result:

$$n = \frac{(z_{\alpha/2})^2 \sigma^2}{E^2}$$

The only difference between this formula and the one for proportions is that we have σ^2 in place of $\hat{p}\hat{q}$. Just as we would use our best estimate for $\hat{p}\hat{q}$, we would use our best estimate for σ^2, say s^2 from a prior sample. If no prior sample is available, sometimes $\sigma \approx \approx \dfrac{\text{Range}}{4}$ is used as a very rough estimate for σ^2, but good enough for this purpose (especially if nothing better is available). Notice we used two wiggly equal signs to show how rough an approximation this is.

If we use s^2 as an estimate for σ^2, you might wonder whether we should use $t_{\alpha/2}$ instead of $z_{\alpha/2}$; that is, use a T-distribution rather than standard normal, just as we did in the previous section. The problem is that the number of degrees of freedom depends on the sample size. Since the sample size is exactly what we're trying to find out, it just makes things too complicated. We would generally only use this formula when we expect n to be at least 30 anyway, so critical values from a standard normal distribution should be close enough.

The results are summed up in the following box.

Finding the Sample Size Needed to Estimate μ

The sample size needed to estimate μ to within a margin of error E with $100(1 - \alpha)\%$ confidence is:

$$n = \frac{(z_{\alpha/2})^2 \sigma^2}{E^2}$$

- Use the best available estimate for σ^2, say s^2 from a prior sample. If no prior sample is available, use $\sigma \approx \approx \frac{Range}{4}$ as a very rough estimate.
- Round n *up* to the next whole number.

Focus on Understanding

In order to comply with the No Child Left Behind Act (see www.nclb.gov), states must create ways of assessing student achievement, make goals for what those assessments should show, and achieve those goals. There can be severe penalties for not achieving the goals they set. It is essential for states to be able to accurately predict how students will do on the tests that they devise. Suppose your state's department of education would like to estimate the mean score on a 100-point math test being designed for all of the state's eighth graders.

1. How large a sample would be needed to estimate this mean to within 2 points with 95% confidence, assuming that data from similar tests used in recent years indicated that $s \approx 15$?
2. Answer the same question, assuming that no prior data is available.

You should now have a basic understanding of how to estimate means and proportions with confidence intervals. You are now ready to learn how to use sample data to decide between two competing claims about a population. That is the subject of the next chapter.

Chapter 9 Summary

This chapter developed ideas about confidence intervals for means and proportions. When we find such a confidence interval, we are using data from a random sample to predict the location of the mean or proportion of the population we sampled. Because information from random samples varies from sample to sample, some confidence intervals make true predictions (the population mean or population proportion is actually in the interval), while others do not. The probability that a random sample would result in a true prediction about the population is expressed as the confidence level for the interval.

You've seen that, in some situations, confidence intervals are based on a normal distribution. However, when estimating μ with σ unknown, we needed to consider a different type of distribution, the T-distribution. You've seen that T-distributions come in different versions, determined by its number of degrees of freedom. To find confidence intervals in different situations, it is important to know which distribution is appropriate.

In addition to finding confidence intervals for estimating means and proportions, we also considered the question of sample size. If we know the level of accuracy that we'd like (as measured by the confidence level and error bound), we can determine how large a sample we need to select. That is exactly what you did in Section 9.3.

In the real world, decision makers in almost any field depend on using samples to make predictions about populations. In order to make informed decisions, they need to understand how accurately the information from random samples reflects the population they come from. They need to understand confidence intervals and be able to interpret them correctly.

The key terms and ideas from this chapter are listed below:

confidence interval   267
confidence level   267
error bound   267
margin of error   267
upper tail area   275
lower tail area   275

total tail area   275
critical value   276
standard error   281
T-distribution   286
tail-heavy   286
degrees of freedom   286

Assessment is an integral part of every curriculum from the elementary school all the way through college. The question always arises—*what is it that students should be able to do after completing this lesson/unit/chapter?* We have included here our intended learning goals for Chapter 9. Students who have a good grasp of the concepts developed in this chapter should be successful in responding to these items:

- Explain, describe, or give an example of what is meant by each of the terms in the vocabulary list.
- Determine whether a given sample size is large enough to use a normal distribution to find a confidence interval for p.
- Find and interpret confidence intervals for estimating population proportions.
- Find and interpret confidence intervals for estimating population means.
- Determine whether it is necessary to assume the population has a normal distribution to use a confidence interval to estimate μ.
- Interpret what the confidence level means when estimating p or μ with a margin of error.
- Identify situations where a T-distribution should be used, rather than standard normal, and use the correct number of degrees of freedom.
- Find the sample size needed to estimate p or μ to within a given margin of error with a given level of confidence.

Your course instructor may have additional or different assessable outcomes for your class. As teachers (or future teachers) you should think about the assessment outcomes and learning goals for each chapter as you work through them.

EXERCISES FOR CHAPTER 9

1. Using the *Data Analysis and Probability Standards for Grades 6–8* from the NCTM's *Principles and Standards for School Mathematics* found at www.nctm.org, identify the middle school objectives that are found in Chapter 9.
2. Using the state standards for mathematics for the content area of data analysis and probability for your state, identify the middle school objectives that are found in Chapter 9. The following website may be useful: www.doe.state.in.us. This website will allow you to access the web pages of the state departments of education for the 50 states. From the state web pages you should be able to find the state's mathematical standards.
3. Nicky is a 60% free throw shooter, but what about her friend Vicki? Vicki just tried out for the basketball team. In the first week of practice, she attempted 50 free throws and hit 34 of them.
   a. Is a sample of size 50 large enough to use a normal distribution to find a confidence interval for p? (Show how you determine this.)
   b. Find a 95% confidence interval for estimating Vicki's free throw percentage.
   c. Based on the interval you found in (b), would you be 95% confident in saying that Vicki is a better free throw shooter than Nicky? Explain.
   d. Would a 90% confidence interval be wider or narrower than the 95% confidence interval you found in (c)? Explain why you think so.
   e. Find a 90% confidence interval and check to see if it confirms your answer for (d).
4. In the game *Pass the Pigs* (distributed by Milton-Bradley), a pair of pig-shaped dice are rolled. Players score points based on the position that the pigs land in. Double razorbacks (both pigs on their backs) are worth 20 points. In a sample of 200 pig rolls, 13 of the rolls are double razorbacks.
   a. Find the experimental probability of rolling double razorbacks.
   b. Find a 90% confidence interval for the proportion of rolls that are double razorbacks.
   c. Would a 95% confidence interval be wider or narrower than the 90% confidence interval you found in (b)? Explain why you think so.
   d. Find a 95% confidence interval and check to see if it confirms your answer for (c).
   e. Double trotters (both pigs on their feet) are also worth 20 points each. Suppose only 2 of the 200 rolls were double trotters. Explain why we can use the sample of 200 pig rolls to find a confidence interval for the proportion of double razorbacks, but not for double trotters.
5. In a random sample of 80 statistics students, 52 are male and 28 are female.
   a. Find a 95% confidence interval for the proportion of statistics students who are male.
   b. Assuming you did all of the computations correctly for part (a), can you be sure that the true proportion of male statistics students lies within your confidence interval? Explain the meaning of the phrase "95% confidence."
   c. Find a 95% confidence interval for the proportion of statistics students who are female.

300 Chapter 9 Estimating with Confidence

    d. How do the error bounds for the confidence intervals in (a) and (c) compare to each other? Why?

6. The weights of chocolate candies produced by a machine do not vary much from piece to piece ($\sigma = 1.23$ grams). A random sample of 120 candies had a mean weight of 14.75 grams. Find a 95% confidence interval for the mean weight of one of these candies.

7. In the game *Pass the Pigs* (see #4 above), a sample of 200 rolls had an average score of 4.55 with a standard deviation of 5.25.
    a. Find a 90% confidence interval for the mean number of points per roll.
    b. Is it necessary to assume that the distribution of the population of scores is approximately normal? Why or why not?

8. In racquetball, the first person to get 15 points (and be at least 2 points ahead) wins. Mike frequently plays racquetball with the same opponent and loses most of the time. In a random sample of 20 games, he averaged 7.75 points per game with a standard deviation of 2.07.
    a. To find a confidence interval for the average number of points he scores per game, should you use a standard normal distribution or a T-distribution? Why?
    b. Find a 95% confidence interval for the mean number of points he scores per game.
    c. Is it necessary to assume that the distribution of Mike's scores is close to normal? Why or why not?

9. In a vehicle safety test, five cars of the same model are crashed into a steel guard rail at 15 miles per hour. The table at the right lists the repair costs for each car.
    a. Find the mean and standard deviation for the sample.
    b. Find a 90% confidence interval for the mean repair cost in this type of accident.
    c. Why do you think that they used such a small sample for this test?

Car 1	$2,022
Car 2	$2,233
Car 3	$2,111
Car 4	$1,854
Car 5	$1,917

10. In #3 above, Vicki hit 34 out of 50 free throws.
    a. Based on the results of this sample, how large a sample would be needed to estimate Vicki's free throw percentage to within 6% with 95% confidence?
    b. Do you think a good way to collect the data would be to have Vicki take all of these shots one after another? Explain.

11. In #4 above, 200 rolls resulted in 13 double razorbacks. Based on the results of this sample, how large a sample would be needed to estimate the probability of double razorbacks to within 0.02 with 99% confidence?

12. Suppose you'd like to estimate the proportion of people who watched a new television show. With no prior knowledge of this proportion, how large a sample would you need if you want to be 95% confident that your estimate is within 3% of the true value?

13. Based on the sample in #6 above, how many candies would need to be weighed to estimate the mean weight to within 0.1 grams with 95% confidence?

14. Based on the sample in #7 above, how many rolls would be needed to estimate the average score per roll to within two-tenths of a point with 95% confidence?

15. A sociologist studying poverty in the United States is focusing on those whose annual income is $10,000 per year or less. She'd like to estimate the average income for this group to within $500 with 99% confidence. How large a sample does she need?

Testing Hypotheses

CHAPTER

10

10.1 WHAT IS A HYPOTHESIS TEST?
10.2 TESTS ABOUT PROPORTIONS
10.3 THE P-VALUE FOR A TEST
10.4 TESTS ABOUT MEANS

In Chapter 8, you learned about probability distributions for sample proportions and sample means. In Chapter 9, you used this knowledge to find reliable estimates for population means and population proportions. You are now ready to use information from a sample to decide between two competing claims about a population. This type of decision making is called hypothesis testing. In Section 10.1, we will give a general introduction to hypothesis testing and explore some of the ideas behind it. In the rest of the chapter, we will discuss methods for testing hypotheses about population proportions and population means.

10.1 WHAT IS A HYPOTHESIS TEST?

We'll start with a story that we will use to illustrate some of the ideas involved in hypothesis testing. We'll call this story *Satellite Phones*.

Satellite Phones

My wife is shopping at a store in the mall, and I'm there waiting for her. I'm standing there, and a salesman comes up to me. "I'd like to demonstrate our new satellite cell phone technology for you." he says. I have nothing better to do; I'm just waiting around, so I say "Okay."

"You may have heard some people claim that our satellite cell phones are not dependable, but I tell you that there is no evidence to support such a claim. Wouldn't you agree that it is wrong to believe such an accusation if you have no evidence to support it? I mean, would you come out and call me a liar without any evidence?"

I agree in principle, and he continues. "Using our satellite cell phones, you can make a call from anywhere in the world, and our satellites will be in position to receive your signal and connect you with the person you are calling. It's a miracle of modern technology! Now, I will admit that, at the present time, it is possible that some calls fail to go through, but I claim that this happens only 1% of the time, maybe even less."

Perhaps I look skeptical, because he says, "You don't believe me? Here's a phone. Dial any number you want, and see if the call goes through."

I dial a number, but the phone's display reads "Currently out of service area." The salesman hesitates for a moment. Then he grins and says, "That was just the one time in one hundred that a call will fail. Try another number!"

I probably should have walked away right then, but I try calling someone else. The result is the same. "Hey, where are you going?" the salesman asks. "Come back! This phone will take pictures, too. Let me show you!"

Classroom Exploration 10.1

In *Satellite Phones*, the salesman claims: **Only 1% of calls fail to go through.**

1. Suppose the salesman's claim is true. What is the probability that two calls in a row would fail? (Assume that the calls are independent.)
2. Suppose we do not know the true proportion of calls that fail. If two calls in a row fail, is it reasonable for us to believe that only 1% of all calls fail? Explain, based on the probability in #1.
3. Suppose the salesman's claim is false and the truth is that **20% of all calls fail**. Would you want such a phone? What is the probability that two successive calls both go through?
4. If two calls in a row had both gone through, should we be convinced that the salesman's claim (only 1% fail) is true? Explain, based on the probability in #3.

Satellite Phones can be thought of as an informal hypothesis test. A **hypothesis test** is a method for making a decision about a population based on evidence from a sample. Hypothesis tests usually follow a very formal step-by-step procedure. A basic outline of this procedure might look like the list in the following box.

> **Steps in a Hypothesis Test**
> 1. Formulating the hypotheses
> 2. Selecting an appropriate test statistic
> 3. Determining the decision rule
> 4. Collecting the data
> 5. Evaluating the test statistic
> 6. Arriving at a conclusion

Satellite Phones didn't really do that, at least not explicitly. It is also unusual because the decision is based on a very small sample: two attempted phone calls. Most hypothesis tests require much larger samples to reach reliable decisions. Despite these differences, we can still use it to illustrate some of the basic ideas behind hypothesis testing.

The order of steps in the previous box is what we would follow in an ideal situation. In practice, it doesn't always work that way. In particular, it is not usually feasible for students to collect data as a part of textbook problems. Therefore, it may appear in most problems as if the data are collected before the hypotheses are ever formulated. While we can't say that this never happens in real life, it is not the best procedure. Ideally, you should decide what hypotheses are to be tested and how the data will be analyzed before collecting the data.

This section will focus on the beginning and end of the process outlined: formulating the hypotheses and arriving at a conclusion. We'll leave the rest for Section 10.2.

Formulating Hypotheses

A **hypothesis** is a claim or statement about a characteristic of a population. In *Satellite Phones*, the population consists of all calls attempted on these new phones. The population characteristic that concerns us is the proportion of all attempted calls that fail to go through; call this proportion p. In this situation, a hypothesis would be some statement about p, such as:

$$p = 0.10 \quad p > 0.25 \quad p \leq 0.4 \quad \text{or} \quad p \neq 0.3$$

Students sometimes try to formulate a hypothesis about $\hat{p}$, such as $\hat{p} \leq 0.5$. This is not a hypothesis, because $\hat{p}$ is not a population characteristic. (It's a sample statistic.) A hypothesis must be about a population characteristic.

Any hypothesis test involves two competing claims, the **null hypothesis** H_0 and the **alternative hypothesis** H_1 (also called the **research hypothesis**). Some important ideas about these hypotheses follow.

H_1: The Alternative Hypothesis or Research Hypothesis

- Evidence (in the form of sample data) is collected for the purpose of supporting the alternative hypothesis.

- Possible forms for the alternative hypothesis:

    H_1: *population characteristic* $\neq$ *specified value*
    H_1: *population characteristic* $>$ *specified value*
    H_1: *population characteristic* $<$ *specified value*

- Note that H_1 is always a strict inequality; it *excludes* the "specified value."

H_0: The Null Hypothesis

- Initially assumed to be true, but only to determine whether or not the sample provides convincing evidence against it.
- Possible forms of the null hypothesis:

    H_0: *population characteristic* $=$ *specified value*
    H_0: *population characteristic* $\leq$ *specified value*
    H_0: *population characteristic* $\geq$ *specified value*

- Note H_0 always permits equality; it *includes* the "specified value."

Note: In a hypothesis test, both H_0 and H_1 must be about the same population characteristic and the same specified value.

In *Satellite Phones*, the two competing claims are:

Claim 1: *The proportion of calls that fail to go through is 1% (or maybe less).*
Claim 2: *The proportion of calls that fail to go through is more than 1%.*

The first claim is stated by the salesman. The second is not explicitly stated in the story, but is implied by my skeptical look and is, in fact, what I believe to be true in the end. We need to identify which of these claims is the null hypothesis and which is the alternative.

Evidence, in the form of sample data, is collected for the purpose of supporting the alternative hypothesis. In the story, I collected data by making two phone calls. Was I doing this because I wanted to support the salesman's claim or because I doubted his claim and wanted to support Claim 2? I think my skeptical look gave away the answer to that question. I was trying to support Claim 2. Claim 2 is the alternative hypothesis. Another clue in Claim 2 is the phrase "more than 1%." Claim 2 is a strict inequality; it excludes the possibility that $p = 0.01$. The alternative hypothesis is always a strict inequality. Symbolically, we would write Claim 2 as $H_1 : p > 0.01$.

The null hypothesis is initially assumed to be true, but only to determine whether or not there is convincing evidence against it. In the story, I agreed in principle not to call the salesman a liar unless I had evidence that his claim was false. I was willing to treat his claim as truth, but only until I had evidence against it. The salesman's claim is the null hypothesis.

There are two different ways we could formulate the null hypothesis. We could interpret the salesman's claim as either $H_0 : p = 0.01$ or $H_0 : p \leq 0.01$. Both of these contradict the alternative hypothesis, and that is the main requirement for H_0. In testing a hypothesis, we will need to choose one particular value to use for

p when H_0 is true. In the first case, $H_0 : p = 0.01$, it is obvious what the value of p will be. It will always be obvious what value to choose if the null hypothesis has the form "population characteristic = specified value." This is exactly what we did in #1 in Exploration 10.1. We assumed that the null hypothesis was true; we assumed $H_0 : p = 0.01$.

If we formulated the null hypothesis as $H_0 : p \leq 0.01$, it is not so obvious what value of p we would choose. It turns out that the best choice is always the value of p closest to the values in the alternative hypothesis. (We'll explain why in the next section; for now, just accept it.) Since the alternative hypothesis is $p > 0.01$, the closest value would again be $p = 0.01$. In other words, whether we formulate the null hypothesis as $H_0 : p = 0.01$ or $H_0 : p \leq 0.01$, we would still choose the same value for p, $p = 0.01$. For this reason, some people always formulate the null hypothesis as an equation. They always use the form "*population characteristic = specified value.*" They never use $\leq$ or $\geq$ in the null hypothesis. Let's go along with this approach for now and formulate the null hypothesis as $H_0 : p = 0.01$.

Regardless of whether the null hypothesis is formulated as *population characteristic = specified value, population characteristic $\leq$ specified value*, or *population characteristic $\geq$ specified value*, always choose the "*specified value*" when assuming H_0 is true. (We'll explain why in the next section.)

Arriving at a Conclusion

A hypothesis test is similar to an indirect proof (a proof by contradiction) in geometry. For example, suppose we wanted to prove this statement: *If $c^2 = a^2 + b^2$ and a, b, and c are the sides of a triangle, it must be a right triangle.* To prove this by contradiction, we would assume the opposite. That is, we would assume that there could be a triangle with $c^2 = a^2 + b^2$ that was *not* a right triangle. We would then search for evidence that this is a *false* assumption. In geometry (and other branches of mathematics), the evidence of a false assumption would be a contradiction, something that is logically impossible. If we arrive at a logical impossibility through correct reasoning, we would conclude that our assumption was false, and therefore the opposite is true: If a triangle has sides a, b, and c with $c^2 = a^2 + b^2$, it must be a right triangle. That is the outline of a typical indirect proof.

In statistics, the two hypotheses are opposites; they contradict each other. The research hypothesis H_1 is the one we are trying to gather evidence to support. To try to show that H_1 is true, we start by assuming the opposite; we assume that the null hypothesis H_0 is true. We then look for evidence that this is a false assumption. In statistics, the evidence is not something that is logically impossible, but merely something highly unlikely. If the sample shows results that would be highly unlikely if H_0 is true (and more likely if H_1 is true), we conclude that H_0 is false and H_1 is true. This is called **rejecting the null hypothesis in favor of the alternative**.

In #1 in Exploration 10.1, you assumed that the salesman's claim was true; that is, you assumed $H_0 : p = 0.01$. Based on that assumption, you should have determined the probability of two failed calls in a row to be 0.0001. It would be highly unlikely to get two failed calls in a row, but that is exactly what happened. The sample of two calls showed results that would be highly unlikely if H_0 is true.

Therefore, we reject H_0 in favor of H_1. We decide that the true proportion of failed calls is more than 1% and walk away, confident that we have evidence (in the form of data) that the salesman's claim is false.

Errors in Hypothesis Testing

Is it possible that we made the wrong decision? Yes, it is. If the salesman's claim is true, it is still possible to get two failed calls in a row, it is just not very likely. It would be very unusual for that to happen. Whenever we make decisions based on the outcome of a random process (like choosing a random sample), it is possible to make the wrong decision because we might be basing that decision on an unusual sample.

Focus on Understanding

The table below shows four different possibilities, based on the real, actual truth about H_0 and what we might decide about H_0. For each possibility, decide whether we would be making a correct decision or an error.

		Real, actual truth about H_0	
		H_0 is true	H_0 is false
Our decision about H_0	Reject H_0	1. Correct or Error?	2. Correct or Error?
	Do not reject H_0	3. Correct or Error?	4. Correct or Error?

There are two different types of errors illustrated in the table above, and the two types have different probabilities. A **Type 1 error** occurs in situation (1) in the table above: H_0 is true, but we decide to reject H_0. When H_0 is true, we would like the probability of rejecting H_0 to be small. This Type 1 error probability is called α. It is no coincidence that we use the same symbol to stand for the total tail area (more on that in the next section).

A **Type 2 error** occurs in situation (4) in the table above: H_0 is false, but we decide not to reject H_0. When H_0 is false, we would like the probability of not rejecting H_0 to be small. This Type 2 error probability is called β.

In situations involving chance, it is generally impossible to create a decision rule that eliminates both types of errors. If we try to make Type 1 errors next to impossible, we end up with lots of Type 2 errors. If we change our decision rule to cut down on Type 2 errors, we increase the probability of a Type 1 error. Decreasing α increases β and vice versa.

In *Satellite Phones*, part of my rule for making a decision appeared to be: If two calls in a row fail, reject the salesman's claim. If the salesman's claim was the actual honest truth, the probability that my decision rule would lead me to make a Type 1 error is exactly the probability that you computed in #1 in Exploration 10.1 ($\alpha = 0.0001$). We have a very small probability for a Type 1 error.

However, suppose that my decision rule had also included: If two calls in a row both go through, accept the salesman's claim. In #3 in Exploration 10.1, you showed that, if I followed this decision rule, the salesman's claim could be false, and I could still have a very high probability of accepting it as true ($\beta = 0.64$). With a sample of only two calls, the probability of making a Type 2 error would have been extremely high. I'm glad that didn't happen; I might have ended up with a pretty lousy phone.

This is part of the reason we say "Do not reject H_0." rather than "Accept H_0." Think back to the idea of the indirect proof in geometry. We are trying to prove that a triangle with $c^2 = a^2 + b^2$ must be a right triangle. We assume the opposite, that there is a triangle with $c^2 = a^2 + b^2$ that is not a right triangle. Now, suppose that we cannot find a contradiction. Have we shown that what we started out to prove is false? No, we haven't proved anything. Just because we have not found a contradiction does not mean there isn't one. It might be that we were just not persistent enough to find it.

If I fail to find evidence against the salesman's claim, it does not necessarily mean that I should believe that the salesman's claim is true. It could be that I was not persistent enough to find the evidence; I did not select a large enough sample to detect the difference between the salesman's claim and the truth. Therefore, I do not necessarily accept the salesman's claim as true. I can withhold judgment until more data is collected. I do not reject H_0, but I don't accept it either.

Another way to think about this is like a criminal trial. If you've seen courtroom dramas on television, you might have noticed that the jury does not choose between *guilty* and *innocent*, but between *guilty* and *not guilty*. The difference between *innocent* and *not guilty* comes from the fact that the burden of proof is on the prosecution. The prosecution must present convincing evidence that the defendant is guilty. If they do not provide evidence that convinces the jury of the defendant's guilt, the defendant goes free. The defendant does not need to prove his innocence, but neither does the jury find him innocent.

The decision we face in a hypothesis test is similar. We do not need to choose between believing H_1 or believing H_0. Instead, we must decide whether we have found convincing evidence of H_1 or we haven't. Just because we have not found convincing evidence in favor of H_1 does not make H_0 true. We might not have sufficient evidence to reject H_0, but that does not mean that we must believe H_0 is true.

We can decide to accept H_0 as the result of a hypothesis test, but we would usually only do so if we had found that the probability of a Type 2 error was very small. It is generally more difficult to compute β than α. Therefore, our strategy will be to control α, keeping the probability of a Type 1 error reasonably small. We will usually not compute β, so we will not accept the null hypothesis; we will either reject it or not reject it.

Here's another story to use in the next *Focus on Understanding*. We'll call this one *The Good Seeds*.

The Good Seeds

I walk into a garden store, looking for some seeds. I'm trying to decide which kind to buy, and a salesman walks over to help me. I tell him, "I bought some seeds once before and most of them didn't come up, but I can't remember what kind they were."

He says, "This package says that more than 90% of these seeds should germinate." I look puzzled. "That means that more than 90% of them should come up."

"Oh," I said, "I don't know. I don't always believe what it says on the package."

"Look," he says, "I trust this company and I want you to be convinced that what it says on this package is true. I'll give you 10 of the seeds for free. You go home and plant them. If all 10 of them germin ... come up, you can come back and buy some."

"Okay," I said, "If all 10 of them come up, I'll be convinced. Thanks for the free *sample*—ha ha!" The salesman just walked away. Some people just don't appreciate a good joke.

Focus on Understanding

In *The Good Seeds*, the package claims that more than 90% of the seeds will germinate. My decision rule is: I will be convinced that this claim is true if the 10 selected seeds all germinate.

1. In this case, is the claim on the package the null hypothesis or the alternative? Formulate it symbolically.
2. Formulate the other competing hypothesis.
3. Suppose that the true probability of a seed germinating is not more than 90%, but exactly 90%. Even so, all 10 seeds could germinate. Find the probability of this event. If this happens, my decision rule would lead me to make an error. What type of error would it be? Is the probability you just found α or β? Explain.
4. Suppose the package is correct. In fact, the true probability of a seed germinating is 0.95. Find the probability that fewer than 10 seeds germinate. If this happens, my decision would be incorrect. What type of error would I be making? Is the probability you just found α or β? Explain.
5. Based on the error probabilities you found above, do you think that an error would be highly unlikely or fairly common in this situation? Do you think that a sample of size 10 is large enough to confidently decide whether or not to believe the package's claim? Explain.

10.2 TESTS ABOUT PROPORTIONS

The Good Seeds is an informal hypothesis test based on a sample of 10 seeds. The decision rule was given in the problem, and you used it to find both Type 1 and Type 2 error probabilities. Unfortunately, the error probabilities are very high: $\alpha = 0.349$ and $\beta = 0.401$. A sample of size 10 is not large enough to do this test and be confident in the conclusions. In this section, we will be looking at hypothesis tests involving larger samples. Just as in *The Good Seeds*, these tests will be about proportions. Unlike *The Good Seeds*, the decision rule will not be given. Instead, we will show how statisticians determine the decision rule from the Type 1 error probability rather than the other way around.

Return of the Good Seeds

Suppose we really want to determine the truth about those seeds. We now realize that we need a much larger sample, so we choose a random sample of 400 seeds. We're still trying to find evidence that supports the claim on the package, so the null and alternative hypotheses have not changed. They are still $H_0 : p = 0.9$ and $H_1 : p > 0.9$.

As we discussed in 10.1, a hypothesis test is like a proof by contradiction. If we are trying to show that H_1 is true, we start by assuming H_0. So, assuming $H_0 : p = 0.9$, we would be choosing a random sample of 400 seeds from a population that is 90% good seeds, seeds that germinate. We are going to be concerned with how the sample proportion of good seeds will vary from sample to sample, so that is the subject of Exploration 10.2.

Classroom Exploration 10.2

1. In the situation in the previous paragraph, is the sample size large enough for the distribution of $\hat{p}$ to be close to normal? (Show how you check this.)
2. What is the mean of the distribution of $\hat{p}$?
3. Compute the standard deviation of the distribution of $\hat{p}$.
4. Describe what a Type 1 error would be in this situation. What (if anything) might be the consequences of making a Type 1 error?
5. Describe what a Type 2 error would be in this situation. What (if anything) might be the consequences of making a Type 2 error?

Selecting an Appropriate Test Statistic

We are going to be looking at the sample of 400 seeds to try to decide whether there is evidence to support $H_1 : p > 0.9$. What kind of sample statistic would be helpful in making this decision? The obvious answer is $\hat{p}$, the proportion of good seeds in the sample:

$$\hat{p} = \frac{\text{number of sample seeds that germinate}}{\text{number of seeds in sample}}$$

However, we would like the value of the sample statistic to tell us (without much additional work), whether or not this sample would be unusual if the null hypothesis was true. By itself, the sample proportion $\hat{p}$ does not do this. What we really want to know is where the value of $\hat{p}$ falls within the distribution of possible sample proportions. Therefore, the appropriate **test statistic** for this situation is the z-score for the value of $\hat{p}$. In Exploration 10.2, you found that if we assume that H_0 is true, then $\mu_{\hat{p}} = 0.9$ and $\sigma_{\hat{p}} = 0.015$. Therefore, the z-score for $\hat{p}$ would be $z = \frac{\hat{p} - \mu_{\hat{p}}}{\sigma_{\hat{p}}} = \frac{\hat{p} - 0.9}{0.015}$. In general, in tests about p, the appropriate test statistic is $z = \frac{\hat{p} - \mu_{\hat{p}}}{\sigma_{\hat{p}}} = \frac{\hat{p} - p_0}{\sqrt{\frac{p_0 q_0}{n}}}$, where p_0 is the value of p we get by assuming that the null hypothesis is true.

Determining the Decision Rule

The decision rule is based on the probability of a Type 1 error, rejecting the null hypothesis even though it is true. So, we must ask ourselves how large a Type 1 error probability we are willing to accept. Why should we be willing to accept any Type 1 errors at all? Remember that reducing α makes β larger (and vice versa). If we try to make Type 1 errors next to impossible, we will find that the probability of a Type 2 error is very large. We want α to be small, but we don't want to be unreasonable about it, or β would be too large.

So, what is a reasonable level to set for α? It depends on the reasons we have for doing the test and the consequences of making an error. In this case, our motivation for doing this test is mostly just curiosity; we just want to know whether or not we should believe the claim on the package.

What about the consequences of a Type 1 error? A Type 1 error would be rejecting the null hypothesis when it is actually true. In this case, a Type 1 error would lead us to believe that, for this type of seed, more than 90% germinate, when this is actually false. At worst, the consequences might be that we buy a few packages of seeds, thinking that they are good when they actually aren't. Unless the seeds are extremely expensive, this is not very important. We'll use $\alpha = 0.05$, accepting a 5% probability of a Type 1 error. If the consequences of a Type 1 error were more severe, we would want α to be smaller. There are cases where the consequences of a Type 1 error can be very significant, costing companies millions of dollars or risking people's lives. In such cases, we would want α to be very small (maybe $\alpha = 0.0001$ or even smaller). Notice that the value we would choose for α depends on the importance or significance of a Type 1 error—the greater the significance, the smaller we'd like α to be. For this reason, α is sometimes referred to as the **level of significance** for the test.

In general, choosing the appropriate value of α for a given situation can require a great deal of experience and familiarity with the consequences of an error. Even experts can disagree about the appropriate choice for α in a given situation. Therefore, in problems in this book, the value of α will usually be given. In cases where we don't give α, use $\alpha = 0.05$.

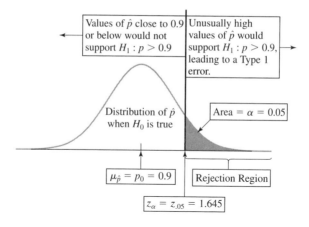

FIGURE 10.2.1 Using the distribution of $\hat{p}$ to determine the decision rule

Now that we have a value for α, we are ready to determine the **decision rule**. Figure 10.2.1 shows the distribution of $\hat{p}$ when H_0 is true. Even if H_0 is true, an unusual sample might still convince us to reject H_0, resulting in a Type 1 error. We would tend to believe $H_1 : p > 0.9$ if the proportion of good seeds in the sample was well above 0.9. Values of $\hat{p}$ that are either close to 0.9 or below 0.9 would not convince us to reject $H_0 : p = 0.9$. Since Figure 10.2.1 represents the situation where H_0 is true, rejecting H_0 would be a Type 1 error. The upper tail area in this figure is α, the probability of a Type 1 error. In this case, $\alpha = 0.05$. Using Table 2, we can find the z-value with this upper tail area, $z_\alpha = z_{.05} = 1.645$. If we reject H_0 when the z-score for the value of $\hat{p}$ places it in the highest 5% of this distribution, the probability of a Type 1 error will be 0.05, exactly as planned. Therefore, the decision rule that has the correct Type 1 error probability is: Reject H_0 if $z > 1.645$, where z is the z-score of the sample proportion. The z-scores that lead us to reject H_0 are said to lie in the **rejection region**. In this case, the rejection region is $z > 1.645$.

Collecting the Data and Evaluating the Test Statistic

We have formulated the hypotheses, selected a test statistic, and determined a decision rule. Collecting the data would come next. Don't worry, we're not going to make you plant 400 seeds and wait for them to sprout. Let's suppose we did and 370 of the 400 seeds sprouted. We would compute $\hat{p}$ and use it to evaluate the test statistic.

$$\hat{p} = \frac{\text{number of sample seeds that germinate}}{\text{number of seeds in sample}} = \frac{370}{400} = 0.925$$

$$z = \frac{\hat{p} - p_0}{\sqrt{\frac{p_0 q_0}{n}}} = \frac{\hat{p} - 0.9}{0.015} = \frac{0.925 - 0.9}{0.015} = 1.67$$

Arriving at a Conclusion

To arrive at a conclusion, we just need to compare the value of the test statistic to the rejection region. In this case, since $z = 1.67 > 1.645$, z lies in the rejection region, so we would reject H_0 in favor of H_1. We would conclude that there is convincing evidence to believe that, for seeds of this type, more than 90% germinate.

Are we 100% certain of this conclusion? No. We have found that the value of $\hat{p}$ we obtained from the sample would be in the highest 5% of the distribution of $\hat{p}$ if the null hypothesis was true; it is in the shaded area in Figure 10.2.1. This would certainly not be impossible, but it would be unusual; its probability is only 0.05. Getting a value of $\hat{p}$ that would be unusually high if H_0 was true convinces us to reject H_0 and believe $H_1 : p > 0.9$.

Would we reach the same conclusion if we changed the level of significance? Suppose we used $\alpha = 0.01$; in other words, if we were only willing to accept a Type 1 error probability of 1%? If you look back at what we did, you will see that we used $\alpha = 0.05$ to determine the decision rule. If $\alpha = 0.01$, the upper tail area in Figure 10.2.1 would be 0.01 instead of 0.05. From Table 2, $z_{.01} = 2.326$, so the rejection region is now $z > 2.326$. The z-value from the data, 1.667, is now outside the rejection region. We would not reject H_0. We would not have sufficient evidence to conclude that more than 90% of these seeds germinate.

Notice that the conclusion we reach depends not only on the data, but also on α, the level of significance. We would reject H_0 at the 0.05 level of significance, but not at the 0.01 level. In general, the smaller the value of α, the more difficult it is to reject the null hypothesis. To reject H_0 at the 0.05 level, we need the value of $\hat{p}$ to be in the highest 5% of the distribution of $\hat{p}$ (when H_0 is true). To reject H_0 at the 0.01 level, we need the value of $\hat{p}$ to be in the highest 1% of this distribution. To emphasize that the decision depends on the level of significance, statisticians sometimes describe the results of a test in terms of α. In the situation we have been dealing with, they would say that the results are **statistically significant** at the 0.05 level, but not at the 0.01 level.

Classroom Connection

You've seen the excerpt in Figure 10.2.2 before. It is from *Mathematics in Context: Great Expectations* (page 51). Use it in the *Focus on Understanding* that follows. ◆

> In the small town of Adorp (population 2,000), a journalist from the local newspaper went to the mall and surveyed 100 people about building a new mall on the other side of town. Twenty-five people said they would like a new mall. The next day the journalist wrote an article about this issue with the headline:
>
> "A Large Majority of the People in Adorp Do NOT Want a New Mall."
>
> 1. Is this headline a fair statement? Explain your answer.
>
> A television reporter wanted to conduct her own survey. She called a sample of 20 people, whose names she randomly selected from the Adorp telephone directory. Sixty percent said they were in favor of the new shopping mall.
>
> 2. Is it reasonable to say that 60% of all the people in Adorp are in favor of the new mall? Explain why or why not.

FIGURE 10.2.2 Drawing conclusions from samples

Focus on Understanding

The television reporter understands that she cannot conclude that 60% of the people in Adorp favor the new mall. She wonders, however, whether her sample of size 20 provides sufficient evidence to conclude that more that half of the people in Adorp favor the new mall. You are going to be conducting a hypothesis test to answer this question. Use $\alpha = 0.05$.

1. Formulate the null and alternative hypotheses for the reporter's test. Remember that these are based not on the data, but on the claim that the reporter is trying to support.
2. If the null hypothesis is true, is $n = 20$ large enough for the distribution of $\hat{p}$ to be close to normal? (Show how you check.)
3. What z-values would lead you to reject the null hypothesis? In other words, what is the rejection region?
4. Evaluate the test statistic, using the reporter's sample of size 20.
5. Should you reject H_0? Does the sample provide evidence that a majority of people in Adorp favor the new mall?
6. Suppose that the reporter selects a larger random sample of 100 people. Fifty-nine of these people favor the new mall. Use this new sample to redo the same test. Whether you reject H_0 or not, state what your conclusion means in this situation.

Summarizing the Procedure: Tests About Proportions

There are three different types of hypothesis tests: **upper-tailed**, **lower-tailed**, and **two-tailed**. The form of the research hypothesis determines which type the test will be. The tests we have just done were upper-tailed tests, since the alternate hypothesis had the form:

$$H_1 : population\ characteristic > specified\ number$$

When the alternative hypothesis has this form, the rejection region is in the upper tail of the distribution, hence the name. Similarly, the rejection region will be in the lower tail when the alternate hypothesis has the form:

$$H_1 : population\ characteristic < specified\ number$$

Lastly, the rejection region will be split, with an area of $\frac{\alpha}{2}$ in each tail, when the alternate hypothesis has the form:

$$H_1 : population\ characteristic \neq specified\ number$$

The following box summarizes some basic ideas about the tests for proportions.

Hypothesis Tests about Proportions

	Upper-Tailed	Lower-Tailed	Two-Tailed
Null Hypothesis	$H_0 : p = p_0$ or $H_0 : p \leq p_0$	$H_0 : p = p_0$ or $H_0 : p \geq p_0$	$H_0 : p = p_0$
Alternative Hypothesis	$H_1 : p > p_0$	$H_1 : p < p_0$	$H_1 : p \neq p_0$

Caution: In order for this test to be valid:

- The sample must be *random*.
- n must be large enough for $\hat{p}$ to have a normal distribution when H_0 is true. To verify this, check that $n \cdot \min(p_0, q_0) > 5$.

Test Statistic	$z = \dfrac{\hat{p} - p_0}{\sqrt{\dfrac{p_0 q_0}{n}}}$	Same	Same
Rejection Region	$z > z_\alpha$	$z < -z_\alpha$	$z < -z_{\alpha/2}$ or $z > z_{\alpha/2}$

Evaluate the test statistic z based on the sample data.

Decision Rule:

- If z is in the rejection region, reject H_0. There is sufficient evidence to conclude H_1 is true.
- If z is not in the rejection region, do not reject H_0. There is not sufficient evidence to conclude H_1 is true.

A Lower-Tailed Test

Suppose we wanted to show that less than 4% of all mayflies live 11 hours or more. We are going to be using a random sample of 1,000 mayflies to conduct a hypothesis test at the 0.01 level of significance. Since we are looking for evidence to support this statement, this is the research hypothesis: $H_1 : p < 0.04$. We would test this against the null hypothesis, which could be formulated as either $H_0 : p = 0.04$ or $H_0 : p \geq 0.04$. Either way we formulate H_0, we are going to be assuming $p = 0.04$ when we assume that H_0 is true.

The sample size ($n = 1,000$) seems large enough for $\hat{p}$ to have a normal distribution, but we'll check anyway.

$$n \cdot \min(p_0, q_0) = 1,000(.04) = 40 > 5 \quad \text{Okay!}$$

This guarantees that the test statistic will have a standard normal distribution if H_0 is true. Therefore, we can use a standard normal distribution to find the rejection region. (See Figure 10.2.3.)

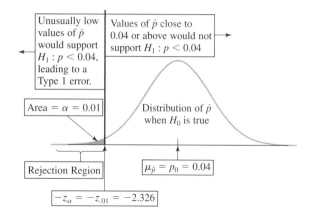

FIGURE 10.2.3 Using the distribution of $\hat{p}$ for a lower-tailed test

This time, because the alternative hypothesis is $H_1 : p < 0.04$, small sample proportions support H_1, rather than large ones. The rejection region ends up in the lower tail rather than the upper. Since we are conducting this test at the 0.01 level of significance, the lower tail area is $\alpha = 0.01$. If we look at Table 2 to find the critical z-value, we have a slight problem. Table 2 only gives upper tail areas. No problem! Using an upper tail area of 0.01, we find $z_\alpha = z_{.01} = 2.326$. Because the standard normal curve is symmetrical about $z = 0$, the critical value we need is exactly the same, but negative: $-z_\alpha = -z_{.01} = -2.326$. The rejection region is $z < -2.326$.

The sample we are using comes from *Mathematics in Context: Great Expectations* (page 31). You have seen the data before, in Figure 9.1.9. In this sample, only 20 of the 1,000 mayflies lived 11 hours or more. Computing the sample proportion and evaluating the test statistic, we get:

$$\hat{p} = \frac{20}{1,000} = 0.02 \quad z = \frac{\hat{p} - p_0}{\sqrt{\frac{p_0 q_0}{n}}} = \frac{0.02 - 0.04}{\sqrt{\frac{(0.04)(0.96)}{1,000}}} = -3.23$$

Notice that the denominator of the formula above uses p_0, not $\hat{p}$. This is because we are initially assuming that H_0 is true. The value of the test statistic ($z = -3.23$) lies in the rejection region, so we reject H_0 in favor of H_1. There is sufficient evidence to conclude that less than 4% of all mayflies live 11 hours or more (in laboratory conditions).

We'll now let you try a two-tailed test. Note that, in a two-tailed test, half of the Type 1 error probability goes in each tail.

Focus on Understanding

Here's a question: Are coin tosses really fair? Because the imprints on the two sides are different, maybe the images stamped on the two sides of a coin are different enough to affect the probability. You will be using a sample of 500 coin tosses to look for evidence that the coin tosses are unfair.

1. Let p stand for the probability of heads. If coin tosses are fair, what does that say about p? Is this the null hypothesis or the alternative? Formulate this hypothesis.
2. If coin tosses are unfair, what does that say about p? Formulate this hypothesis.
3. Suppose a sample of 500 tosses resulted in 238 heads. Use this sample to test your hypotheses at the 0.05 level of significance. State what your conclusion means in this situation.
4. Figure 10.2.4 shows the results of 50 different random samples of 500 coin tosses each, sorted according to the value of the test statistic z from the sample. Explain why you should not expect all samples to lead to the same conclusion. How many samples would lead us to reject H_0? Is this number about what you would expect? Why isn't it exactly what you would expect?

Number of heads	Proportion of heads	Value of test statistic	Number of heads	Proportion of heads	Value of test statistic
226	0.452	−2.147	252	0.504	0.179
226	0.452	−2.147	252	0.504	0.179
231	0.462	−1.699	252	0.504	0.179
233	0.466	−1.521	253	0.506	0.268
234	0.468	−1.431	255	0.510	0.447
235	0.470	−1.342	256	0.512	0.537
236	0.472	−1.252	256	0.512	0.537
237	0.474	−1.163	256	0.512	0.537
238	0.476	−1.073	257	0.514	0.626
239	0.478	−0.984	258	0.516	0.716
239	0.478	−0.984	258	0.516	0.716
239	0.478	−0.984	258	0.516	0.716
243	0.486	−0.626	259	0.518	0.805
244	0.488	−0.537	259	0.518	0.805
246	0.492	−0.358	261	0.522	0.984
246	0.492	−0.358	261	0.522	0.984
246	0.492	−0.358	262	0.524	1.073
247	0.494	−0.268	262	0.524	1.073
247	0.494	−0.268	263	0.526	1.163
247	0.494	−0.268	263	0.526	1.163
247	0.494	−0.268	264	0.528	1.252
248	0.496	−0.179	266	0.532	1.431
248	0.496	−0.179	266	0.532	1.431
251	0.502	0.089	279	0.558	2.594
251	0.502	0.089	280	0.560	2.683

FIGURE 10.2.4 Results from 50 random samples of 500 coin tosses each

Why Do We Choose $p = p_0$?

Now that you have seen some hypothesis tests and are a little more familiar with how they work, we are finally ready to explain why, when we assume that H_0 is true, we always assume that $p = p_0$, even if we have formulated the null hypothesis as $p \leq p_0$ or $p \geq p_0$. Suppose we go back to *Return of the Good Seeds* at the beginning of this section. We are still using a random sample of 400 seeds to test $H_1 : p > 0.9$. This time, let's suppose that we had formulated the null hypothesis as $H_0 : p \leq 0.9$ and, contrary to the advice we'd been given, we decided to choose $p = 0.8$ when we assumed that H_0 was true, rather than $p = 0.9$ as before. After all, $0.8 \leq 0.9$, so why not?

Figure 10.2.5 shows the distribution we would be using to find the rejection region. Notice that, since we are assuming $p = 0.8$, $\mu_{\hat{p}}$ is now 0.8, rather than 0.9. In the same way, the standard deviation of the distribution is now $\sigma_{\hat{p}} = \sqrt{\frac{(0.8)(0.2)}{400}} = 0.02$, rather than 0.015 as before.

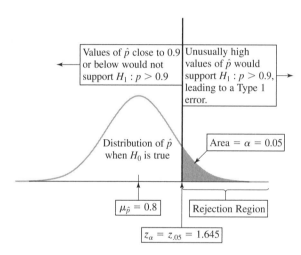

FIGURE 10.2.5 Why do we choose $p = p_0$?

At first glance, we appear to end up with the same rejection region as before, but remember that the z-score of $\hat{p}$ will now be calculated as if the mean of the distribution was $\mu_{\hat{p}} = 0.8$. The key question is now, "What values of $\hat{p}$ are in this new rejection region?" Since the rejection region is $z > 1.645$ and $z = \frac{\hat{p} - \mu_{\hat{p}}}{\sigma_{\hat{p}}} = \frac{\hat{p} - 0.8}{0.02}$, we can substitute $z = \frac{\hat{p} - 0.8}{0.02}$ in place of z in the inequality and solve for $\hat{p}$.

$$z > 1.645$$
$$\frac{\hat{p} - 0.8}{0.02} > 1.645 \quad \text{(Substituting)}$$
$$\hat{p} - 0.8 > (0.02)(1.645) \quad \text{(Multiplying both sides by 0.02)}$$
$$\hat{p} - 0.8 > .0329$$
$$\hat{p} > 0.8329 \quad \text{(Adding 0.8 to both sides)}$$

So, we would be rejecting the null hypothesis in favor of the alternative if the sample proportion was above 0.8329. Something seems definitely odd about that. Let's think about what we are trying to do. We are trying to find convincing evidence that supports $H_1 : p > 0.9$ as opposed to $H_0 : p \leq 0.9$. Do you think that a sample proportion of 84% should convince us that the population proportion is more than 90%? No way! But this is exactly what we would be doing if we chose $p = 0.8$, rather than $p = 0.9$ as the value of p when H_0 is true. In effect, we are doing the test as if we were testing $H_1 : p > 0.8$, rather than $H_1 : p > 0.9$. We would be accepting evidence that more than 80% of the seeds are good as if it supported the conclusion that more than 90% of the seeds are good. This is the reason we always choose $p = p_0$ when we assume H_0 is true, even if H_0 is formulated as $p \leq p_0$ or $p \geq p_0$. This is also why some people always formulate H_0 as an equation and never use $\leq$ or $\geq$ in the null hypothesis.

In this section, you saw how to use a Type 1 error probability to find a rule for reaching a decision in a hypothesis test. The decision rule is based on identifying a rejection region, a range of possible values for the test statistic that would lead us to reject H_0. This method for reaching a decision is called the **classical approach**. In the next section, you will see a different way to reach a decision in a hypothesis test, sometimes called the **modern approach**.

10.3 THE P-VALUE FOR A TEST

In this section, we will show how to find the **P-value** (also called the **observed significance level**) for a hypothesis test. The P-value is not the same as the population proportion p or the sample proportion $\hat{p}$; this is something different. We will explain how to use the P-value to determine the conclusion of the hypothesis test, as an alternative to using the rejection region as we did in the previous section.

Finding a P-value

Classroom Exploration 10.3

In *Return of the Good Seeds*, we were testing $H_0 : p = 0.9$ against the alternative $H_1 : p > 0.9$. We had selected a random sample of 400 seeds and found that the proportion of good seeds in the sample was $\hat{p} = 0.925$.

1. Suppose that the null hypothesis is true. Based on this assumption, find the probability of selecting a random sample in which the sample proportion is at least as large as the one we saw. In other words, find $P(\hat{p} \geq 0.925)$.
2. Based on the probability in #1, would it be unusual or fairly common to get a sample proportion as high as 0.925, if the null hypothesis is true?

The probability that you have just found is the **P-value** of the test. What is a P-value? The box below gives a general definition.

Definition of P-value

The **P-value** for a hypothesis test is the probability of choosing a random sample that supports H_1 at least as much as the sample we actually chose, assuming H_0 is true.

The definition is quite a mouthful, but it's not so bad if you break it into small bites. Let's see how it applies to the situation in Exploration 10.3. First, notice that the definition implies that we have chosen a random sample already, which we have. The sample we chose had $\hat{p} = 0.925$.

Second, we are concerned with samples that support H_1 at least as much as the sample we actually chose. Which samples would those be? Since this is an upper-tailed test, sample proportions in the upper tail of the distribution of $\hat{p}$ support H_1. The higher the value of $\hat{p}$, the more we would be inclined to believe $H_1 : p > 0.9$. So, the samples that support H_1 at least as much as the sample we actually chose (with $\hat{p} = 0.925$) are those with $\hat{p} \geq 0.925$, farther into the upper tail than $\hat{p} = 0.925$. (See Figure 10.3.1.)

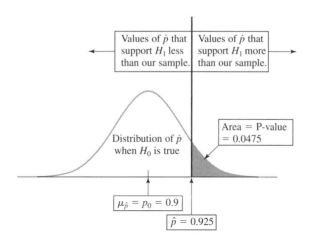

FIGURE 10.3.1 The P-value of a hypothesis test

The P-value is the probability of choosing one of these high-proportion samples, assuming that the null hypothesis is true. This probability is the shaded area in Figure 10.3.1 and is exactly what you were looking for in Exploration 10.3. You should have gotten 0.0475.

Notice that, just as we state the rejection region terms of z rather than $\hat{p}$, we can do the same thing with the P-value. Since the z-score for $\hat{p} = 0.925$ was $z = 1.67$, we could just as easily have said that the P-value was $P(z \geq 1.67)$. As a matter of fact, that is exactly how you computed it.

320 Chapter 10   Testing Hypotheses

We have discussed finding the P-value in this particular example. The following box gives a general procedure for finding the P-value of any hypothesis test.

Finding the P-value for a Hypothesis Test

1. Use the sample data to compute the value of the test statistic.
2. Determine which values of the test statistic would support H_1 at least as much as the value from the sample.
3. Assuming that the null hypothesis is true, find the probability that a random sample would result in one of the values you found in step 2. This is the P-value of the test.

Interpreting the P-value

In Section 10.2 we used the significance level $\alpha = 0.05$ to find the rejection region for the upper-tailed test in *Return of the Good Seeds*. We were determining how high a sample proportion needed to be in order for it to be in the highest 5% of the distribution of $\hat{p}$. Without determining this rejection region, but judging by the P-value 0.0475 that we just found, can we tell whether $\hat{p} = 0.925$ is in the highest 5% of the distribution $\hat{p}$? Sure we can! Look at Figure 10.3.1. The sample proportion $\hat{p} = 0.925$ is (just barely) in the highest 4.75% of the distribution, so it must be in the highest 5% as well. We can tell from the P-value that $\hat{p} = 0.925$ is high enough to reject the null hypothesis without finding the rejection region!

Figure 10.3.2 focuses on the upper tail of the distribution of $\hat{p}$ in *Return of the Good Seeds*, with three z-values and their corresponding upper tail areas marked. Just as the rejection region $z > 1.645$ allows us to determine whether or not a

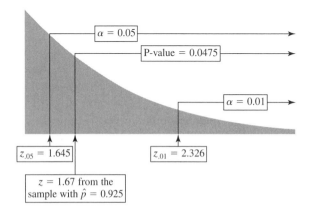

FIGURE 10.3.2 Relating the P-value to α

particular sample proportion lies in the highest 5% of the distribution of $\hat{p}$, the rejection region $z > 2.326$ allows us to determine whether it is in the highest 1%. The P-value addresses the same issue. Knowing that the proportion from our sample has a P-value of 0.0475 tells us that it is just barely in the highest 4.75% of the distribution. Therefore, it is inside the highest 5%, but outside the highest 1%. It is high enough to reject H_0 at the 0.05 level, but not the 0.01 level. This is exactly what we decided in the previous section. In general, the P-value and the rejection region lead to exactly the same conclusions.

In a hypothesis test, we initially assume that the null hypothesis is true. We then search for evidence to cast doubt on this assumption. In *Return of the Good Seeds*, the evidence is a sample proportion that would be unusually high if the null hypothesis was true ($p = 0.9$). The question addressed by the rejection region is: How high does the sample proportion need to be to convince us to reject H_0? The answer depends on α. If $\alpha = 0.05$, we need a sample proportion high enough for its z-score to be greater than 1.645. If $\alpha = 0.01$, we need $z > 2.326$.

The question addressed by the P-value is different. We are still in the same situation as above, knowing that an unusually high sample proportion would cast doubt on the assumption that H_0 was true. But now suppose we collect the data before finding the rejection region. We have chosen a random sample and found that the sample proportion is $\hat{p} = 0.925$. We can now ask ourselves: How unusual is it for a sample proportion to be this high when H_0 is true? That is the question addressed by the P-value. When you found that $P(\hat{p} \geq 0.925) = 0.0475$, you found that only 4.75% of random samples have proportions that high.

One of the advantages of using P-values to make decisions is that the rejection rule is always exactly the same, regardless of whether the test is upper-tailed, lower-tailed, or two-tailed. Here it is:

Rejection Rule Based on P-values

- If *P-value* $\leq \alpha$, reject H_0. There is sufficient evidence to conclude H_1 is true.
- If *P-value* $> \alpha$, do not reject H_0. There is not sufficient evidence to conclude H_1 is true.

In the example we've been considering, the P-value 0.0475 is smaller than $\alpha = 0.05$. Following the rule above, we would reject the null hypothesis at the 0.05 level. In the same way, $0.0475 > \alpha = 0.01$, so we would not reject the null hypothesis at the 0.01 level.

Focus on Understanding

In the previous section, you used two different samples to test the hypothesis that more than half of the people in Adorp favor the new mall. In both cases, the hypotheses were $H_0: p = 0.5$ and $H_1: p > 0.5$. Both tests were done using $\alpha = 0.05$.

1. Sixty percent of the first sample ($n = 20$) favored the new mall. For this sample, you should have computed the value of the test statistic as $z = 0.89$.

   a. Find the P-value for this test.
   b. Based on the P-value, would you reject or not reject H_0 at the 0.05 level? Is this the same decision that you made in Section 10.2?

2. Fifty-nine percent of the second sample ($n = 100$) favored the new mall. For this sample, you should have computed the value of the test statistic as $z = 1.80$.

   a. Find the P-value for this test.
   b. Based on the P-value, would you reject or not reject H_0 at the 0.05 level? Is this the same decision that you made in Section 10.2?

Historically, using the rejection region to reach a decision came first. This method is referred to as the **classical approach**. The method based on P-values is called the **modern approach**. The table below compares the two methods.

The Classical Approach	The Modern Approach
• Start with α. • Use α to determine the rejection region. • Reject H_0 if the value of the test statistic lies in the rejection region.	• Compute the value of the test statistic. • Use the value of the test statistic to find the P-value. • Reject H_0 if the P-value $\leq \alpha$.

We've already mentioned that one advantage of using the modern approach is that the rejection rule is always the same. Another advantage of the P-value approach is that, in published studies, it allows the reader to choose the level of significance. By comparing the chosen α to the published P-value, the reader can decide whether or not to reject the null hypothesis. The modern approach allows the author of a study to report results while leaving the choice of α to the reader.

The P-value for a Two-Tailed Test

Students sometimes find the P-value for a two-tailed test to be confusing and difficult to understand, so we'll go through an example illustrating this situation. In Section 10.2, you used a sample of 500 coin tosses to test the hypothesis that the coins were unfair. The 500 tosses resulted in 238 heads. The hypotheses were H_0: $p = 0.5$ and $H_1: p \neq 0.5$. You should have found that:

$$\hat{p} = \frac{238}{500} = 0.476 \text{ and}$$

$$z = \frac{\hat{p} - p_0}{\sqrt{\frac{p_0 q_0}{n}}} = \frac{0.476 - 0.5}{\sqrt{\frac{(0.5)(0.5)}{500}}} = -1.07$$

We now must determine what values of z would support the alternative hypothesis at least as much as $z = -1.07$. Figure 10.3.3 may be helpful. Since the alternative hypothesis is $H_1: p \neq 0.5$, values of $\hat{p}$ in both tails support H_1. If a sample proportion with a z-score of -1.07 convinced us to reject the null hypothesis, a sample proportion with $z = +1.07$ would be equally convincing. The z-values farther out in the tails would support H_1 even more.

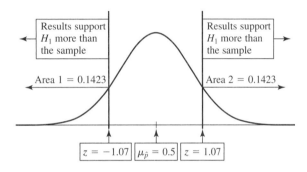

FIGURE 10.3.3 Finding the P-value for a two-tailed test

We use Table 1 to find the area to the left of $z = -1.07$. We get 0.1423. Because of the symmetry of the standard normal curve, the area to the right of $z = 1.07$ must be the same. The P-value of this test is: $P(z < -1.07 \text{ or } z > 1.07) = $ Area1 + Area2 = 0.2846.

In general, if the P-value is greater than α, we do not reject H_0. Since the P-value = $0.2846 > \alpha = 0.05$, we would not reject H_0. We do not have sufficient evidence to conclude that the coins are unfair. This should be the same decision you reached in Section 10.2.

In a two-tailed test like the one we have been discussing, both high and low sample proportions would support the alternative hypothesis; either an unusually high proportion of heads or an unusually low proportion of heads could convince us that the coins were unfair. If we use $\alpha = 0.05$ to find the rejection region, we are determining the location of the most extreme 5% of the distribution (the highest 2.5% and the lowest 2.5%). Is the value of the test statistic from our sample in the most extreme 5% of the distribution? No! The P-value tells us that it is barely in the most extreme 28.5%. It is not nearly extreme enough to convince us that coin tosses are unfair. In general, if the P-value is greater than α, we do not reject H_0.

Focus on Understanding

We suspect that a number cube made by the Sneaky Pete Manufacturing Company may be unfair. (Maybe the name makes us suspicious.) We could check the probability for all of the sides, but we'll just focus on sixes.

1. Let p stand for the probability of rolling a six. If the number cube was fair, what is p? Formulate this as a hypothesis.
2. Formulate the competing hypothesis.
3. We roll the cube 60 times and get 16 sixes. Use this sample to find the P-value for the test.
4. Based on the P-value, would you reject the null hypothesis at the 0.05 level of significance? Interpret what your conclusion means in this situation.
5. Would you reject the null hypothesis at the 0.01 level of significance? Interpret what your conclusion means in this situation.

Using the TI-83 plus

Calculators and computer programs that do hypothesis testing often require an understanding of P-values to reach a conclusion. The TI-83 plus is no exception. To use it to do a hypothesis test about a proportion, hit STAT and use the right arrow key to go to TESTS. We want a one-proportion test based on a standard normal distribution, so we select 5: 1-PropZTest. As an example, we'll redo the test from *Return of the Good Seeds*. Remember that we were testing $H_0: p \leq 0.9$ versus $H_1: p > 0.9$ by using a sample of 400 seeds, 370 of which were good.

The value of p that we use when we assume the null hypothesis is true is p_0. In this case, $p_0 = 0.9$. The calculator uses x to stand for the number of "successes" in the sample, and n is the sample size, so enter 370 for x and 400 for n.

The next line uses the alternative hypothesis to indicate whether we want a lower-tailed, an upper-tailed, or a two-tailed test. Since the alternative hypothesis says that the population proportion is greater than p_0, use the right arrow key to move to $>p_0$ and hit enter. Your screen should look like this:

```
1-PropZTest
  p0:.9
  x:370
  n:400
  prop≠p0  <p0  >p0
  Calculate  Draw
```

Use the down arrow key to move to Calculate and hit ENTER. You should now see:

```
1-PropZTest
  prop>.9
  z=1.666666667
  p=.0477903304
  p̂=.925
  n=400
```

The calculator has done a one-proportion z-test of $H_1: p > p_0$. The value of the test statistic was $z = 1.666666667$. The P-value was 0.0477903304. The reason the P-value is not exactly the same as ours is that Table 1 forced us to round z to two decimal places. The calculator uses more decimal places. Aside from that, the results are the same as what we got earlier. The calculator does not state a conclusion. It is up to us to compare the P-value to α and draw the conclusion for ourselves.

Hit STAT, TESTS, and 5: 1-PropZTest again. Leave all the settings the same, but go to Draw instead of Calculate. Hit ENTER. You should see a graph of the standard normal curve with the area representing the P-value shaded. (See Figure 10.3.4.)

Focus on Understanding

Use the TI-83 plus to redo the test about Sneaky Pete's number cube from the previous *Focus on Understanding*. Are your results consistent with what you got before?

In Sections 10.2 and 10.3, you have seen how to test hypotheses about proportions, using both the classical approach (using rejection regions) and the modern approach (using P-values). Both approaches lead to exactly the same conclusions. You should

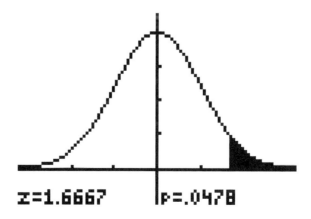

FIGURE 10.3.4 A one-proportion z-test on the TI-83 plus

be able to use either approach to decide whether there is sufficient evidence to support the alternative hypothesis. In the next section, we will shift our attention from proportions to means. You will see that many of the ideas from tests about proportions carry over into tests about means as well.

10.4 TESTS ABOUT MEANS

So far, this chapter has dealt with hypothesis tests about proportions. In a test about p, we use information from a random sample ($\hat{p}$) to make a decision about the population proportion. In a similar way, we can conduct a hypothesis test about μ, using information from a random sample ($\bar{x}$) to make a decision about the population mean. We'll start with an example we'll call *The Drink Dispenser*.

The Drink Dispenser

A fast food restaurant has a machine that dispenses soft drinks in preset amounts for small, medium, and large drinks. For drinks of the same size, the machine is very consistent; the amount of liquid varies very little from drink to drink ($\sigma = 0.24$ ounces). However, it is possible for the machine to be set incorrectly, so that it consistently dispenses too much or too little. Too little is bad because customers start to complain; too much is also bad because soda goes down the drain or ends up on the floor behind the counter. It is supposed to be dispensing an average of 12 ounces for small drinks, 20 ounces for medium drinks, and 32 ounces for large drinks. We are going to be conducting hypothesis tests to determine whether there is evidence that the machine has been set incorrectly. We'll start with the medium drinks.

Classroom Exploration 10.4

Let μ = *the mean amount the machine dispenses for medium drinks*.

1. If the machine is set correctly, what is the value of μ? Formulate this as a hypothesis.
2. Formulate the competing hypothesis.
   You will be conducting a hypothesis test using $\alpha = 0.05$ and a random sample of 36 medium drinks. Think back to Section 8.2.
3. Assuming that the null hypothesis is true, find $\mu_{\bar{x}}$, the mean of the distribution of $\bar{x}$.
4. Find $\sigma_{\bar{x}}$, the standard deviation of the distribution of $\bar{x}$.
5. What type of distribution does $\bar{x}$ have? What information in the problem is needed to determine this?
6. In tests about proportions, the test statistic was the z-score for the sample proportion: $z = \dfrac{\hat{p} - \mu_{\hat{p}}}{\sigma_{\hat{p}}} = \dfrac{\hat{p} - p_0}{\sqrt{\dfrac{p_0 q_0}{n}}}$. Based on your answers to #3, 4, and 5 above, take a guess at the test statistic we will be using for this test about μ.

Let's see how well you did in Exploration 10.4. The hypotheses should be:
$$H_0 : \mu = 20 \quad \text{and} \quad H_1 : \mu \neq 20$$
Assuming H_0 is true, the mean of the distribution of $\bar{x}$ is $\mu_{\bar{x}} = \mu = 20$. The standard deviation of $\bar{x}$ is $\sigma_{\bar{x}} = \frac{\sigma}{\sqrt{n}} = \frac{0.24}{\sqrt{36}} = \frac{0.24}{6} = 0.04$. The distribution of $\bar{x}$ will be close to normal, since the sample size $n = 36 \geq 30$. Lastly, a good guess for the test statistic would be the z-score for the sample mean:
$$z = \frac{\bar{x} - \mu_{\bar{x}}}{\sigma_{\bar{x}}} = \frac{\bar{x} - \mu_0}{(\sigma/\sqrt{n})} = \frac{\bar{x} - 20}{0.04}.$$
Not only is this a good guess, it is exactly right.

Let's go ahead and talk through the rest of this test. We are following the general procedure that we gave in Section 10.1. (See the box at the right.) We have completed the first two steps and are ready for *Step 3: Determining the decision rule*. We now have a choice about this. We can use the classical approach (using the rejection region)

Steps in a Hypothesis Test
1. Formulating the hypotheses
2. Selecting an appropriate test statistic
3. Determining the decision rule
4. Collecting the data
5. Evaluating the test statistic
6. Arriving at a conclusion

or the modern approach (using the P-value). We'll do it both ways, just to show that we get the same results, but we'll start with the modern approach.

In the modern approach, we collect the data before computing the P-value. Suppose we selected a random sample of 36 medium drinks and determined that the mean size for the sample was 20.09 ounces. The value of the test statistic is:
$$z = \frac{\bar{x} - 20}{0.04} = \frac{20.09 - 20}{0.04} = \frac{0.09}{0.04} = 2.25$$
We need to determine which values of the test statistic would support the alternative hypothesis at least as much as the value from our sample. The alternative hypothesis is $H_1: \mu \neq 20$. Because this is a two-tailed test, values in both tails support the alternative hypothesis. Sample means close to 20 won't convince us to reject H_0, but sample means that are either much higher than 20 or much lower than 20 would support H_1. A z-value of -2.25 would be just as convincing as 2.25, the value from our sample. Values further out in the tails would be even more convincing. Therefore, the P-value will be $P(z \leq -2.25 \text{ or } z \geq 2.25)$. (See Figure 10.4.1.)

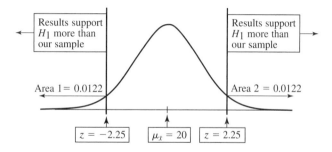

FIGURE 10.4.1 The P-value for a two-tailed test about μ

We are going to find this probability using Table 1, the table of standard normal curve areas. However, I know someone will read this and wonder why we aren't using a T-distribution, instead of standard normal. The reason is that the population standard deviation was given in the problem. Remember that we get a T-distribution when we use $(s/\sqrt{n})$ in place of $(\sigma/\sqrt{n})$ when we compute the z-score for $\bar{x}$. We'll be doing that later, but not now.

So, using Table 1, we find that the area to the left of $z = -2.25$ is 0.0122; the area to the right of 2.25 is the same. Therefore, the P-value is: $0.0122 + 0.0122 = 0.0244$.

We are ready to make a decision. Looking way back at Exploration 10.4, we see that $\alpha = 0.05$. In other words, we would reject H_0 if the mean from our sample was in the most extreme 5% of the distribution of $\bar{x}$. Is $\bar{x} = 20.09$ extreme enough? Yes. The P-value tells us that, if H_0 was true, only 2.44% of all random samples will have means that are as extreme as ours. The results from our sample would be highly unusual if H_0 was true. We reject H_0; we always reject H_0 if the P-value $\leq \alpha$. There is convincing evidence that the machine is not dispensing the correct amount for medium drinks.

We'll also look at the rejection region, just to demonstrate that we reach the same conclusion. For a two-tailed test with $\alpha = 0.05$, we want $\frac{\alpha}{2} = 0.025$ in each tail. Finding this upper tail area in Table 2, the critical z-value is $z_{.025} = 1.96$. Because of the symmetry of the standard normal curve, the critical value in the lower tail is $-z_{.025} = -1.96$. Therefore, we would reject H_0 if either $z < -1.96$ or $z > 1.96$. Since $\bar{x} = 20.09$ results in $z = 2.25 > 1.96$, our sample is in the upper tail area of the rejection region. We reject H_0, just as we decided from the P-value.

We have just done a hypothesis test about a population mean in a situation where the population standard deviation was known. In realistic situations it is more common for σ to be unknown, but we'll get to that a little later. First let's summarize the tests about μ when σ is known.

Hypothesis Tests about μ when σ is Known

	Upper-Tailed	Lower-Tailed	Two-Tailed
Null Hypothesis	$H_0 : \mu = \mu_0$ or $H_0 : \mu \leq \mu_0$	$H_0 : \mu = \mu_0$ or $H_0 : \mu \geq \mu_0$	$H_0 : \mu = \mu_0$
Alternative Hypothesis	$H_1 : \mu > \mu_0$	$H_1 : \mu < \mu_0$	$H_1 : \mu \neq \mu_0$
Test Statistic	$z = \dfrac{\bar{x} - \mu_0}{\left(\dfrac{\sigma}{\sqrt{n}}\right)}$	Same	Same
Rejection Region	$z > z_\alpha$	$z < -z_\alpha$	$z < -z_{\alpha/2}$ or $z > z_{\alpha/2}$

Caution: In order for this test to be valid:

- The sample must be *random*.
- The distribution of $\bar{x}$ must be close to normal, so either the sample size must be large ($n \geq 30$) or we must assume that the population has a normal distribution.

Evaluate the test statistic z based on the sample data.

Decision Rule:

- If z is in the rejection region, reject H_0. There is sufficient evidence to conclude H_1 is true.
- If z is not in the rejection region, do not reject H_0. There is not sufficient evidence to conclude H_1 is true.

Or, instead of finding the rejection region, compute the P-value.

- If *P-value* $\leq \alpha$, reject H_0. There is sufficient evidence to conclude H_1 is true.
- If *P-value* $> \alpha$, do not reject H_0. There is not sufficient evidence to conclude H_1 is true.

Focus on Understanding

You are going to use a sample of 64 small drinks to test the hypothesis that the drink dispenser is not set correctly for small (12-ounce) drinks. Before you start, the manager says. "I know the machine is not under-filling small drinks. The people who buy small drinks would be the first ones to complain, and we haven't had any complaints. Just check to see if it is over-filling."

1. Formulate the hypothesis, taking the manager's instructions into account.
2. You collect a sample of 64 small drinks and find that $\bar{x} = 12.042$ ounces. Find the P-value for this test. Assume that $\sigma = 0.24$, just as for the medium drinks.
3. Based on the P-value, what would you conclude at the 0.05 level of significance? Interpret what your conclusion means in this situation.
4. Find the rejection region and see if it leads to the same conclusion.

We'll leave the large drinks for an exercise. For now, we have another important issue to discuss.

What if σ is Unknown?

The tests with σ known depend upon $z = \frac{\bar{x}-\mu}{(\sigma/\sqrt{n})}$ having a standard normal distribution. This will be true if either n is large ($n \geq 30$) or the population we're sampling has a normal distribution. As we discussed in Section 9.2, if we replace

330   Chapter 10   Testing Hypotheses

the population standard deviation σ with the sample standard deviation s, we add some additional variability to the situation. This additional variability results in a T-distribution, instead of standard normal. The following box summarizes the changes that are needed in the hypothesis tests about μ when σ is unknown.

Hypothesis Tests about μ when σ is Unknown

- **Test Statistic:** $t = \frac{\bar{x} - \mu_0}{(s/\sqrt{n})}$. The sample standard deviation s replaces σ.
- **Rejection Region:** Now based on a T-distribution with $n - 1$ degrees of freedom, rather than standard normal. For upper-tailed tests: $t > t_\alpha$. For lower-tailed tests: $t < t_\alpha$. For two-tailed tests: $t < -t_{\alpha/2}$ or $t > t_{\alpha/2}$.
- **P-value:** Also based on a T-distribution with $n - 1$ degrees of freedom, not standard normal.

In Section 8.1, we introduced the Grade 8 Database from *Connected Mathematics* (*Samples and Populations*, page 39). We have selected a random sample of 10 students from the database and listed their daily sleep hours below:

$$6.2 \quad 8.5 \quad 5.0 \quad 7.0 \quad 8.5 \quad 7.0 \quad 7.3 \quad 8.5 \quad 9.0 \quad 7.5$$

Doctors recommend at least 8 hours of sleep per night. Does this sample provide evidence that the average student in the Grade 8 Database gets less than the recommended 8 hours of sleep? We'll be conducting a hypothesis test at the .05 level to answer this question.

Formulating the Hypotheses: We are looking for evidence that the average student gets less than 8 hours of sleep, so that is the alternative hypothesis: $H_1 : \mu < 8$. The null hypothesis could be formulated either as $H_0 : \mu \geq 8$ or $H_0 : \mu = 8$. Either way is fine.

Caution: We do have a random sample, but because the sample size is small ($n = 10$), we cannot count on the Central Limit Theorem to make the distribution of $\bar{x}$ normal. We must assume that the sample is being selected from a population whose distribution is close to normal.

The Rejection Region: Because σ is unknown, the test statistic is $t = \frac{\bar{x} - \mu_0}{(s/\sqrt{n})}$. Since we have replaced σ with s, the test statistic will have a T-distribution with $n - 1 = 9$ degrees of freedom. This is a lower-tailed test with $\alpha = 0.05$, so the rejection region will have a lower tail area of 0.05. (See Figure 10.4.2.) Looking at Table 2, we see that it does not list lower tail areas. No problem! (See Figure 10.4.3.)

Using the column for an upper tail area of .05 and the row for 9 degrees of freedom, we find the critical value $t_{.05} = 1.833$. Because of the symmetry of the T-distribution, the critical value we want is just the opposite: $-t_{.05} = -1.833$. Therefore, the rejection region is $t < -1.833$.

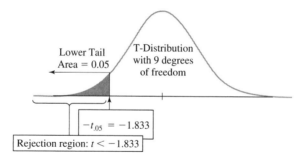

FIGURE 10.4.2 The rejection region in a T-distribution

Area to Left of t_0	0.90	0.95	0.975	0.99	0.995	0.999	0.9995
Upper Tail Area	0.10	0.05	0.025	0.01	0.005	0.001	0.0005
Central Area	0.80	0.90	0.950	0.98	0.990	0.998	0.9990
Standard Normal	1.282	1.645	1.960	2.326	2.576	3.090	3.291
Degrees of Freedom for T							
1	3.078	6.314	12.706	31.821	63.657	318.309	636.619
2	1.886	2.920	4.303	6.965	9.925	22.327	31.599
3	1.638	2.353	3.182	4.541	5.841	10.215	12.924
4	1.533	2.132	2.776	3.747	4.604	7.173	8.610
5	1.476	2.015	2.571	3.365	4.032	5.893	6.869
6	1.440	1.943	2.447	3.143	3.707	5.208	5.959
7	1.415	1.895	2.365	2.998	3.499	4.785	5.408
8	1.397	1.860	2.306	2.896	3.355	4.501	5.041
9	1.383	1.833	2.262	2.821	3.250	4.297	4.781

FIGURE 10.4.3 Finding a critical value from a T-distribution

Evaluating the Test Statistic: To compute $t = \frac{\bar{x} - \mu_0}{(s/\sqrt{n})}$, we first need $\bar{x}$ and s.

$$\bar{x} = frac\sum xn = \frac{74.5}{10} = 7.45$$

$$s = \sqrt{\frac{n\sum x^2 - (\sum x)^2}{n(n-1)}} = \sqrt{\frac{10(568.73) - (74.5)^2}{10(9)}} = 1.234$$

$$t = \frac{\bar{x} - \mu_0}{(s/\sqrt{n})} = \frac{7.45 - 8}{(1.234/\sqrt{10})} = -1.409$$

The Conclusion: Since $t = -1.409 > -1.833$, t is not in the rejection region. We do not reject H_0. The sample does not provide sufficient evidence to conclude that the average student in the Grade 8 Database gets less than 8 hours sleep.

In fact, the mean number of sleep hours in the Grade 8 Database is 7.962, which is less than 8. The alternative hypothesis was true, but we failed to reject H_0. We committed a Type 2 error. A random sample (especially a small sample) will sometimes lead to a Type 2 error. The small sample did not provide enough evidence to reject H_0, even though it was false. A larger sample may have provided enough information to detect the difference between $H_0 : \mu = 8$ and the truth, $\mu = 7.962$, but this sample was too small to detect this difference.

Finding the P-value: Like the rejection region, the P-value comes from the distribution of the test statistic if the null hypothesis is true, in this case a T-distribution with 9 degrees of freedom. Unlike the rejection region, which is based on α, the P-value is based on the value of the test statistic, $t = -1.409$. In a lower-tailed test such as this one, lower t-values support H_1 more, so the P-value of this test is $P(t \leq -1.409)$, the area to the left of $t = -1.409$. (See Figure 10.4.4.)

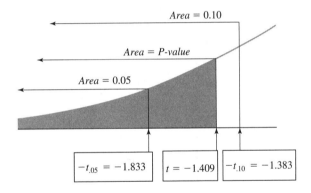

FIGURE 10.4.4 The P-value for a T-test

Since we are dealing with a T-distribution, rather than a normal distribution, we need to use Table 2. However, using Table 2 to find the P-value poses two problems. First, as you saw when finding the rejection region, Table 2 gives only upper tail areas and therefore only positive t-values. This is not really a problem because, as before, we know that the symmetry of the T-distribution makes the critical values in the lower tail the same as those for the upper tail, except negative.

The second problem is more serious: Table 2 does not list nearly as many values as Table 1. Ideally, we'd be looking for $t = 1.409$ in the row for 9 degrees of freedom. (See Figure 10.4.3 or Table 2.) This particular value of t is not listed; the closest ones are 1.383 and 1.833, with tail areas of 0.10 and 0.05. The most we can determine from Table 2 is that, since $t = 1.409$ is between 1.383 and 1.833, the P-value must be between those two areas; that is, $0.05 < P\text{-value} < 0.10$. Typically, we can't use Table 2 to find the P-value exactly; we can only determine two tail areas that it must lie between. This is called **bounding the P-value**.

Even though we cannot find the P-value exactly, we can still use it to draw a conclusion about the hypotheses. We always reject the null hypothesis if the P-value $\leq \alpha$. Since the P-value is greater than $\alpha = 0.05$, we would not reject H_0 at the 0.05 level. There is not sufficient evidence to conclude that the average student in the database gets less than 8 hours of sleep. We reach the same conclusion from the P-value as we did from the rejection region.

Notice that the P-value is less than 0.10. Therefore, we would have reached the opposite conclusion if we had used $\alpha = 0.10$. In this case, we would have avoided a Type 2 error. We would have concluded correctly that the average number of sleep hours in the database is below 8. This should not be too much of a surprise. Increasing α decreases β and vice versa. By increasing α, we would decrease the probability of a Type 2 error, avoiding an error in this case. The problem is that, in practice, we don't know what the actual truth is. You only know that $\mu = 7.962$ because we told you it was. In a real situation, we wouldn't be doing this test if we knew the truth. We wouldn't know we were making a Type 2 error. We would only know how great a risk we were taking for a Type 1 error. By increasing α to 0.10, we would have twice the chance of a Type 1 error as we would with $\alpha = 0.05$. In practice, we would have to weigh the consequences of each type of error and decide how large a Type 1 error probability we are willing to accept.

Classroom Connection

Figure 10.4.5 comes from *Mathematics in Context: Dealing with Data* (page 29). Question 14(a) in the figure asks students to compare the annual snowfall in the area

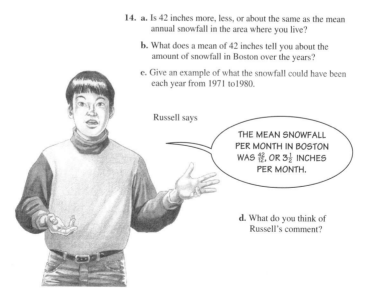

FIGURE 10.4.5 Mean annual snowfall

where they live to Boston's mean of 42 inches. Suppose Russell suspects that his town's average annual snowfall is less than 42 inches, but he would like to collect data to test this. Luckily, his local newspaper publishes a year-end summary that includes the total snowfall for the year. He finds the information for the last 6 years; it is listed in the table at the right. Is this a random sample? Maybe not, but it seems to be just as random as the sample the meteorologist used, so we will treat it as if it is.

Year	Snowfall
1999	39.5
2000	40.5
2001	32.7
2002	42.5
2003	37.7
2004	28.3

Focus on Understanding

You are going to help Russell determine whether his sample provides evidence (at the 0.05 level) that his town averages less than 42 inches of snow per year.

1. Formulate the hypotheses for the test.
2. Find the rejection region.
3. Use Russell's data to find the sample mean and standard deviation.
4. Find the value of the test statistic.
5. What do you conclude from your test? Interpret what your conclusion means in this situation.
6. Use Table 2 to bound the P-value as closely as possible.
7. Using $\alpha = 0.05$, does the P-value lead to the same conclusion as the rejection region?
8. Would you have reached a different conclusion if $\alpha = 0.01$? Describe the advantages and disadvantages of using $\alpha = 0.05$ and $\alpha = 0.01$.

Using the TI-83 plus

The TI-83 plus can perform hypothesis tests about μ in situations where σ is known (z-tests) and those where σ is unknown (T-tests). For an example with σ known, let's return to *The Drink Dispenser*. We used a sample of 36 medium drinks to test $H_0 : \mu = 20$ against the alternative $H_1 : \mu \neq 20$. Our sample had a mean of 20.09 ounces and we were told that the machine was constructed so that the population standard deviation was 0.24 ounces. We conducted the test at the 0.05 level of significance.

To conduct this test with the TI-83 plus, hit STAT and move to TESTS. Select 1: Z-Test. In this case, we want to input sample statistics rather than raw data, so we choose Stats, rather than Data under Inpt. The value of μ_0 comes from the hypotheses: $\mu_0 = 20$. The value of σ was given as 0.24. The sample mean $\bar{x}$ is 20.09; the sample size n is 36. Lastly, we want a two-tailed test, so we select "$\neq \mu_0$" for the alternative hypothesis. Your screen should look like this:

```
      Z-Test
       Inpt:Data Stats
       µ0:20
       σ:.24
       x̄:20.09
       n:36
       µ:≠µ0 <µ0 >µ0
       Calculate Draw
```

Move down to Calculate and hit ENTER. You should see:

```
      Z-Test
       µ≠20
       z=2.25
       p=.0244488668
       x̄=20.09
       n=36
```

The P-value 0.0244488668 is consistent with what we got earlier; the calculator just shows more decimal places. Just as we saw in the previous section, the calculator does not reach a conclusion, it just reports the P-value. It is up to us to compare the P-value to α and arrive at the appropriate conclusion. Since the P-value $< \alpha = 0.05$, we would reject the null hypothesis and conclude that the machine was not filling medium drinks correctly.

Hit STAT again and move to TESTS. Choose 1: Z-Tests again, but this time choose Draw instead of Calculate. You should see a standard normal curve with the area representing the P-value shaded. Draw also displays the value of the test statistic ($z = 2.25$) and the P-value ($P = 0.0244$). You can use either Calculate or Draw to reach the same conclusion.

For an example with σ unknown, look at the test we did on the average sleep hours in the Grade 8 Database. Remember that we were testing $H_0 : \mu = 8$ or $H_0 : \mu \geq 8$ against $H_1 : \mu < 8$, using $\alpha = 0.05$. We had selected the sample below:

6.2   8.5   5.0   7.0   8.5   7.0   7.3   8.5   9.0   7.5

To do this test with the TI-83 plus, we start by entering the sample data into a list. Hit STAT and choose 1: Edit. Enter the data into L1. Hit STAT again and go to Tests. Since σ is unknown, we need to use a T-distribution, rather than standard normal, so choose 2: T-Test. This time, we have raw data, rather than sample statistics, so choose Inpt: Data. The value of μ_0 comes from the hypotheses: $\mu_0 = 8$. The data is in list L1; each value has a frequency of 1. Lastly, we are doing a lower-tailed test. If you enter all of this information correctly, the screen should look like this:

```
T-Test
  Inpt:DATA Stats
  μ0:8
  List:L1
  Freq:1
  μ:≠μ0  <μ0  >μ0
  Calculate Draw
```

Select Calculate and you should see this:

```
T-Test
  μ<8
  t=-1.409432963
  p=.0961557977
  x̄=7.45
  Sx=1.234008824
  n=10
```

The results are consistent with what we got before. The calculator uses more decimal places and is not limited by a T-table with only a few different values. We had decided that the P-value was between 0.05 and 0.10. The calculator gives the P-value with much more accuracy, but it is between 0.05 and 0.10, just as we decided. Since the P-value $> \alpha = 0.05$, we would not reject H_0. This sample does not provide evidence that the average student in the database gets less than 8 hours of sleep per night.

We could also have used Draw instead of Calculate. We would not get as much information, but we would see a pretty picture of a T-distribution with the area representing the P-value shaded.

Focus on Understanding

1. Earlier, you used a sample of 64 small drinks to test $H_1 : \mu > 12$. Your sample had a mean of 12.042 ounces. Just as with the medium drinks, the population standard deviation was 0.24 ounces. Redo this test with the TI-83 plus. Using $\alpha = 0.05$ as before, do you reach the same conclusion?
2. Use the TI-83 plus and Russell's data on the annual snowfall in his town to test the hypothesis that, on average, his town gets less than 42 inches of snow. Using $\alpha = 0.05$ as before, do you reach the same conclusion? Does the calculator's P-value fall between the two areas you had determined earlier?

Chapter 10 Summary

This chapter introduced the ideas behind hypothesis testing, using information from a random sample to decide between two competing claims about a population. The logical argument in a hypothesis test is similar to the idea of indirect proof. To see if there is evidence to support one claim (the alternative hypothesis), we start by assuming the opposite (the null hypothesis). We determine whether the results from the data would be highly unlikely if the null hypothesis is true. If so, we reject the null hypothesis and decide in favor of the alternative.

You've tested hypotheses about both population proportions and population means, and you've used two different ways to judge whether sample results are unlikely enough to justify rejecting the null hypothesis. Both approaches are based on α, the Type 1 error probability. In the classical approach, we use α to find the rejection region and reject H_0 if the value of the test statistic lies in this region. In the modern approach, we use the value of the test statistic to find the P-value, and reject H_0 if the P-value is less than α.

Confidence intervals (Chapter 9) and hypothesis tests (Chapter 10) are both forms of inference, using information from samples to draw conclusions about populations. In the real world, inferential statistics is extremely important. It shows up whenever we want to know something about a large group, whether it's how long light bulbs will last or what proportion of people will benefit from a new medication.

The key terms and ideas from this chapter are listed below:

hypothesis testing   302
hypothesis   303
null hypothesis   303
alternative hypothesis   303
research hypothesis   303
Type 1 error   306
Type 2 error   306
test statistic   310
level of significance   310
decision rule   311

rejection region   311
statistically significant   312
lower-tailed test   313
upper-tailed test   313
two-tailed test   313
P-value   318
observed significance level   318
classical approach   322
modern approach   322
bounding the P-value   332

Assessment is an integral part of every curriculum from the elementary school all the way through college. The question always arises—*what is it that students should be able to do after completing this lesson/unit/chapter?* We have included here our intended learning goals for Chapter 10. Students who have a good grasp of the concepts developed in this chapter should be successful in responding to these items:

- Explain, describe, or give an example of what is meant by each of the terms in the vocabulary list.
- Formulate the null and alternative hypotheses for a test in a given situation.
- Identify or describe Type 1 and Type 2 errors in a given situation.
- Given a decision rule for a hypothesis test, find the probability of Type 1 and Type 2 errors.
- List the six steps in hypothesis testing.
- Decide whether the sample size is large enough to use a normal distribution in a test about p.
- Select the appropriate test statistic for a test about p or μ.

- Use a given level of significance to find the correct rejection region for a test about p or μ.
- Evaluate the test statistic and use its value to arrive at the appropriate conclusion for a hypothesis test.
- Use the standard normal distribution (Table 1) to find the P-value for a test about p or a test about μ when σ is known.
- Use a T-distribution (Table 2) to bound the P-value for a test about μ when σ is unknown.
- Use the P-value to reach a conclusion for a hypothesis test.
- Combine the abilities listed above to conduct hypothesis tests about p and μ.
- Use the TI-83 plus to conduct hypothesis tests about p and μ.

Your course instructor may have additional or different assessable outcomes for your class. As teachers (or future teachers) you should think about the assessment outcomes and learning goals for each chapter as you work through them.

EXERCISES FOR CHAPTER 10

1. Using the *Data Analysis and Probability Standards for Grades 6–8* from the NCTM's *Principles and Standards for School Mathematics* found at www.nctm.org, identify the middle school objectives that are found in Chapter 10.

2. Using the state standards for mathematics for the content area of data analysis and probability for your state, identify the middle school objectives that are found in Chapter 10. The following website may be useful: www.doe.state.in.us. This website will allow you to access the web pages of the state departments of education for the 50 states. From the state web pages you should be able to find the state's mathematical standards.

3. A college's mathematics department must create a policy for placing students in College Algebra or Intermediate Algebra. They will treat this decision like a hypothesis test between two competing claims:

   Claim 1: *The student is ready for College Algebra.*
   Claim 2: *The student is not ready for College Algebra.*

   They will believe Claim 1 only if it is supported by sufficient evidence from the student's ACT scores, high school grades, or a score on the college's own math placement test. Otherwise, they will believe Claim 2 and place the student in Intermediate Algebra.

   a. Which claim are they treating as the null hypothesis, and which is the alternative? Explain why you think so.
   b. What would be a Type 1 error in this situation? What are the consequences of this type of error?
   c. What would be a Type 2 error in this situation? What are the consequences of this type of error?

4. Billy's mother asks him, "Shouldn't you be studying for your spelling test instead of watching television?" Billy replies, "I already studied."

   "You know all of the words on this list?" his mother asked skeptically. "Well, maybe not all of them," he replies, "but I bet I know over 90% of them."

   "I am going to pick four words from this list at random. If you spell all four of them correctly, you can watch television. If not, you have to study for an hour."

"Okay, it's a deal!"

Think of this situation as a hypothesis test about the proportion of words on the list that Billy can spell. Call this proportion p.

    a. Billy claims that he can spell over 90% of the words. Is this the null hypothesis or the alternative? Formulate this hypothesis.

    b. Formulate the competing hypothesis.

    c. Suppose that Billy's claim is false, and he can really only spell 75% of the words. He could still spell all four words correctly, leading his mother to make an incorrect decision. What type of error would this be? What is the probability that he spells all four words correctly?

    d. Suppose that Billy can spell exactly 92% of the words (more than 90%, just as he claimed). He could still miss at least one of the four words and have to study for an hour anyway. What type of error would this be? What is the probability that he misses at least one word?

    e. Give a reason why the error probabilities are so large in this situation.

5. Last year, Nicky was a 60% free throw shooter. She went to basketball camp this summer, and now she claims that her free throw percentage has improved. To support her claim, she takes 50 shots from the free throw line and gets 35 of them. You are going to conduct a hypothesis test at the .10 level to decide whether there is sufficient evidence that Nicky has improved her free throw percentage.

    a. Formulate the hypotheses.

    b. Is the sample large enough to use a normal distribution for this test? Show how you check this.

    c. Find the rejection region.

    d. Find the value of the test statistic.

    e. What is your conclusion? Interpret what your conclusion means in this situation.

6. In 2001 and 2002, the company that makes M&M's allowed people from all over the world to vote to choose a new color for M&M's. The winner was purple with 41% of the vote (http://us.mms.com/us/). Tastes change from year to year. Suppose they decided to conduct a smaller poll in 2005 to decide whether there is evidence that this percentage has changed since 2002. In a random sample of 1,000 people, 375 people voted for purple. You are going to be conducting a hypothesis test at the 0.05 level of significance.

    a. Formulate the hypotheses for testing whether the proportion of people who prefer purple has changed since 2002.

    b. Find the rejection region.

    c. Find the value of the test statistic.

    d. What is your conclusion? Interpret what your conclusion means in this situation.

7. To raise money for a class trip, the eighth graders plan to hire a band for a dance and charge admission. The plan will work only if more than 40% of the middle school students go to the dance. They survey a random sample of 20 students and find that 11 of them plan to attend. You are going to conduct a hypothesis test to determine whether this sample provides evidence that their plan will work. Use $\alpha = .05$.

    a. Formulate the hypotheses.

    b. Is this sample large enough to use a normal distribution for this test? Show how you check this.

    c. Find the rejection region.

d. Find the value of the test statistic.
e. What is your conclusion? Interpret what your conclusion means in this situation.

8. In Section 7.6, we introduced Kalifa, who had tossed a tack 100 times. Fifty-eight of the tosses were "point up." You are going to conduct a hypothesis test to see if this sample provides evidence that "point up" and "point down" are not equally likely. Use the .05 level of significance.
   a. Formulate the hypotheses.
   b. Find the rejection region.
   c. Find the value of the test statistic.
   d. What is your conclusion? Interpret what your conclusion means in this situation.

9. "I don't think this game is fair!" says Lucy. "If it was fair, I'd win half the time. I've played it a dozen times, and I've only won four times. I think it is rigged so that I will lose most of the time." You are going to conduct a hypothesis test to test Lucy's claim that the game is rigged against her. Use the .10 level of significance.
   a. Formulate the hypotheses.
   b. Is this sample large enough to use a normal distribution for this test? Show how you check this.
   c. Find the rejection region.
   d. Find the value of the test statistic.
   e. What is your conclusion? Interpret what your conclusion means in this situation.

10. Find the P-value for the test in #4 and check to see whether it leads to the same conclusion. Would you reach the same conclusion if $\alpha = .05$?

11. Find the P-value for the test in #5 and check to see whether it leads to the same conclusion. Would you reach the same conclusion if $\alpha = .10$? What about $\alpha = .01$?

12. Find the P-value for the test in #6 and check to see whether it leads to the same conclusion. Would you reach the same conclusion if $\alpha = .10$? What about $\alpha = .01$?

13. Find the P-value for the test in #8 and check to see whether it leads to the same conclusion. Would you reach the same conclusion if $\alpha = .05$?

14. Remember *The Drink Dispenser* from Section 10.4? We were told that $\sigma = .24$. If the machine is set correctly, large drinks should average 32 ounces. Suppose you plan to use a random sample of 25 large drinks to see whether there is evidence that the machine is set incorrectly for large drinks.
    a. Formulate the hypotheses.
    b. For this test to be valid, what do we need to assume about the population of large drinks? Why do we need this assumption in this situation?
    c. Suppose drinks in the sample averaged 31.91 ounces. Find the P-value for the test.
    d. Are your results statistically significant at the .05 level? Use the P-value to reach a conclusion. Interpret what your conclusion means in this situation.
    e. Find the rejection region and check to see whether it leads to the same conclusion.

15. A package of light bulbs of a particular brand says "Average life: 1,000 hours." A consumer protection agency routinely conducts tests to see is there is evidence of false advertising. They select a random sample of 36 of these light bulbs and find that the sample mean is 982 hours with a standard deviation of 75.3 hours. You

are going to conduct a hypothesis test to determine whether this sample provides evidence that this brand of light bulb has a shorter life than the package says. Use the .05 level of significance.
   a. Formulate the hypotheses for this test.
   b. What, if anything, do we need to assume about the population or the sample for this test to be valid?
   c. Find the rejection region.
   d. Find the value of the test statistic.
   e. What is your conclusion? Interpret what your conclusion means in this situation.
   f. Bound the P-value as closely as possible and check to see whether it leads to the same conclusion.

16. In Section 9.2, a group of students were studying the life of a mayfly. For a sample of 1,000 mayflies, they found the average life span to be 8.25 hours with a standard deviation of 1.23. You are going to conduct a hypothesis test to determine whether this sample provides evidence that the average mayfly lives more than 8 hours. Use the .01 level of significance.
   a. Formulate the hypotheses for this test.
   b. What, if anything, do we need to assume about the population or the sample for this test to be valid?
   c. Find the rejection region.
   d. Find the value of the test statistic.
   e. What is your conclusion? Interpret what your conclusion means in this situation.

17. A company that markets an online speed reading course claims that people who complete their course read at an average speed of 1,200 words per minute with an 80% comprehension rate. A random sample of 12 people who completed the course had an average reading speed of 1,126 words per minute with a standard deviation of 132. Use the .05 level of significance to test the hypothesis that the reading speed of people who complete the course is actually less than what the company claims.
   a. Formulate the hypotheses.
   b. What, if anything, do we need to assume about the population or the sample for this test to be valid?
   c. Find the rejection region.
   d. Find the value of the test statistic.
   e. What is your conclusion? Interpret what your conclusion means in this situation.
   f. Bound the P-value as closely as possible and check to see whether it leads to the same conclusion.

18. The sample of 12 people in #14 had an average comprehension rate of 78.5% with a standard deviation of 6.1%. Use the .05 level of significance to test the hypothesis that the comprehension rate of people who complete the course is actually less than the 80% the company claims. Interpret what your conclusion means in this situation.

Some Final Thoughts

When we write a final exam for a course, we always try to write it so that the average student will be able to answer most of the questions correctly. At the same time, we try to ask questions that students could not possibly have answered correctly before taking the course. We want students to feel that the time they spent on the course was worthwhile and that they now know a lot that they didn't know before. We hope you feel that way right now.

If you've been through the whole book, you should have learned quite a bit about statistics, from collecting, organizing, and displaying data through probability and counting techniques to confidence intervals and hypothesis testing. Is that all there is to statistics? A glance at the menus on the TI-83 plus will tell you that there is a lot more that we didn't get to: two-sample tests for comparing two populations, tests about variance, tests about the least squares line, etc. We'll leave those for another book. Our goal was to include enough for a one-semester course for preparing future middle school teachers. We felt strongly that we should get far enough to introduce hypothesis testing, but we could not see getting much further while still doing a thorough job on the topics that are closer to the middle school curricula. If this course has sparked your interest to study statistics further, great! There are other statistics courses and plenty more to learn from them.

Throughout this book, you have seen many examples taken from NSF-funded middle school curricula. In these examples, you have seen the kinds of things that middle school teachers are teaching these days. You should also be comfortable in knowing that what we have asked you to do goes well beyond the middle school level. You should be able to handle almost any question a precocious eighth grader can throw at you.

Lastly, we tried to be entertaining. If you came to this book thinking that statistics is a deadly dull, dry subject that is about as interesting as timing how long it takes paint to dry, we hope we've changed your mind. We have always felt that making a subject interesting is part of a teacher's job. We hope you feel the same way; you may be teaching our grandchildren someday.

Best of luck for your future in teaching!

Mike and Debbie Perkowski

Glossary

alternative hypothesis or research hypothesis: A claim about a population characteristic for which data is collected to support.

back-to-back stem-and-leaf plot: A type of stem-and-leaf plot in which two data sets are represented using one plot. For one data set the leaves go to the right of the stems and for the other the leaves go to the left of the stems. This type of graph is useful for comparing two numerical data sets.

bar graph: A visual way of displaying **categorical** or **nominal data**. Bar graphs may be vertical (up and down) or horizontal (back and forth). One axis of the graph displays the classes or categories while the other axis displays the frequency scale. Bar graphs are useful for comparing classes or categories to each other, but harder to compare classes or categories to the whole than other types of graphs.

binomial probability distribution: The probability distribution associated with a binomial experiment in which the number of trials is fixed, each of the trials is independent, and the probability of a success is the same from trial to trial.

binomial random variable: The number of successes associated with n trials in a binomial experiment. The number of trials must be fixed, each of the trials must be independent, and the probability of a success must be the same from trial to trial in order for the experiment to be binomial.

bivariate data: A data set in which two variables are considered or studied.

bounding the P-value: When it is not possible to find the exact area corresponding to the P-value for a hypothesis test, we find two areas, one that is smaller than the P-value (the lower bound) and one that is larger (the upper bound).

box-and-whisker plot/box plot: A visual way to display numerical data showing the distribution of the data. Box plots use the upper and lower quartiles, the median, the minimum value, and the maximum value of the data set in their construction.

categorical data: Also referred to as **qualitative data**. Data can be sorted into mutually exclusive groups like eye color, political party affiliation, or year in school. Qualitative data can be represented by numbers such as gender being recorded as a "0" for male and a "1" for female.

Central Limit Theorem: Regardless of the distribution of the population being sampled, the distribution of sample means taken from random samples of size n is approximately normal when n is large.

chance experiment: A situation with more than one possible outcome.

circle graph: A visual way of displaying categorical or nominal data. Circle graphs use the size of each sector to show the relative size of each class or category to the whole.

class boundaries: The numerical value halfway between one numerical interval and the next. For example, the class boundary between the class 0–4 and the class 5–9 would be 4.5. Class boundaries are typically used in the construction of histograms.

class frequency graph: Also referred to as a dot plot or a line plot.

class mark: The middle of a class in a frequency distribution. For example, the class mark for the class interval 0–6 is 3.

class width: The number of possible items in a class for discrete data or the difference between successive class marks in a frequency distribution table.

classical approach: The traditional method of reaching a conclusion in a hypothesis test. We reject the null hypothesis if the value of the test statistic lies in the rejection region.

cluster sampling: A sampling technique where the population is divided into sections and then a few of those sections are randomly selected. All the members within the chosen sections are then included in the study.

combinations: One of the three basic counting techniques for determining how many outcomes are possible in an experiment or chance situation. Combinations are used when order does not make a difference. In general, the number of combinations possible from n distinct objects taken r at a time is $_nC_r = \binom{n}{r} = \frac{n!}{(n-r)! \cdot r!}$.

complementary events: The events which are "opposite" in nature. For example, if rolling a two is the event, then **not** rolling a two would be the complementary event. The probability of the complement is 1 − probability of the event or $P(\overline{A}) = 1 - P(A)$.

conditional probability: A probability where the likelihood of a particular outcome is dependent upon the results of a prior outcome. $P(A|B) = \frac{P(A \cap B)}{P(B)}$.

confidence interval: A range of values predicting the location of a population characteristic. Confidence intervals typically have the form (estimate) ± (margin of error).

confidence level: The probability that a confidence interval will actually contain the population characteristic being estimated. Equivalently, the confidence level is the probability that the distance between an estimate and the true value will be less than the margin of error.

contingency table: A way of representing the outcomes of an experiment using rows and columns.

continuity correction: Adding and/or subtracting 0.5 to the values of a discrete (usually binomial) random variable when approximating probabilities with a (continuous) normal curve.

continuous data: Numerical data (such as temperature, time, weight, height, and so on) that can take on any value in a finite or infinite range. There are uncountably many possible values within any given range.

convenience sampling: A sampling technique in which participants are selected based solely on convenience or availability.

correlation: A measure of the strength and direction of the linear relationship between two variables.

critical value: A cutoff value determined by the level of significance desired (or confidence level) and the sample size.

data: Pieces of information.

data set: The collection or group of data on all members of a study or survey.

decile: Similar to a percentile, deciles are a measure of the position of a data item among all the other items in the set. Deciles are every tenth percentile rank (so the tenth percentile would be the first decile and so on). The second decile would correspond to the data value separating the bottom 20 percent of the data from the top 80 percent of the data.

decision rule: A rule for determining whether or not to reject the null hypothesis in a hypothesis test.

degrees of freedom: A number that distinguishes between different T-distributions. The higher the number of degrees of freedom, the closer the T-distribution gets to a standard normal distribution. When T-distributions are used for hypothesis tests and confidence intervals in this book, the number of degrees of freedom is $n - 1$.

descriptive statistics: A branch of statistical study that deals with ways of collecting, organizing, and describing data.

deviation: The difference between an individual data value and the mean.

discrete data: Data (typically numerical) in which the possible values are distinct and separate, like the number of tornadoes in a given year. There are countably many possible values within any given range.

dot plot: A dot plot is essentially the same type of visual display as a line plot only dots are used instead of "x's" to plot the frequencies.

Empirical Rule: A rule of thumb that states that roughly 68% of all of the data values within a data set will lie within one standard deviation of the mean and close to 100% of the items will lie within three standard deviations of the mean. The Empirical Rule applies when the distribution is approximately normal.

error bound: Also called the **margin of error**, the maximum distance we would expect between an estimate and the true value. The probability of getting an estimate within this distance of the true value is the confidence level.

estimator: A statistic computed from a sample and used to estimate the value of a population characteristic.

event: A subset of the sample space of an experiment.

expected value: Also referred to as the mean of a random variable. The expected value is found by summing the product of each of the possible values in a chance experiment and its associated probability or $E(x) = \mu = x_1 \cdot P(x_1) + x_2 \cdot P(x_2) + \cdots + x_n \cdot P(x_n)$.

experiment: A situation where some element of chance is associated with the possible outcomes.

experimental probability: A probability based on the outcomes obtained from conducting or simulating an experiment.

factorial notation: Factorial notation is a shorthand way of denoting a whole series of multiplications where each factor is one less than the previous factor. $n!$ means $n \cdot (n - 1) \cdot (n - 2) \ldots \cdot 1$ where n is a whole number. $0!$ is defined to be 1.

fitted line: A line used to describe the data in a bivariate data set and to predict values of one variable from the other.

five-number summary: Five-number summaries are the five measures associated with a data set used to construct a box-and-whisker plot. Five-number summaries consist of the median, the upper and lower quartiles, the minimum, and the maximum of a data set.

frequency: The number of items in a particular class or category. Frequency is usually denoted by the letter f.

frequency table: A table that organizes and shows the number of data items that fall into each class or category.

Fundamental Counting Principle: Also referred to as the multiplication of choices principle. One of the three basic counting techniques for determining how many possible outcomes there are for an experiment or counting situation. If there are n_1 choices for the first selection, n_2 choices for the second, n_3 choices for the third, and so on, then the total number of ways a person can choose one from each of the selections is $n_1 \cdot n_2 \cdot n_3 \cdot \ldots$

general addition rule: A way of computing the probability of the union of two events. $P(A \cup B) = P(A) + P(B) - P(A \cap B)$.

general multiplication rule: A way of computing the intersection of dependent events. $P(A \cap B) = P(A) \cdot P(B|A)$.

geometric probability: Probabilities determined by using geometric models (typically areas).

histogram: A visual way of displaying **continuous data** using bars to represent the frequency of numerical values in a specified interval. The intervals all have equal width and the bars of the graph touch each other.

hypothesis: A claim about a population characteristic.

hypothesis testing: A formal statistical method for deciding between two competing claims about a population characteristic.

independent events: Two events are said to be independent if the likelihood of one event is not influenced by knowing that the other has occurred.

inferential statistics: A branch of statistical study that uses information obtained from samples to make estimates or draw conclusions about a population.

interquartile range: The difference between the upper quartile and the lower quartile values.

interval data: A type of numerical data for which differences between categories on any part of the scale reflect equal differences in the characteristic being measured. Temperature is the classical example of interval data. The difference between 2 degrees and 4 degrees is the same as the difference between 70 degrees and 72 degrees. For interval data ratios do not make sense. For example, 4 degrees is not twice as hot as 2 degrees. The value of zero is simply another point on the scale and does not represent the lack of the particular characteristic being measured.

Law of Large Numbers: As an experiment is conducted over and over again and the results pooled, the experiment probability will get closer and closer to the theoretical probability.

least squares line: A common type of fitted line (sometimes referred to as the line of best fit). The "squares" are the squared vertical distances between the data points and a line. The least squares line is the line for which the total of these squared vertical distances is as small as possible.

level of significance: The largest Type 1 error probability we are willing to accept in a hypothesis test. The level of significance is the total tail area above the rejection region and is symbolized by the Greek letter α (alpha).

line graph: A visual way of displaying data values *over time*. Continuity is implied since all the points on the graph are connected.

line plot: A visual way of displaying **numerical data** using either a vertical or horizontal scale (horizontal is more typical) including both the smallest and largest value in the data set. Frequencies are plotted either using "x's" or dots above the values on the scale. *Occasionally* you will see a line plot used with categorical data, though bar graphs are far more common.

lower class limits: The smallest value in each class in a frequency distribution table.

lower fence: A value used to determine data values that are considered outliers—much larger or much smaller than other values in the set. The lower fence is found by subtracting 1.5 times the interquartile range from the first quartile.

lower quartile: The value for a data set that partitions the lower 25 percent of the data from the upper 75 percent of the data.

lower tail area: An area at the left end of a continuous probability distribution such as a normal or T-distribution.

lower-tailed test: A hypothesis test in which the rejection region is in the lower tail of the distribution of the test statistic. We get a lower-tailed test when the alternative hypothesis has the form: H_1: (population characteristic) < (specified value).

margin of error: Also called the **error bound,** the maximum distance we would expect between an estimate and the true value. The probability of getting an estimate within this distance of the true value is the confidence level.

mean: The mean is sometimes referred to as the arithmetic mean. It is a measure of central tendency found by adding all of the values in the data set together and then dividing by the total number of items in the set. The mean of a sample is represented with the symbol $\bar{x}$, while the mean of a population is represented using the Greek letter μ.

mean deviation: The sum of all of the absolute values of the deviations from the mean divided by the total number of items in the data set or $\frac{\sum |x - \bar{x}|}{n}$. This measurement gives the average value that each data item varies from the mean or the average distance that each value is from the mean.

mean of a binomial random variable: If a random variable is binomial, then the mean may be found by multiplying the number of trials by the probability of a success on any given trial. $\mu = n \cdot p$

mean of a random variable: Also referred to as the expected value. The mean of a random variable is found by summing the product of each of the possible values in a chance experiment and its associated probability or

$$E(x) = \mu = x_1 \cdot P(x_1) + x_2 \cdot P(x_2) + \cdots + x_n \cdot P(x_n).$$

mean of $\hat{p}$: The average or expected value of a sample proportion, symbolized by $\mu_{\hat{p}}$ or $E(\hat{p})$. If the sampling is done at random, $E(\hat{p}) = p$, the corresponding proportion from the population being sampled.

mean of $\bar{x}$: The average or expected value of a sample mean, symbolized by $\mu_{\bar{x}}$ or $E(\bar{x})$. If the sampling is done at random, $E(\bar{x}) = \mu$, the mean of the population being sampled.

median: The physical middle of a data set. If the items are all ordered from high to low (or low to high) the middle value is the median. If the data set has an even number of items, then the median is the value that falls halfway between the two middle values of the set.

median-median line: A type of fitted line found by splitting the data into three roughly equal groups and then using the median points from these groups to find the fitted line.

misleading graph: A type of graph (bar, circle, histogram, line graph, whatever) that may mislead the reader by use of inconsistent scales, shortening of scales, inaccurate portrayal of relative sizes, using different width bars, etc.

mode: The item or class with the highest/largest frequency.

modern approach: Using the P-value to reach a conclusion in a hypothesis test. We reject the null hypothesis if the P-value $\leq \alpha$.

multiplication of choices principle: One of the three primary counting techniques used for determining how many ways are possible involving sets of two or more choices. If there are n_1 choices for the first selection, n_2 choices for the second, n_3 choices for the third, and so on, then the total number of ways a person can choose one from each of the selections is $n_1 \cdot n_2 \cdot n_3 \cdot \ldots$

mutually exclusive events: Events that may not occur at the same time. For example, rolling a two with one toss of a fair die AND rolling a four at the same time are mutually exclusive events.

negative correlation: An increase in one variable of a bivariate data set leads to a decrease in the other variable.

nominal data: A type of categorical data. Data can be sorted into distinct categories based on some defined characteristic such as eye color. Nominal data may be numbers, but the numbers are just used as identifiers and not to quantify a particular characteristic. Telephone numbers are an example of nominal data. Nominal data may not be added, subtracted, multiplied, or divided as, for example, you may not find the average telephone number of all your friends or the average eye color of your class.

null hypothesis: The null hypothesis is the claim against which a test is conducted to determine if there is enough evidence to support or reject the claim. The null hypothesis is assumed to be true until convincing sample evidence is given to reject it.

numerical data: Also referred to as **quantitative data.** Data for which the desired characteristic of the data set can be measured and represented by a number such as height.

observed significance level: Also referred to as the P-value. The probability of selecting a random sample that supports the alternative hypothesis at least as much as the sample actually chosen, assuming the null hypothesis is true.

ordinal data: Data (either categorical or numerical) that can be placed in some type of logical order like music preferences or Likert Scale-type responses of strongly agree, agree, etc.

outcomes: Possible results of an experiment. For example, when flipping a fair coin (the experiment) there are two possible outcomes—heads or tails.

outliers: Extreme values that are much larger or much smaller than other values in the data set. Upper and lower fences are used to determine the outliers in a data set.

P-value: Also referred to as the **observed significance level**. The probability of selecting a random sample that supports the alternative hypothesis at least as much as the sample actually chosen, assuming the null hypothesis is true.

Pearson's Correlation Coefficient: A formula for measuring the strength and direction of the linear relationship between two quantitative variables.

percent: The relative frequency, $\frac{f}{n}$, multiplied by 100 and represented numerically as a decimal such as 13.6%.

percentile: A measure of position for an item in a data set based on its rank order position among all the other items in the set. The percentile rank indicates the percentage of data below that particular value in the data set.

permutations: One of the three basic counting techniques for determining how many outcomes are possible in an experiment or chance situation. Permutations are used when order **makes a difference**. In general, the number of permutations possible from n distinct objects taken r at a time is $_nP_r = \frac{n!}{(n-r)!}$.

population: The complete set of people or objects being studied from which a more manageable sample is typically drawn. For example, rather than studying all of the teachers in a particular state (the population), a sample of 1,000 could be randomly selected for study.

population mean: One of the measures of central tendency for a population. In general, the population mean is the sum of the numerical data items divided by the total number of items. Population means are usually denoted by the Greek letter mu, μ.

population parameters: The specific characteristics of a population. Population parameters are generally represented using Greek letters or capital letters such as μ (mu) for mean or N for population.

population proportion: The ratio of the number of items in a population that have some special characteristic (like blue eyes) to the total number of items in the population.

population standard deviation: The standard deviation for a population rather than a sample. The average amount each item in the data set varies from the mean in a population. For population standard deviations the divisor is n rather than $n - 1$.

positive correlation: A increase in one variable in a bivariate data set leads to an increase in the other variable.

probability: The ratio of the number of favorable outcomes obtained from either a sample space or from a simulation to the total number of possible outcomes or total number of obtained outcomes.

probability density function for a binomial random variable: A probability density function (or pdf for short) is a rule for finding the probabilities associated with the

number of successes (the random variable) for a binomial experiment.
$P(x) = \binom{n}{x} p^x q^{n-x}$

probability distribution: A rule or table for finding the probabilities for values of a random variable.

qualitative data: Also referred to as **categorical data**. Data can be sorted into mutually exclusive groups like eye color, political party affiliation, or year in school. Qualitative data can be represented by numbers such as a freshman being recorded as year "9" in school.

quantitative data: Data for which the desired characteristic can be measured and represented by a number such as height. Also known as **numerical data**.

quartile: Similar to percentiles and deciles, quartiles indicate the data has been split into quarters or fourths. The second quartile corresponds to the fifth decile, the fiftieth percentile, or the median of a data set. Quartiles are used in the construction of box-and-whisker plots.

random sample: A sample chosen from the population in such a way that every possible element of the population or every possible sample of the same size is equally likely to be chosen.

random variable: A quantitative variable whose value is determined by the outcome of a chance experiment.

range: The range is a measure of the spread of a data set. It is the difference between the largest value of the data set and the smallest value.

ratio data: A type of numerical data for which it makes sense to compute ratios or quotients of characteristics possessed by two objects, such as weight. It makes sense to say that a father weighs twice as much as his son. Zero indicates a total absence of the characteristic.

raw data: Bits of information about the elements of a population or sample like eye color, annual salary, age, and so on.

rejection region: In a hypothesis test, the values of the test statistic that are extreme enough (high enough or low enough) to support the alternative hypothesis. The area above the rejection region is α, the Type 1 error probability.

relative frequency: The ratio or quotient of the frequency of a particular class, f, to the total number of items in the survey or study, n.

S_{xx}: One of the three "S's" used to compute Pearson's Correlation Coefficient. This value is the sum of the squares of the deviations from the mean of the independent variable in a bivariate relationship. In other words, S_{xx} is $\Sigma(x - \bar{x})^2$.

S_{xy}: One of the three "S's" used to compute Pearson's Correlation Coefficient. This value is the sum of the products of the deviations from the mean of both the independent and the dependent variable in a bivariate relationship. In other words, S_{xy} is $\Sigma(x - \bar{x})(y - \bar{y})$.

S_{yy}: One of the three "S's" used to compute Pearson's Correlation Coefficient. This value is the sum of the squares of the deviations from the mean of the dependent variable in a bivariate relationship. In other words, S_{yy} is $\Sigma(y - \bar{y})^2$.

sample: A subset of a population from which data are obtained.

sample mean: One of the measures of central tendency for a sample. In general, the mean of a sample is the sum of the numerical data items in the sample divided by the total number of items in the sample. Sample means are denoted using $\bar{x}$.

sample proportion: The ratio of the number of items in a sample taken from a population that have some special characteristic (like the number of students with blue eyes) to the total number of items in the sample. Sample proportions are denoted by the symbol p-hat, $\hat{p}$.

sample space: The set of all possible outcomes from an experiment.

sample standard deviation: The standard deviation associated with a sample rather than a population. The average amount each item in the data set varies from the mean in a sample. For sample standard deviations the divisor is $n - 1$ rather than n.

sample statistics: The characteristics of a sample found by consolidating or summarizing the raw data from the sample.

scatter plot: A plot displaying bivariate data where one variable represents the independent variable in the plot and the other variable represents the dependent variable.

simple random sampling: A sampling technique where every element of the population or every sample of the same size has an equal chance of being selected. Computers are often used to determine which elements are selected.

slope: The change in the dependent variable that corresponds to a one-unit increase in the independent variable or the ratio of the change in the dependent variable to the change in the independent variable.

standard deviation: The square root of the variance; taking the square root makes the units on standard deviation the same as the units for the data. Standard deviation can be interpreted as a typical or standard distance between data values and the mean. In general, the larger the standard deviation, the more "spread out" the values in the data set.

standard deviation of a binomial random variable: A way of measuring the spread or variability associated with a random variable from a binomial distribution. $\sigma = \sqrt{npq}$

standard deviation of $\hat{p}$: A measure of spread or variability for the distribution of sample proportions, symbolized by $\sigma_{\hat{p}}$. When samples are chosen at random, $\sigma_{\hat{p}} = \sqrt{\frac{pq}{n}} \approx \sqrt{\frac{\hat{p}\hat{q}}{n}}$. Also referred to as the standard error of the mean.

standard deviation of a random variable: A way of measuring the spread or variability of a random variable. $\sigma = \sqrt{\sum_1^n [(x_i - \mu)^2 \cdot P(x_i)]} = \sqrt{\sum_1^n [x_i^2 \cdot P(x_i)] - \mu^2}$

standard deviation of $\bar{x}$: A measure of spread or variability for the distribution of sample proportions, symbolized by $\sigma_{\bar{x}}$. When samples are chosen at random, $\sigma_{\bar{x}} = \frac{\sigma}{\sqrt{n}} \approx \frac{s}{\sqrt{n}}$. Also referred to as the standard error of the mean.

standard error: The standard deviation of the probability distribution of an estimator. The standard error of the mean is $\frac{\sigma}{\sqrt{n}} \approx \frac{s}{\sqrt{n}}$. The standard error of a proportion is $\sqrt{\frac{pq}{n}} \approx \sqrt{\frac{\hat{p}\hat{q}}{n}}$.

standard normal curve: The density curve for the normal distribution whose mean is 0 and standard deviation is 1. Sometimes referred to as the "bell" curve, its equation is $f(z) = \frac{1}{\sqrt{2\pi}} \cdot e^{-\left(\frac{z^2}{2}\right)}$.

statistically significant: When the data provides sufficient evidence to reject the null hypothesis at a certain level of significance, the results are said to be statistically significant at that level.

statistics: A branch of mathematical sciences focused on how to collect, organize, describe, and interpret numbers and other information about some topic and how to use that information to make predications and draw conclusions about some topic.

stem-and-leaf plot: A visual way of displaying numerical data. A vertical line separates the "stem" from the "leaf" part of a given number. The stem is typically the highest place-value digit of the number and the leaf is the lowest. For example, in the number 72, the 7 would be the stem (the highest place-value digit) and the 2 would be the leaf (the lowest place-value digit).

stratified sampling: A sampling technique where the population is partitioned or divided into several layers or strata and then a sample is drawn from each layer. For example, socioeconomic status divides our society into several layers. If a random sample were taken from each of these layers the result would be a stratified sample.

systematic sampling: A sampling technique where every n^{th} element of the population is selected for study.

T-distribution: A continuous probability distribution similar to the standard normal distribution, but more tail-heavy than standard normal. The distribution of $z = \frac{\bar{x} - \mu}{(\sigma/\sqrt{n})}$ is standard normal, but replacing the population standard deviation σ with the sample standard deviation s to get $t = \frac{\bar{x} - \mu}{(s/\sqrt{n})}$ results in a T-distribution with $n - 1$ degrees of freedom. As the number of degrees of freedom increases, the T-distribution approaches standard normal.

tail heavy: A term indicating that a mound-shaped distribution has more area in the tail than the standard normal curve.

test statistic: The value (a z-score or t-score for tests of means and proportions) that is compared to the critical value to determine whether to accept or reject the null hypothesis in a hypothesis test. The value of the test statistic is also used to find the P-value for the test.

theoretical probability: A probability based on the set of all possible outcomes for an experiment.

total tail area: The probability associated with the area under the normal curve for the sum of the areas for the extreme leftmost and rightmost regions of the curve.

two-tailed test: A hypothesis test in which the rejection region is equally divided between the lower and upper tails of the distribution. We get a two-tailed test when the alternative hypothesis has the form: H_1: (population characteristic) $\neq$ (specified value).

Type 1 error: Rejecting a null hypothesis that is actually true.

Type 2 error: Not rejecting a null hypothesis that is actually false.

unbiased estimator: A statistic which, on average, neither overestimates nor underestimates the true value of the population characteristic. The mean of the distribution of the estimator is exactly equal to the true value to be estimated.

univariate data: A data set in which only one variable is considered or studied.

upper class limits: The largest value in each class of a frequency distribution table.

upper fence: A value used to determine data values that are considered outliers—much larger or much smaller than other values in the set. The upper fence is found by adding the third quartile to the 1.5 times of the interquartile range.

upper quartile: The value that partitions the lower 75 percent of a data set from the upper 25 percent of a data set.

upper tail area: An area at the right end of a continuous probability distribution such as a normal or T-distribution.

upper-tailed test: A hypothesis test in which the rejection region is in the lower tail of the distribution. We get a lower-tailed test when the alternative hypothesis has the form: H_1: (population characteristic) > (specified value).

variable: Characteristics of individuals within the population that vary from individual to individual, like eye color or height.

variance: A measure of spread or variability in a data set. The larger the variance, the more spread out the data is. The variance is the average (or predicted average) of the squared distances between data values and the mean. For a sample, the variance is $s^2 = \frac{\Sigma(x-\bar{x})^2}{n-1} = \frac{n\Sigma x^2 - (\Sigma x)^2}{n(n-1)}$; for a population, the variance is $\sigma^2 = \frac{\Sigma(x-\mu)^2}{N} = \frac{N\Sigma x^2 - (\Sigma x)^2}{N^2}$. Variance is also the square of the standard deviation of a data set.

variance of a binomial random variable: A measure of spread or variability for the random variable in a binomial experiment. For binomial random variables, we can get the same results as the usual variance formula by using a simpler formula: $\sigma^2 = npq$.

variance of $\hat{p}$: A measure of spread or variability for the distribution of sample proportions, symbolized by $\sigma_{\hat{p}}^2$ or $Var(\hat{p})$. When samples are chosen at random, $Var(\hat{p}) = \frac{pq}{n} \approx \frac{\hat{p}\hat{q}}{n}$.

variance of a random variable: A measure of spread or variability for the distribution of a random variable. $\sigma^2 = \Sigma_1^n[(x_i - \mu)^2 \cdot P(x_i)] = \Sigma_1^n[x_i^2 \cdot P(x_i)] - \mu^2$

variance of $\bar{x}$: A measure of spread or variability for the distribution of sample proportions, symbolized by $\sigma_{\bar{x}}^2$ or $Var(\bar{x})$. When samples are chosen at random, $Var(\bar{x}) = \frac{\sigma^2}{n} \approx \frac{s^2}{n}$.

Venn diagram: A diagram in which sets are represented by circles within a rectangle representing the universal set.

weighted mean: Another way of finding the center of a data set where certain values have more weight than others. Grade point averages are a typical example of a weighted mean. Weighted means are found by multiplying each value by its weight and then dividing the sum of those products by the total sum of the weights.

with replacement: To conduct an experiment with replacement means that after each item is drawn or selected from a group (like a cube from a bag), it is returned to the group and may be selected again and again.

without replacement: To conduct an experiment without replacement means that after an item is drawn or selected from a group (like a cube from a bag), it is NOT returned to the group and may NOT be selected again.

***y*-intercept:** The value of the dependent variable that corresponds to a value of 0 for the independent variable. On a graph, where the graph intersects the y-axis.

***z*-score:** Probably the most commonly used measure of position for an item in a data set. The z-score measures how many standard deviations an individual item is from the mean of a data sets. The sign of the z-score indicates whether the item is above or below the mean. Z-scores are used to scale the axes of normal curves.

References

Connected Mathematics Project, Prentice Hall, 2004.
Mathematics in Context, Encyclopedia Britannica, 1998.
MathScape, Creative Publications, 1998.
MathThematics, McDougal-Littell/Houghton Mifflin, 1999.
Peck, R., Olsen, C., & Devore, J. *Introduction to Statistics and Data Analysis*, Duxbury, Pacific Grove, CA, 2001.
Principles and Standard for School Mathematics, National Council of Teachers of Mathematics, Reston, VA, 2000.

Answers and Hints for Odd Numbered Exercises

CHAPTER 1

1. Answers may vary on this exercise. The point of this particular exercise (along with exercise 1.2) is to become familiar with the middle school standards from the NCTM and your own state. Since nearly every state has adopted state standards based on the NCTM's *Principles and Standards for School Mathematics* (or the older book, *Standards for School Mathematics*) there will probably be quite a bit of overlap between your responses for exercises 1.1 and 1.2.

   From the NCTM PSSM e-standards online: Data Analysis and Probability Standard
   Formulate questions that can be addressed with data and collect, organize, and display relevant data to answer them. (General benchmark)

   Specifically:
   In grades 6–8 all students should formulate questions, design studies, and collect data about a characteristic shared by two populations or different characteristics within one population.

3. a. i. The population would be all sixth graders. The question can then be raised of what does "all sixth graders" actually mean? Is it just the sixth graders in that particular school? That district? That region of the state? That state? The nation? The world?
   ii. The sample would be the class of 25 students.
   iii. The variable would be the time they spend watching television.
   iv. The raw data would be the 25 values of time in hours recorded by the students.
   v. A sample statistic would be the average of 3.15 hours per day.
   vi. A population parameter would be the national, regional, district, or whatever was agreed upon as the population average.
   b. The teacher used convenience sampling. One of the advantages of this method is that the teacher had easy access to the sample—they were at hand. One of the disadvantages is that her sample may not be representative of whatever the intended population was.
   c. The variable in this case would be of the ratio data type.

5. a. Number of points scored in a basketball game—discrete
   b. Average number of points per game for a basketball player—continuous
   c. Number of eggs in a bird's nest—discrete
   d. Weight of eggs in a bird's nest—continuous
   e. Length of eggs in a bird's nest—continuous

f. Number of hurricanes in a given three-month period—discrete
g. Scores on your most recent statistics quiz—probably discrete
h. Length of time spent by students riding the bus to and from school—continuous

7. Answers will vary. Here are some sample answers:
   a. discrete—number of tornadoes in a year
   b. nominal—numbers on football jerseys
   c. continuous—weight of football players
   d. ordinal—Likert scale items like disagree, neutral, agree
   e. ratio—weight of football players
   f. ratio, discrete—number of tornadoes in a year
   g. ratio, continuous—weight of football players
   h. interval, discrete—IQs
   i. interval, continuous—temperature (in any measurement system)
   j. nominal, continuous—not possible
   k. nominal, discrete—jersey numbers for football players
   l. ordinal, discrete—Likert scale items where each choice of agree, disagree, etc, is assigned a numerical value
   m. categorical—eye color
   n. quantitative—height

CHAPTER 2

1. Chapter 2 supports, from the NCTM's *Principles and Standards for School Mathematics*, "select, create, and use appropriate graphical representations of data, including histograms, box plots, and scatter plots."

3. a.

Color	Frequency	Relative Frequency
Blue	8	0.4
Yellow	2	0.1
Green	5	0.3
Red	4	0.2

b.

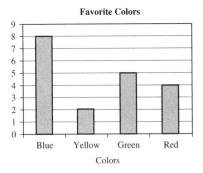

c.

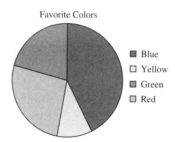

Favorite Colors
- Blue
- Yellow
- Green
- Red

5. a.

Weight	Frequency	Relative Frequency
1 to 10	2	0.08
11 to 20	3	0.12
21 to 30	2	0.08
31 to 40	2	0.08
41 to 50	5	0.20
51 to 60	2	0.08
61 to 70	4	0.16
71 to 80	2	0.08
81 to 90	1	0.04
91 to 100	2	0.08

b. class width = 10
c. class marks $\{5.5, 15.5, 25.5, 35.5, \ldots, 95.5\}$
d. class boundaries $\{0.5, 10.5, 20.5, \ldots, 100.5\}$
e.

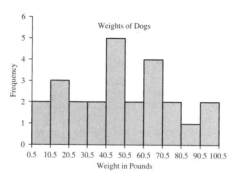

362 Answers and Hints for Odd Numbered Exercises

f.

```
0 | 6 8
1 | 2 5 8
2 | 2 6
3 | 5 6
4 | 3 4 5 5 8
5 | 7
6 | 0 1 2 5 7
7 | 2 8
8 | 4
9 | 1 2
```

g. A line plot would not be good since there are too many individual values and very few duplicates.

7. a.

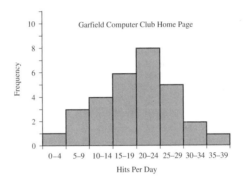

b.

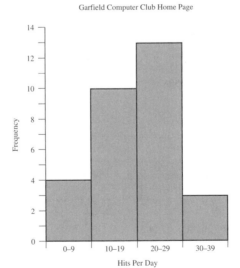

c.

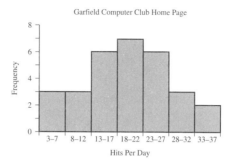

d. All three graphs show a high center, no gaps, and a slight heaviness on the left side. Graph c looks the most symmetrical, while graph b shows the least symmetry. Graph a shows more stretch on the left side of the center.

e. Graph a shows more information and also goes all the way to zero on the independent variable axis. All three graphs accurately depict the data.

f. Once the original raw data is grouped into classes, it cannot be recovered because the frequencies represent the class, not the individual independent variable values.

9. The line graph is inappropriate because the data is categorical and there is no reason to connect the data points. In fact, a bar graph is much more appropriate. The headline also seems to indicate some type of time progression, where the independent variable (x-axis) should be time, which it is not.

CHAPTER 3

1. From the NCTM's *Principles and Standards for School Mathematics* (page 248): Students should be able to "find, use, and interpret measures of center and spread, including mean and interquartile range."

3.

	Mean	Median	Mode	Range	Sample standard deviation	Sample variance
a.	68.88	70.5	74	54	16.34	266.92
b.	11.95	12	12	18	4.56	20.83
c.	33.81	33	none	32	9.63	92.83
d.	13	12	10	13	4.18	17.5

5. a. The mean is 72, the median is 74, the modes are 73 and 82, and the sample standard deviation is 13.70.

b. James scored the old median of 74, which is higher than the old mean.

c. Any score 76 or above would make the median 75 since the two scores closest to the middle are already 74 and 76.

d. For the mean to be 75, the total of all scores would need to be 16(75) = 1200. The sum of the 15 scores is 1080, so James would have to score 1200 − 1080 = 120. Either a lot of extra credit is needed, the test must have at least 120 points, or Ms. Hanson is out of luck.

7. a. Sharla's GPA will be (4(3) + 2(3) + 3(3) + 4(1) + 2(4) + 1(1))/15 = 2.6̄ or 2.67.
  b. Sharla's GPA will be higher since she increased her grade in a class with 3 hours and lowered her grade in a class with only 1 hour.
9. a. The minimum is 34, Q1 is 78, median is 120, Q3 is 141, and the maximum is 254.
  b.

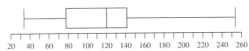

  c. The interquartile range is 141 − 78 = 63.
  d. There is one outlier at 254.

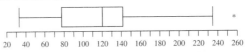

11. The five-number summary includes 16, 18, 20, 24, and 26. You could start with just those five numbers. The minimum, median, and maximum are correct. Add a number between 18 and 20 to get Q1 = 18, and add a number between 20 and 24 to get Q2 = 24. {16, 18, 20, 23, 24, 26} works.
13. Using Fathom™ or similar software, you should find that State 1 has much lower utility bills. If the cost of utilities was the most important factor, live in State 1.
15. To find the original scores, use the formula $x = sz + \bar{x}$
  a. 1.56(12) + 73.5 = 92.22
  b. −2.08(12) + 73.5 = 48.54
  c. 2.35(12) + 73.5 = 101.7
  d. −0.25(12) + 73.5 = 70.5

CHAPTER 4

1. The middle school objectives from the NCTM's *Principles and Standards for School Mathematics* that are addressed in this chapter are:
Make conjectures about possible relationships between two characteristics of a sample on the basis of scatter plots of the data and approximate lines of fit (page 248).
3. As the temperature increases, the number of cups of coffee decreases, so there is a negative correlation.

5. a. Negative correlation
  b. The point (48, 18) might be considered an outlier in the sense that the temperature was significantly higher than most of the data set. However, it still follows the same trend as the rest of the data set.
  c. $S_{xx} = \sum x^2 - \dfrac{(\sum x)^2}{n} = 7850$
  $- \dfrac{248^2}{9} = 1016.22$

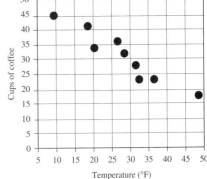

$S_{yy} = \Sigma y^2 - \dfrac{(\Sigma y)^2}{n} = 9348 - \dfrac{280^2}{9} = 636.89$

$S_{xy} = \Sigma xy - \dfrac{(\Sigma x)(\Sigma y)}{n} = 6951 - \dfrac{(248)(280)}{9} = -764.56$

$r = \dfrac{S_{xy}}{\sqrt{S_{xx} \cdot S_{yy}}} = \dfrac{-764.56}{\sqrt{(1016.22)(636.89)}} = -0.950$

This value of r indicates a strong negative correlation between the variables.

d. $m = \dfrac{S_{xy}}{S_{xx}} = \dfrac{-764.56}{1016.22} = -0.752$

$b = \bar{y} - m\bar{x} = \dfrac{280}{9} - (-0.752)\left(\dfrac{248}{9}\right) = 51.83$

Least-squares line: $y = -0.752x + 51.83$

e. 35 cups of coffee (35.291)

f. The equation would predict about 59 cups of coffee (59.35), but it would be inappropriate to use the equation this way because $-10°F$ is too far outside the range of the collected data.

g. The left median point is (18, 41). The middle median point is (28, 32). The right median point is (36, 23).

h. $y = -1x + 59$

i. $y = -1x + 59.33$

j. The equation predicts 37 cups of coffee (37.33) which is very close to 35.

k. See graph at right.

l. The median-median line is steeper and has a higher y-intercept, but the two are similar.

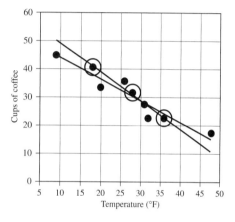

7. a. The scatter plot shows a positive correlation between the variables.

b. The change is acceptable if it is noted in the label that the prices are in thousands of dollars. The change will not affect the correlation coefficient at all. The regression equation will change so that the slope will be in thousands of dollars per square foot. Likewise, the y-intercept will be in thousands of dollars.

c. The point (800, 200) is an outlier.

d. $S_{xx} = \Sigma x^2 - \dfrac{(\Sigma x)^2}{n} = 64{,}181{,}725 - \dfrac{28{,}485^2}{14} = 6{,}224{,}923$

$$S_{yy} = \sum y^2 - \frac{(\sum y)^2}{n} = 313{,}250 - \frac{1{,}946^2}{14} = 42{,}756$$

$$S_{xy} = \sum xy - \frac{(\sum x)(\sum y)}{n} = 4{,}229{,}300 - \frac{(28{,}485)(1{,}946)}{14} = 269{,}885$$

$$r = \frac{S_{xy}}{\sqrt{S_{xx} \cdot S_{yy}}} = \frac{269{,}885}{\sqrt{(6{,}224{,}923)(42{,}756)}} = 0.523$$

This value of r indicates a moderate positive correlation between the variables.

e. $m = \dfrac{S_{xy}}{S_{xx}} = \dfrac{269{,}885}{6{,}224{,}923} = 0.04336$

$b = \bar{y} - m\bar{x} = \dfrac{1{,}946}{14} - (0.04336)\left(\dfrac{28{,}485}{14}\right) = 50.78$

Least-squares line: $y = 0.04336x + 50.78$

f. The equation predicts a cost of about $153,000.

g. The equation predicts a cost of about $224,000, but it is questionable to use this equation to make a prediction so far out of the data's range.

h. The three median points are (1250, 70), (2175, 157.5), and (2650, 175).

i. $m = \dfrac{175 - 70}{2650 - 1250} = 0.075$

$y - 70 = 0.075(x - 1250) \Rightarrow$
$y = 0.075x - 23.75$

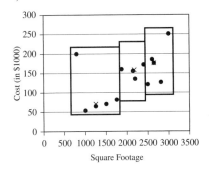

j. Parallel line through M_2:

$$y - 157.5 = 0.075(x - 2175) \Rightarrow y = 0.075x - 5.625$$

y-intercept of median-median line:

$$\frac{(-23.75) + (-23.75) + (-5.625)}{3} = -17.71$$

Median-median line: $y = 0.075x - 17.71$

k. The equation predicts a cost of about $158,500. This is quite a bit higher than the $153,000 predicted by the least-squares line.

l. See graph at right.

m. The slopes and y-intercepts are quite different in this case. Because the outlying data point (800, 200) does not follow the trend of the other points, it "pulls" the left end of the least-squares line upward, but does not affect the median-median line. If we believe that this particular house is unusual, the median-median line seems the better fit.

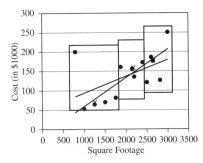

n. Yes. It makes sense to predict the price from the square footage if you are the seller trying to decide on a price. If you are a buyer looking at prices, it makes sense to predict the number of square feet you would get for a given price.

9. **a.** $r = 0.696$; there is a moderate positive relationship between the first and second exams.
   b. $r = 0.730$; there is a moderate positive relationship between the first exam and the final.
   c. $r = 0.977$; there is a strong positive relationship between the second exam and the final.
   d. Yes; as a predictor of performance on the final exam, the second exam is better than the first.
   e. The correlation coefficient would not change. In real life, it makes sense to use the second exam to predict the final exam, not the other way around. The second exam comes first.

CHAPTER 5

1. The middle school objectives from the NCTM's *Principles and Standards for School Mathematics* (page 248) that are found in Chapter 5 are:

    - understand and use appropriate terminology to describe complementary and mutually exclusive events
    - use proportionality and a basic understanding of probability to make and test conjectures about the results of experiments and simulations
    - compute probabilities for simple compound events, using such methods as organized lists, tree diagrams, and area models.

3. The theoretical probability is calculated making assumptions about what would happen in ideal cases or experiments that continued with infinite number of trials. Experimental probability is determined by actually performing some finite number of trials. In practice, there is a very small probability that a coin will land on its edge, which is usually dismissed in the theoretical case. (I have actually seen this happen.)

5. **a.** 6/20 or 0.3   **b.** 1/20 or 0.05   **c.** 3/20 or 0.15
   d. 10/20 or 0.5   **e.** 7/20 or 0.35   **f.** 13/20 or 0.65
   g. 0 since blue and white are mutually exclusive

7. **a.** $85/120 = 17/24$ or 0.7083   **b.** 53/120 or 0.4417
   c. 99/120 or 0.825   **d.** 32/120 or 0.2667
   e. 53/85 or 0.6235   **f.** 21/53 or 0.3962
   g. 53/67 or 0.7910   **h.** 14/67 or 0.2090 (nearest. 0001)

9. $P(\text{room A}) = (1/2)(1/3) + (1/2)(1/3)(1/2) = 1/6 + 1/12 = 3/12 = 1/4$
   $P(\text{room B}) = (1/2) + (1/2)(1/3) + (1/2)(1/3)(1/2) = 1/2 + 1/6 + 1/12$
   $= 9/12 = 3/4$

11. If the dart is thrown randomly and does hit the target, the probability of hitting in a white space is $(3/4)(1/4) = 3/16$, and the $P(\text{shaded region}) = 3(1/4) + (1/4)(1/4) = 13/16$.

13. $P(\text{girl}) = 0.5$, since theoretically the probability of having a girl is 0.5. It does not matter that the other three children are girls, since there is no biological memory involved on the part of the process that determines gender. In other words, each child is an independent event.

15. a. $24/40 = 0.6$ **b.** $36/40 = 0.9$ **c.** $28/40 = 0.7$
d. $8/20 = 0.4$ **e.** $12/16 = 0.75$

CHAPTER 6

1. At first glance it may seem that there are no middle school objectives from the NCTM's *Principles and Standards for School Mathematics* that pertain to Chapter 6; however, counting techniques are embedded in the objectives for probability as well as the techniques of making tree diagrams, organized lists, and so on, that are specifically mentioned.

3. $3 \times 4 = 12$

5. **a.** $\dfrac{6!}{1!3!2!} = 60$ **b.** $\dfrac{9!}{2!1!1!1!1!1!2!1!} = 90{,}720$

   c. $\dfrac{10!}{3!3!1!2!1!} = 50{,}400$ **d.** $\dfrac{8!}{3!2!1!1!1!} = 3360$

7. **a.** $_{15}C_6 = \dfrac{15!}{9!6!} = 5005$ **b.** $_{15}C_4 \cdot {}_5C_2 = 1365 \cdot 10 = 13{,}650$

   c. $_{15}C_1 \cdot {}_5C_5 = 15 \cdot 1 = 15$ **d.** $_{15}C_3 \cdot {}_5C_3 = 455 \cdot 10 = 4550$

   e. $_{20}C_6 = 38{,}760$ **f.** $13{,}650/38{,}760 = 0.3522$ (rounded)

   g. $5005/38{,}760 = 0.1291$ (rounded)

9. $7! = 6040$. While the order makes no difference to us, it certainly would to the drivers who would like to come in first in the race. The car in front has a slight advantage.

11.

Number of Nickels	Probability of the Event
3	$\dfrac{{}_6C_3 \cdot {}_4C_0}{{}_{10}C_3} = \dfrac{20 \cdot 1}{120} = \dfrac{1}{6} \approx 0.1667$
2	$\dfrac{{}_6C_2 \cdot {}_4C_1}{{}_{10}C_3} = \dfrac{15 \cdot 4}{120} = \dfrac{60}{120} = 0.5$
1	$\dfrac{{}_6C_1 \cdot {}_4C_2}{{}_{10}C_3} = \dfrac{6 \cdot 6}{120} = \dfrac{36}{120} = 0.30$
0	$\dfrac{{}_6C_0 \cdot {}_4C_3}{{}_{10}C_3} = \dfrac{1 \cdot 4}{120} = \dfrac{4}{120} \approx 0.0333$

CHAPTER 7

1. The middle school objectives from the NCTM's *Principles and Standards for School Mathematics* that are addressed in this chapter are embedded in the general objective of "develop and evaluate inferences and predictions that are based on data." This chapter sets up the ideas needed to work with sampling distribution from which inferences and predictions will be made about populations.

Answers and Hints for Odd Numbered Exercises

3. a. Since part **a** is based on a simulation, many answers are possible. The table at the right shows one typical example.

Customers Paying Cash	Number in Simulation	Probability
0	4	0.2
1	10	0.5
2	6	0.3

b.

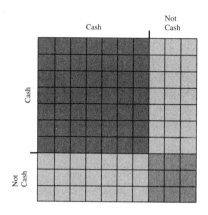

c.

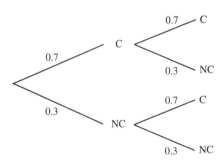

d.

Customers Paying Cash	Probability
0	0.09
1	0.42
2	0.49

e. The tree diagram is probably easier to draw by hand because you don't have to worry about the relative sizes of the boxes. If you are using pre-lined graph paper, the area model may be easier to draw.

f. Area models are limited to two-stage experiments; tree diagrams can be extended to more than 2 stages.

g. This is more like Nikki's situation because the second customer paying cash is not dependent on the first customer's method of payment. The second stage does not depend on the outcome of the first stage.

5. a.

Winnings	Probability
−10	0.10
−5	0.15
0	0.25
10	0.15
25	0.30
50	0.05

  b. $9.75
  c. Variance = 246.19   Standard Deviation = $15.69

7. a. This situation is binomial as the number of trials is fixed ($n = 20$), the probability of success (rolling a number divisible by 3) is the same from trial to trial, and the trials are all independent.
  b. This situation is not binomial. The number of trials is not fixed nor is there a success/failure scenario. Three different events are being monitored at the Dairy Knight.
  c. This situation is binomial as the number of trials is fixed ($n = 10$), the probability of a success (a patient needing fillings) is the same from patient to patient, and the trials are all independent. One could argue that the trials might not be independent if the trials (patients) all came from the same family.
  d. This situation is not binomial. The probability of a success (receiving a single catalog) is not the same from trial (day) to day and the number of trials is not fixed.
  e. This situation is binomial as the number of trials is fixed, the probability of success is the same from trial to trial, and the trials are all independent.
  f. This situation is not binomial. The number of trials is not fixed.

9. a. The mean number of catalogs per day would be:
  $\mu = 1(0.05) + 2(0.10) + 3(0.15) + 4(0.23) + 5(0.28) + 6(0.09) + 7(0.06) + 8(0.04) = 4.3$ catalogs per day.
  b. The variance for the number of catalogs received per day would be:
  $\sigma^2 = \Sigma X^2 \cdot P(X) - \mu^2 = 21.22 - 4.3^2 = 2.73$. The standard deviation would be $\sigma = \sqrt{\sigma^2} = \sqrt{2.73} = 1.652271164$ catalogs per day.

11. Let X be the number of minutes that a customer has to stand in the check-out line. $\mu = 6.4$ minutes, $\sigma = 1.5$ minutes
  a. $P(5 \leq X \leq 8) = 0.6815$ **b.** $P(X < 3) = 0.0116$
  c. $P(X > 9) = 0.0418$ **d.** $P(X < 8) = 0.8577$

13. $\mu = 5.8$ hurricanes per year, $\sigma = 1.45$ hurricanes per year
  a. $P(X = 7) \approx P(y \leq 7.5) - P(y \leq 6.5) = 0.8790 - 0.7794 = 0.0996$
  b. $P(X \leq 3) \approx P(y \leq 3.5) = 0.0559$
  c. $P(X \geq 8) = 1 - P(X \leq 7) \approx 1 - P(y \leq 7.5) = 1 - 0.8790 = 0.1210$

15. $\mu = ?$ minutes, $\sigma = 10.5$ minutes, $X = 150$ minutes, $P(X > 150) = 0.4180$
  We are asked to find μ. If $P(X > 150) = 0.4180$, then $P(X \leq 150) = 0.5820$. This area corresponds to a z-score of 0.21. This gives us $0.21 = \frac{150 - \mu}{10.5}$. Solving for μ yields a mean value of 147.795 minutes to take the state assessment.

17. The mean length of time spent on mathematics instruction is given as 27.3 minutes with $P(X \geq 45) = 0.0885$. If $P(X \geq 45) = 0.0885$, then $P(X < 45) = 1 - 0.0885 = 0.9115$. The z-score that corresponds to an area to the left of

0.9115 is 1.35. Using this information we get $1.35 = \frac{45-27.3}{\sigma}$. Solving for σ we get $\sigma = 13.11$ minutes.

19. We are given that $p = 0.75$ and $n = 8$.
   a. It is not reasonable to approximate this distribution using a normal curve because $n \cdot \min(p,q) = 8 \cdot (0.25) = 2$, which is not greater than 5. We will have to use binomial distribution techniques on this problem.
   b. $P(X \geq 5) = P(5) + P(6) + P(7) + P(8) = 0.2076 + 0.3115 + 0.2670 + 0.1001 = 0.8862$ OR $P(X \geq 5) = 1 - P(X \leq 4) = 1 - 0.1138 = 0.8862$.
   c. $P(3 \leq X \leq 6) = P(3) + P(4) + P(5) + P(6) = 0.0231 + 0.0865 + 0.2076 + 0.3115 = 0.6287$ OR binomcdf$(8, 0.75, 6)$ − binomcdf$(8, 0.75, 2) = 0.6287$.

CHAPTER 8

1. The middle school objectives from the NCTM's *Principles and Standards for School Mathematics* that are addressed in this chapter are embedded in the general objective of "develop and evaluate inferences and predictions that are based on data." This chapter sets up the ideas needed to work with sampling distributions from which inferences and predictions will be made about populations.

3. a. There would be six different samples. The people in each sample are listed in the table at the right.

Sample 1	1	7	13	19	25
Sample 2	2	8	14	20	26
Sample 3	3	9	15	21	27
Sample 4	4	10	16	22	28
Sample 5	5	11	17	23	29
Sample 6	6	12	18	24	30

   b. Person #17 is in only one of the six possible samples. The probability that the sample containing #17 is selected is $\frac{1}{6}$ or 0.1667.
   c. Each person on the list is in only one of the six samples, so each person has the same probability of being selected: $\frac{1}{6}$.
   d. While each person has an equal chance of being chosen, this is not a random sample. We cannot get all possible samples of size 5 this way. For example, it is impossible to get the sample $\{1, 8, 15, 22, 29\}$.
   e. Use randInt(1,30,5) on the TI-83 plus, and if any person is chosen more than once, try again and use the next group of 5. A second method might be to write the numbers 1 to 30 on congruent slips of paper and draw 5 from a container without looking. Using either of these methods, all possible samples of size 5 are equally likely.
   f. $_{30}C_5 = \frac{30!}{25!5!} = 124,506$
   g. Once we choose #17 as being in the set, there are 29 numbers left and we want 4 of them, in any order. There are $_{29}C_4 = \frac{29!}{24!4!} = 23,751$ such groups. So the probability of choosing person #17 is $\frac{23,751}{124,506} = 0.1667$ or $\frac{1}{6}$.
   h. Yes, exactly the same reasoning would apply to any person on the list.

5. a. AB, AC, AD, AE, BC, BD, BE, CD, CE, DE
   b. 4/10 or 0.4
   c. Yes, every prize is in exactly four of the ten samples.

d.

Sample	AB	AC	AD	AE	BC	BD	BE	CD	CE	DE
Values	1,1	1,1	1,2	1,3	1,1	1,2	1,3	1,2	1,3	2,3
Average Value ($\bar{x}$)	1	1	1.5	2	1	1.5	2	1.5	2	2.5

$\bar{x}$	1	1.5	2	2.5
Probability	0.3	0.3	0.3	0.1

e. $\mu = 1\left(\dfrac{3}{5}\right) + 2\left(\dfrac{1}{5}\right) + 3\left(\dfrac{1}{5}\right) = 1.6$ or $\mu = \dfrac{1 + 1 + 1 + 2 + 3}{5} = 1.6$

f. $P(\bar{x} = \mu) = 0$

g. $P(\bar{x} < \mu) = 0.6 \quad P(\bar{x} > \mu) = 0.4$
It is more likely that $\bar{x}$ underestimates μ.

h. $\mu_{\bar{x}} = 1(0.3) + 1.5(0.3) + 2(0.3) + 2.5(0.1) = 1.6$

i. Yes, $\bar{x}$ is an unbiased estimator of μ, since the mean of the distribution of $\bar{x}$ is the same as μ.

7. a. For the sampling distribution, the mean is 3, the variance is 1, and the standard deviation is 1.

b. For the original population, the mean is 3, the variance is 2, and the standard deviation is $\sqrt{2}$, or about 1.414. The formula used for the variance is $\sigma^2 = \dfrac{\Sigma(x-\mu)^2}{N} = \dfrac{10}{5} = 2$

c. The mean of the sampling distribution is exactly the same as the mean of the original distribution.

d. The variance of the sampling distribution is half the variance of the original population. This agrees with the formula $(\sigma_{\bar{x}})^2 = \dfrac{\sigma^2}{n}$, where $n = 2$.

e. If the sample size was increased from 2 to 4 the mean would remain the same, $\mu_{\bar{x}} = 3$. Regardless of the sample size, the sample mean is an unbiased estimator of the population mean: $\mu_{\bar{x}} = \mu$.

f. If the sample size was increased to 4, the variance would be cut in half again to get 0.5, one-fourth the original variance. $Var(\bar{x}) = \dfrac{\sigma^2}{n}$, now with $n = 4$.

9. a. $\mu_{\bar{x}} = \mu = 18.2 \quad \sigma_{\bar{x}} = \dfrac{\sigma}{\sqrt{n}} = \dfrac{4.7}{\sqrt{50}} = 0.6647$

b. The distribution of $\bar{x}$ is (approximately) *normal*. The key information that leads us to this conclusion is the *sample size* (50). According to the Central Limit Theorem, regardless of the distribution of the population, the distribution of $\bar{x}$ will be close to normal if the sample size is large enough (at least 30, by our rule of thumb).

c.

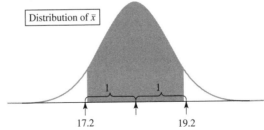

Values of $\bar{x}$ between 17.2 inches and 19.2 inches are within 1 inch of the population mean.

d. $z_1 = \dfrac{17.2 - 18.2}{0.6647} = -1.50 \qquad z_2 = \dfrac{19.2 - 18.2}{0.6647} = 1.50$

$P(-1.50 \le z \le 1.50) = 0.9332 - 0.0668 = 0.8664$

e. The larger sample size should <u>increase</u> the probability that $\bar{x}$ is close to μ.

f. $\sigma_{\bar{x}} = \dfrac{4.7}{\sqrt{100}} = 0.47$

$z_1 = \dfrac{17.2 - 18.2}{0.47} = -2.13 \qquad z_2 = \dfrac{19.2 - 18.2}{0.47} = 2.13$

$P(-2.13 \le z \le 2.13) = 0.9834 - 0.0166 = 0.9668$

The probability increased, just as predicted.

11. a. $\dfrac{1}{2} = 0.5 \qquad P\left(\dfrac{1}{2} \text{ red}\right) = \dfrac{(_2C_1)(_3C_1)}{(_5C_2)} = \dfrac{6}{10} = 0.6$

(You could get this same result by using a tree diagram.)

b. $0, 0.5,$ and 1

c.

$\hat{p}$	Probability
0	0.3
$\dfrac{1}{2}$	0.6
1	0.1

d. $p = \dfrac{2}{5} = 0.4 \qquad P(\hat{p} = p) = 0$

e. $P(\hat{p} < p) = 0.3 \qquad P(\hat{p} > p) = 0.7$

The sample is more likely to overestimate p.

f. $E(\hat{p}) = 0(0.3) + \left(\dfrac{1}{2}\right)(0.6) + 1(0.1) = 0.4$

g. Yes, $\hat{p}$ is an unbiased estimator of p. Although the value of $\hat{p}$ is sometimes larger than p and sometimes smaller, the average or expected value of $\hat{p}$ is 0.4, exactly the same as p.

13. a.

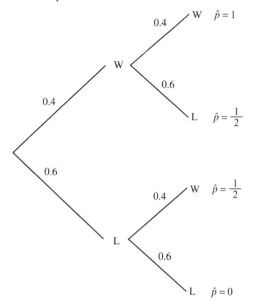

374 Answers and Hints for Odd Numbered Exercises

b.

$\hat{p}$	$P(\hat{p})$
0	0.36
$\frac{1}{2}$	0.48
1	0.16

c. $P(\hat{p} = p) = 0$
d. $P(\hat{p} < p) = 0.36 \qquad P(\hat{p} > p) = 0.64$
The sample is more likely to overestimate p.
e. $E(\hat{p}) = 0(0.36) + \frac{1}{2}(0.48) + 1(0.16) = 0.4$
f. Since the expected value of $\hat{p}$ is 0.4, exactly the same as p, $\hat{p}$ is an unbiased estimator of p.
g. $\sigma_{\hat{p}}^2 = \left[0^2(0.36) + \left(\frac{1}{2}\right)^2(0.48) + 1^2(0.16) \right] - 0.4^2 = 0.12$

$\sigma_{\hat{p}}^2 = \frac{pq}{n} = \frac{(0.4)(0.6)}{2} = 0.12$ The result is the same.

15. a. $n \cdot \min(p,q) = 44(0.2) = 8.8 > 5$
Yes, the sample size is large enough for $\hat{p}$ to have a normal distribution.
b. $\mu_{\hat{p}} = p = 0.8$

$\sigma_{\hat{p}}^2 = \frac{(0.8)(0.2)}{44} = 0.00364$

$\sigma_{\hat{p}} = \sqrt{0.00364} = 0.0603$

c.

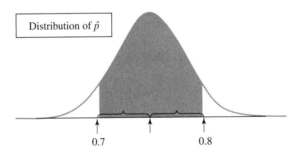

Values of $\hat{p}$ between 0.75 and 0.85 are within 0.05 of p.
d. $z_1 = \frac{0.75 - 0.80}{0.0603} = -0.83 \qquad z_2 = \frac{0.85 - 0.80}{0.0603} = 0.83$
$P(0.75 \leq \hat{p} \leq 0.85) = P(-0.83 \leq z \leq 0.83) = 0.7967 - 0.2033 = 0.5934$
e. The larger sample size should *increase* the probability that $\hat{p}$ is close to p.
f. $\sigma_{\hat{p}} = \sqrt{\frac{pq}{n}} = \sqrt{\frac{(0.8)(0.2)}{88}} = 0.0426$

$z_1 = \frac{0.75 - 0.80}{0.0426} = -1.17 \qquad z_2 = \frac{0.85 - 0.80}{0.0426} = 1.17$
$P(0.75 \leq \hat{p} \leq 0.85) = P(-1.17 \leq z \leq 1.17) = 0.8790 - 0.1210 = 0.7760$
Just as predicted, the probability increased.

CHAPTER 9

1. Chapter 9 extends the ideas present in the NCTM's *Principles and Standards for School Mathematics* at the middle school and high school level regarding statistical inference. While middle school students will probably not discuss confidence intervals, it is an integral part of a college course on elementary statistics.

3. **a.** $\hat{p} = \dfrac{34}{50} = 0.68 \quad \hat{q} = 1 - 0.68 = 0.32$

   $n \cdot \min(\hat{p}, \hat{q}) = 50(0.32) = 16 > 5$

   Yes, the sample size is large enough for the distribution of $\hat{p}$ to be close to normal.

   b. $E = z_{\alpha/2}\sqrt{\dfrac{\hat{p}\hat{q}}{n}} = 1.96\sqrt{\dfrac{(0.68)(0.32)}{50}} = 0.129$

   $\hat{p} \pm E = 0.68 \pm 0.129 = (0.551, 0.809)$

   c. No; based on the confidence interval, we believe that Vicki's shooting percentage could be anywhere from 55.1% to 80.9%. If it is near 55.1%, she would not be better than Nicky.

   d. We would have less confidence that a narrower interval would include p, so a 90% confidence interval would be narrower than a 95% interval.

   e. $E = z_{\alpha/2}\sqrt{\dfrac{\hat{p}\hat{q}}{n}} = 1.645\sqrt{\dfrac{(0.68)(0.32)}{50}} = 0.109$

   $\hat{p} \pm E = 0.68 \pm 0.109 = (0.571, 0.789)$

   As predicted, this interval is narrower than the 95% interval.

5. **a.** $\hat{p} = \dfrac{52}{80} = 0.65 \quad \hat{q} = 1 - 0.65 = 0.35$

   $E = z_{\alpha/2}\sqrt{\dfrac{\hat{p}\hat{q}}{n}} = 1.96\sqrt{\dfrac{(0.65)(0.35)}{80}} = 0.105$

   $\hat{p} \pm E = 0.65 \pm 0.105 = (0.545, 0.755)$

   b. No, we cannot be sure that the true proportion of male students is in the interval. "Ninety-five percent confidence" means that by using the same procedure, 95% of all random samples would result in intervals that included the true proportion. Since we can't be sure our sample was one of the 95% or one of the 5%, we cannot be 100% confident that the true proportion is in this interval.

   c. $\hat{p} = \dfrac{28}{80} = 0.35 \quad \hat{q} = 1 - 0.35 = 0.65$

   $E = z_{\alpha/2}\sqrt{\dfrac{\hat{p}\hat{q}}{n}} = 1.96\sqrt{\dfrac{(0.35)(0.65)}{80}} = 0.105$

   $\hat{p} \pm E = 0.35 \pm 0.105 = (0.245, 0.455)$

   d. The two error bounds are exactly the same. Since there are only two possible outcomes in this situation, changing the one we are looking at just switches $\hat{p}$ and $\hat{q}$.

7. **a.** Since σ is unknown, we should ordinarily use a T-distribution with $n - 1 = 199$ degrees of freedom. But, since Table 2 only goes as high as 100 degrees of freedom, we'll use standard normal. With a sample size this large, there is very

little difference between the T-distribution and standard normal. If you used the TI-83 plus to find this interval, it would be more accurate to use TInterval with $n = 200$, rather than ZInterval.

$$E = 1.645 \left(\frac{5.25}{\sqrt{200}} \right) = 0.611$$

$$\bar{x} \pm E = 4.55 \pm 0.611 = (3.94, 5.16)$$

For comparison, TInterval gives (3.9365, 5.1635), almost the same.

b. No, it is not necessary to assume that the distribution of scores is approximately normal. Since the sample size is large enough ($200 \geq 30$), the distribution of $\bar{x}$ should be close to normal, regardless of the population's distribution.

9. a. $\bar{x} = \dfrac{\sum x}{n} = \dfrac{10{,}137}{5} = 2027.4$

$$s = \sqrt{\frac{N\sum x^2 - (\sum x)^2}{n(n-1)}} = \sqrt{\frac{5(20{,}643{,}299) - 10{,}137^2}{5(4)}} = 151.28$$

b. Using Table 2 with a central area of 0.90 and $n - 1 = 4$ degrees of freedom, $t_{\alpha/2} = 2.132$.

$$E = t_{\alpha/2} \left(\frac{s}{\sqrt{n}} \right) = 2.132 \left(\frac{151.28}{\sqrt{5}} \right) = 144.24$$

$$\bar{x} \pm E = 2027.4 \pm 144.24 = (1883.16, 2171.64)$$

c. This small sample resulted in over $10,000 worth of damage to the vehicles. A larger sample would be even more expensive.

11. $n = \dfrac{(z_{\alpha/2})^2 \hat{p}\hat{q}}{E^2} = \dfrac{(2.576)^2 (0.065)(0.935)}{0.02^2} = 1008.22$
Rounding up to the nearest whole number gives $n = 1009$ rolls.

13. $n = \dfrac{(z_{\alpha/2})^2 \sigma^2}{E^2} = \dfrac{(1.96)^2 1.23^2}{0.1^2} = 581.20$
Rounding up to the nearest whole number gives $n = 582$ candies.

15. With no prior knowledge about σ, the best we can do is use the rough estimate $\sigma \approx \approx \frac{Range}{4}$. Since incomes that are $10,000 or less must be between $0 and $10,000, $\sigma \approx \approx \frac{10{,}000 - 0}{4} = 2{,}500$

$$n = \frac{(z_{\alpha/2})^2 \sigma^2}{E^2} = \frac{(2.576)^2\, 2500^2}{500^2} = 165.89$$

Rounding up to the nearest whole number gives $n = 166$ people.

CHAPTER 10

1. Chapter 10 extends the ideas present in the NCTM's *Principles and Standards for School Mathematics* at the middle school and high school level regarding statistical inference. While middle school students will probably not discuss hypothesis testing, it is an integral part of a college course on elementary statistics.

Answers and Hints for Odd Numbered Exercises

3. a. The mathematics department is treating Claim 2 as the null hypothesis and Claim 1 as the alternative hypothesis. Evidence is being gathered to support Claim 1 through various means, like high school grades, placement tests, and so on.

b. A Type I error would be the situation where the null hypothesis is true, the student is not ready for College Algebra, but it is rejected. In this case a student would be placed in College Algebra even though they were not ready for College Algebra. Chances are the student would not do very well.

c. A Type II error in this situation would be the case where the null hypothesis is actually false—in other words the student is ready for College Algebra—but the null is accepted, meaning that the student is placed in Intermediate Algebra. One possible consequence of such a conclusion could be that the student performs extremely well in Intermediate Algebra since it would be content they were already comfortable with. Another possible consequence of such a conclusion could be that the student gets bored in class since they already know the content, stops coming to class, and ends up failing the class.

5. a. $H_0: p \leq 0.60$ Nicky's free-throw shooting percentage has remained the same (or less) since last year;
$H_1: p > 0.60$ Nicky's free-throw shooting percentage has improved since last year and is now greater than 60%.

b. The sample is large enough to use a normal distribution for this test as $n \cdot \min(p_0, q_0) = 50 \cdot 0.40 = 20 > 5$.

c. The reject region in this case is the area in the right tail region of the normal curve. Since the significance level was given to be $\alpha = 0.10$, the $z_\alpha = 1.28$. If our test statistic is greater than 1.28, we will reject the null hypothesis.

d. $z = \dfrac{\dfrac{35}{50} - 0.60}{\sqrt{\dfrac{(0.60)(0.40)}{50}}} = 1.44$

e. Based on our test statistic of 1.44 and the significance level of 0.10, we would reject the null hypothesis and conclude that Nicky's free-throw shooting ability has indeed significantly improved to more than 60%.

7. a. $H_0: p \leq 0.40$. The proportion of middle school students who plan to attend the dance is not significantly more than 0.40. $H_1: p > 0.40$. The proportion of middle school students who plan to attend the dance is significantly more than 0.40.

b. As $n \cdot \min(p_0, q_0) = 20(0.4) = 8 > 5$, the sample is large enough to use a normal distribution for this test.

c. The reject region would consist of 5% of the tail area on the right of the normal curve. This reject area corresponds to $z_\alpha = 1.645$. If our test statistic is greater than 1.645 we will reject the null hypothesis. Otherwise, we will reserve judgment on the null.

d. $z = \dfrac{\dfrac{11}{20} - 0.40}{\sqrt{\dfrac{(0.40)(0.60)}{20}}} = 1.37$

e. As the test statistic is less than $z_\alpha = 1.645$, we cannot conclude that the middle school students would support the eighth grade plan. We reserve judgment on the null hypothesis.

378 Answers and Hints for Odd Numbered Exercises

9. **a.** $H_0: p = 0.50$. The game is such that it is equally likely that either player will win. Or, the probability of winning is not significantly different from 0.50. H_1: The game is such that Lucy will lose significantly more than 50% of the time. Or, the probability of winning is significantly less than 0.50.
   b. The sample is barely large enough to use a normal distribution for the test as $n \cdot \min(p_0, q_0) = 12(0.5) = 6 > 5$.
   c. The reject region will be on the left of the normal curve this time as the alternative hypothesis is "less than." At the 0.10 level of significance we get $z_\alpha = -1.28$ so if the test statistic is less than -1.28, we will reject the null hypothesis in favor of the alternative hypothesis.
   d. $z = \dfrac{\dfrac{4}{12} - 0.50}{\sqrt{\dfrac{(0.50)(0.50)}{12}}} = -1.15$
   e. The value of the test statistic is not less than $z_\alpha = -1.28$, so we cannot reject the null hypothesis. There is not enough evidence to support Lucy's claim that the game is rigged so that she will lose most of the time.

11. Using the TEST menu and 1-sample proportions from the TI-83 plus, we get a P-value of 0.0244. As this value is less than the given significance value of 0.05, we would still reject the null hypothesis and conclude that the proportion of people who prefer purple has changed since 2002.
   We would have reached the same conclusion using an alpha of 0.10, as the P-value is still less than 0.10. If 0.01 is used, however, we would not be able to reject the null hypothesis.

13. Using the TEST menu and 1-sample proportions from the TI-83 plus, we get a P-value of 0.124. Since this P-value is greater than 0.10, the significance level, we would be unable to reject the null hypothesis. We would reach the same conclusion if $\alpha = 0.05$, as the P-value would still be greater than α.

15. **a.** $H_0: \mu = 1000$ *hours*. The average life of this particular brand of light bulbs is not significantly different from 1000 hours. $H_1: \mu < 1000$ *hours*. The average life of this particular brand of light bulbs is significantly less than 1000 hours.
   b. Since the population standard deviation is unknown, we must assume that the original population from which the sample was taken has a normal distribution.
   c. As we are using the sample standard deviation as an estimate of the population standard deviation a t-test is used instead of a z-test. The degrees of freedom in this case will be $36 - 1 = 35$. The corresponding $t_\alpha = -1.69$, meaning that the reject region of the normal curve is to the left of $t_\alpha = -1.69$.
   d. $t = \dfrac{982 - 1000}{\dfrac{75.3}{\sqrt{36}}} = -1.43$
   e. Since the test statistic is not less than -1.690, we cannot reject the null hypothesis. We cannot conclude that the average life of this particular brand of light bulbs is significantly less than 1000 hours.
   f. Looking at the row for 35 degrees of freedom in Table 2, we see that 1.43 is between the values 1.306 and 1.690. Therefore, the tail area for 1.43 must be between the tail areas for 1.306 and 1.690, between 0.10 and 0.05. The P-value is the area to the left of -1.43, but this area is exactly the same, so $0.05 < P\text{-value} < 0.10$. Since $P\text{-value} > \alpha = 0.05$, we would not reject the null hypothesis. This is the same conclusion we reached in e).

17. a. H_0: $\mu = 1200$ words per minute. The average reading speed of people who complete the course is not significantly different from 1200 words per minute. H_1: $\mu < 1200$ words per minute. The average reading speed of people who complete the course is significantly less than 1200 words per minute.

b. Since the sample size is small and the population standard deviation is not known, we must assume that the population from which the sample was drawn has a normal distribution.

c. For the 0.05 level of significance and $12 - 1 = 11$ degrees of freedom, the rejection region would be the area to the left of $t_\alpha = -1.796$. If the test statistic is less than -1.796, then the null hypothesis will be rejected.

d. $t = \dfrac{1126 - 1200}{\dfrac{132}{\sqrt{12}}} = -1.94$

e. Based on the test statistic, we would reject the null hypothesis at the 0.05 level of significance and conclude that people who participate in the speed reading course have average speeds significantly less than the advertised speed of 1200 words per minute.

f. Looking at the row for 11 degrees of freedom in Table 2, we see that 1.94 is between the values 1.796 and 2.201. Therefore, the tail area for 1.94 must be between the tail areas for 1.796 and 2.201, between 0.05 and 0.025. The P-value is the area to the left of -1.94, but this area is exactly the same, so $0.025 < P\text{-}value < 0.05$. Since $P\text{-}value < \alpha = 0.05$, we would reject the null hypothesis and reach exactly the same conclusion as in e).

Photo Credits

Page 8	From *Connected Mathematics: Samples and Populations* by Glenda Lappan, James T. Fey, William M. Fitzgerald, Susan N. Friel, and Elizabeth Defanis Phillips. © 2004 by Michigan State University. Published by Pearson Education, Inc., publishing as Pearson Prentice Hall. Used by permission.
Page 9	From *Connected Mathematics: Samples and Populations* by Glenda Lappan, James T. Fey, William M. Fitzgerald, Susan N. Friel, and Elizabeth Defanis Phillips. © 2004 by Michigan State University. Published by Pearson Education, Inc., publishing as Pearson Prentice Hall. Used by permission.
Page 10	From *Connected Mathematics: Samples and Populations* by Glenda Lappan, James T. Fey, William M. Fitzgerald, Susan N. Friel, and Elizabeth Defanis Phillips. © 2004 by Michigan State University. Published by Pearson Education, Inc., publishing as Pearson Prentice Hall. Used by permission.
Page 11	Greg Shuck, Columbia, MO. Used by permission.
Page 12	From *Connected Mathematics: Samples and Populations* by Glenda Lappan, James T. Fey, William M. Fitzgerald, Susan N. Friel, and Elizabeth Defanis Phillips. © 2004 by Michigan State University. Published by Pearson Education, Inc., publishing as Pearson Prentice Hall. Used by permission.
Page 13	From *Connected Mathematics: Samples and Populations* by Glenda Lappan, James T. Fey, William M. Fitzgerald, Susan N. Friel, and Elizabeth Defanis Phillips. © 2004 by Michigan State University. Published by Pearson Education, Inc., publishing as Pearson Prentice Hall. Used by permission.
Page 21	Used by permission of McDougal Littell Inc., a division of Houghton Mifflin.
Page 24	Used by permission of McDougal Littell Inc., a division of Houghton Mifflin.
Page 26	Used by permission of McDougal Littell Inc., a division of Houghton Mifflin.
Page 34	Used by permission of McDougal Littell Inc., a division of Houghton Mifflin.
Page 32	From *MathScape: Looking Behind the Numbers*, © 1991, Glencoe/McGraw-Hill. Reprinted by permission.
Page 32	Used by permission of McDougal Littell Inc., a division of Houghton Mifflin.
Page 34	Used by permission of McDougal Littell Inc., a division of Houghton Mifflin.
Page 35	Used by permission of McDougal Littell Inc., a division of Houghton Mifflin.
Page 35	From *MathScape: What Does the Data Say*, © 1991, Glencoe/McGraw-Hill. Reprinted by permission.
Page 36	From *Connected Mathematics: Variables and Patterns* by Glenda Lappan, James T. Fey, William M. Fitzgerald, Susan N. Friel, and Elizabeth Defanis Phillips. © 2004 by Michigan State University. Published by Pearson Education, Inc., publishing as Pearson Prentice Hall. Used by permission.
Page 36	Used by permission of McDougal Littell, Inc., a division of Houghton Mifflin.
Page 39	Used by permission of McDougal Littell, Inc., a division of Houghton Mifflin.
Page 41	Used by permission of McDougal Littell, Inc., a division of Houghton Mifflin.
Page 41	Adapted from www.the-movie-times.com.
Page 43	Copyright © CNN.com. Reprinted by permission.
Page 44	Used by permission of Dustin Jones.
Page 50	Used by permission of McDougal Littell, Inc., a division of Houghton Mifflin.
Page 52	Used by permission of McDougal Littell, Inc., a division of Houghton Mifflin.
Page 52	© Clive Bromhall/OSF/Animals Animals.

Photo Credits

Page 53	From *Connected Mathematics: Data About Us* by Glenda Lappan, James T. Fey, William M. Fitzgerald, Susan N. Friel, and Elizabeth Defanis Phillips. © 2004 by Michigan State University. Published by Pearson Education, Inc., publishing as Pearson Prentice Hall. Used by permission.
Page 54	From *MathScape: What Does the Data Say*, © 1991, Glencoe/McGraw-Hill. Reprinted by permission.
Page 67	Reprinted with permission from *Mathematics in Context: Dealing with Data*, © 1998 by Encyclopaedia Britannica, Inc.
Page 76	From *Connected Mathematics: Samples and Populations* by Glenda Lappan, James T. Fey, William M. Fitzgerald, Susan N. Friel, and Elizabeth Defanis Phillips. © 2004 by Michigan State University. Published by Pearson Education, Inc., publishing as Pearson Prentice Hall. Used by permission.
Page 78	Used by permission of McDougal Littell Inc., a division of Houghton Mifflin.
Page 79	Used by permission of McDougal Littell Inc., a division of Houghton Mifflin.
Page 87	Reprinted with permission from *Mathematics in Context: Statistics and the Environment,* © 1998 by Encyclopaedia Britannica, Inc.
Page 88	From *MathScape: Looking Behind the Numbers*, © 1991, Glencoe/McGraw-Hill. Reprinted by permission.
Page 89	Reprinted with permission from *Mathematics in Context: Insights Into Data*, © 2003 by Encyclopaedia Britannica.
Page 90	Used by permission of McDougal Littell Inc., a division of Houghton Mifflin.
Page 97	Used by permission of McDougal Littell Inc., a division of Houghton Mifflin.
Page 102	Used by permission of McDougal Littell Inc., a division of Houghton Mifflin.
Page 119	From *MathScape: Chance Encounters*, © 1991, Glencoe/McGraw-Hill. Reprinted by permission.
Page 122	Reprinted with permission from *Mathematics in Context: Take a Chance*, © 2003 by Encyclopaedia Britannica, Inc.
Page 127	From *MathScape: Chance Encounters*, © 1991, Glencoe/McGraw-Hill. Reprinted by permission.
Page 128	Reprinted with permission from *Mathematics in Context: Take a Chance,* © 2003 by Encyclopaedia Britannica, Inc.
Page 129	From *Connected Mathematics: How Likely Is It?* by Glenda Lappan, James T. Fey, William M. Fitzgerald, Susan N. Friel, and Elizabeth Defanis Phillips. © 2004 by Michigan State University. Published by Pearson Education, Inc., publishing as Pearson Prentice Hall. Used by permission.
Page 130	From *Connected Mathematics: How Likely Is It?* by Glenda Lappan, James T. Fey, William M. Fitzgerald, Susan N. Friel, and Elizabeth Defanis Phillips. © 2004 by Michigan State University. Published by Pearson Education, Inc., publishing as Pearson Prentice Hall. Used by permission.
Page 140	From *Connected Mathematics: What Do You Expect?* by Glenda Lappan, James T. Fey, William M. Fitzgerald, Susan N. Friel, and Elizabeth Defanis Phillips. © 2004 by Michigan State University. Published by Pearson Education, Inc., publishing as Pearson Prentice Hall. Used by permission.
Page 141	From *Connected Mathematics: What Do You Expect?* by Glenda Lappan, James T. Fey, William M. Fitzgerald, Susan N. Friel, and Elizabeth Defanis Phillips. © 2004 by Michigan State University. Published by Pearson Education, Inc., publishing as Pearson Prentice Hall. Used by permission.
Page 143	From *MathScape: Chance Encounters*, © 1991, Glencoe/McGraw-Hill. Reprinted by permission.
Page 145	From *Connected Mathematics: How Likely Is It?* by Glenda Lappan, James T. Fey, William M. Fitzgerald, Susan N. Friel, and Elizabeth Defanis Phillips. © 2004 by Michigan State University. Published by Pearson Education, Inc., publishing as Pearson Prentice Hall. Used by permission.
Page 146	Used by permission of McDougal Littell Inc., a division of Houghton Mifflin.

Photo Credits

Page 154	From *MathScape: Looking Behind the Numbers*, © 1991, Glencoe/McGraw-Hill. Reprinted by permission.
Page 155	Reprinted with permission from *Mathematics in Context: Take a Chance*, © 2003 by Encyclopaedia Britannica, Inc.
Page 159	From *Connected Mathematics: Clever Counting* by Glenda Lappan, James T. Fey, William M. Fitzgerald, Susan N. Friel, and Elizabeth Defanis Phillips. © 2004 by Michigan State University. Published by Pearson Education, Inc., publishing as Pearson Prentice Hall. Used by permission.
Page 160	From *MathScape: Looking Behind the Numbers* © 1991, Glencoe/McGraw-Hill. Reprinted by permission.
Page 161	From *Connected Mathematics: Clever Counting* by Glenda Lappan, James T. Fey, William M. Fitzgerald, Susan N. Friel, and Elizabeth Defanis Phillips. © 2004 by Michigan State University. Published by Pearson Education, Inc., publishing as Pearson Prentice Hall. Used by permission.
Page 165	From *MathScape: Looking Behind the Numbers* © 1991, Glencoe/McGraw-Hill. Reprinted by permission.
Page 167	From *Connected Mathematics: Clever Counting* © 2004 by Michigan State University. Glenda Lappan, James T. Fey, William M. Fitzgerald, Susan N. Friel, & Elizabeth Defanis Phillips. Published by Pearson Education, Inc., publishing as Pearson Prentice Hall. Used by permission.
Page 170	Used by permission of McDougal Littell Inc., a division of Houghton Mifflin.
Page 170	Used by permission of McDougal Littell Inc., a division of Houghton Mifflin.
Page 174	From *MathScape: Looking Behind the Numbers*, © 1991, Glencoe/McGraw-Hill. Reprinted by permission.
Page 184	From *Connected Mathematics: What Do You Expect?* by Glenda Lappan, James T. Fey, William M. Fitzgerald, Susan N. Friel, and Elizabeth Defanis Phillips. © 2004 by Michigan State University. Published by Pearson Education, Inc., publishing as Pearson Prentice Hall. Used by permission.
Page 187	Used by permission of McDougal Littell Inc., a division of Houghton Mifflin.
Page 188	From *MathScape: Chance Encounters*, © 1991, Glencoe/McGraw-Hill. Reprinted by permission.
Page 189	From *Connected Mathematics: What Do You Expect?* by Glenda Lappan, James T. Fey, William M. Fitzgerald, Susan N. Friel, and Elizabeth Defanis Phillips. © 2004 by Michigan State University. Published by Pearson Education, Inc., publishing as Pearson Prentice Hall. Used by permission.
Page 190	From *Connected Mathematics: What Do You Expect?* by Glenda Lappan, James T. Fey, William M. Fitzgerald, Susan N. Friel, and Elizabeth Defanis Phillips. © 2004 by Michigan State University. Published by Pearson Education, Inc., publishing as Pearson Prentice Hall. Used by permission.
Page 192	From *Connected Mathematics: What Do You Expect?* by Glenda Lappan, James T. Fey, William M. Fitzgerald, Susan N. Friel, and Elizabeth Defanis Phillips. © 2004 by Michigan State University. Published by Pearson Education, Inc., publishing as Pearson Prentice Hall. Used by permission.
Page 205	From *Connected Mathematics: What Do You Expect?* by Glenda Lappan, James T. Fey, William M. Fitzgerald, Susan N. Friel, and Elizabeth Defanis Phillips. © 2004 by Michigan State University. Published by Pearson Education, Inc., publishing as Pearson Prentice Hall. Used by permission.
Page 227	From *Connected Mathematics: What Do You Expect?* by Glenda Lappan, James T. Fey, William M. Fitzgerald, Susan N. Friel, and Elizabeth Defanis Phillips. © 2004 by Michigan State University. Published by Pearson Education, Inc., publishing as Pearson Prentice Hall. Used by permission.
Page 236	From *Connected Mathematics: Samples and Populations* by Glenda Lappan, James T. Fey, William M. Fitzgerald, Susan N. Friel, and Elizabeth Defanis Phillips. © 2004 by Michigan State University. Published by Pearson Education, Inc., publishing as Pearson Prentice Hall. Used by permission.

Photo Credits **383**

Page 238 — From *Connected Mathematics: Samples and Populations* by Glenda Lappan, James T. Fey, William M. Fitzgerald, Susan N. Friel, and Elizabeth Defanis Phillips. © 2004 by Michigan State University. Published by Pearson Education, Inc., publishing as Pearson Prentice Hall. Used by permission.

Page 239 — From *Connected Mathematics: Samples and Populations* by Glenda Lappan, James T. Fey, William M. Fitzgerald, Susan N. Friel, and Elizabeth Defanis Phillips. © 2004 by Michigan State University. Published by Pearson Education, Inc., publishing as Pearson Prentice Hall. Used by permission.

Page 255 — From *Connected Mathematics: How Likely Is It?* by Glenda Lappan, James T. Fey, William M. Fitzgerald, Susan N. Friel, and Elizabeth Defanis Phillips. © 2004 by Michigan State University. Published by Pearson Education, Inc., publishing as Pearson Prentice Hall. Used by permission.

Page 279 — Reprinted with permission from *Mathematics in Context: Great Expectations*, © 2003 by Encyclopaedia Britannica, Inc.

Page 284 — Reprinted with permission from *Mathematics in Context: Great Expectations*, © 2003 by Encyclopaedia Britannica, Inc.

Page 289 — From *Connected Mathematics: Samples and Populations* by Glenda Lappan, James T. Fey, William M. Fitzgerald, Susan N. Friel, and Elizabeth Defanis Phillips. © 2004 by Michigan State University. Published by Pearson Education, Inc., publishing as Pearson Prentice Hall. Used by permission.

Page 297 — Reprinted with permission from *Mathematics in Context: Great Expectations*, © 2003 by Encyclopaedia Britannica, Inc.

Page 312 — Reprinted with permission from *Mathematics in Context: Great Expectations*, © 2003 by Encyclopaedia Britannica, Inc.

Page 334 — Reprinted with permission from *Mathematics in Context: Dealing with Data*, © 1998 by Encyclopaedia Britannica, Inc.

Index

A

Absolute deviation, 58–59
Alternative hypothesis, 304–306, 314–315
    in tests about proportions, 315
    in tests about means, 329

B

Back-to-back stem-and-leaf plot, 21
Bar graph, 11–12, 23–24
    contrasted with histogram, 32–33
Binomial
    experiment, 202–204
    probability density function (pdf), 203
    probability distribution, 203
    random variable, 200–207, 219–227
Bivariate data, 85
Bounding the P-value, 333
Box-and-whiskers plot, 27–31
Box plot, 27–31

C

Categorical data, 11–12
    displaying categorical data, 20–25
Central Limit Theorem, 249
Chance experiment, 184
Circle graph, 23–24
    for numerical data, 33
Class
    boundaries, 27–33
    frequency graph, 33
    limits, 27–30
    mark, 27–30
    width, 27–30
Classical approach, 319
    contrasted with modern approach, 323
Cluster sampling, 6–7
Combinations, 168–173
Complementary events, 130
Conditional probability, 133–137
Confidence interval
    definition, 267
    for estimating a mean, 280–291
    for estimating a proportion, 267–279
    interpretation, 292–293
Confidence level, 252–253, 267–276
    definition, 267
    interpretation, 292–293
    relation to error bound, 268–276
Contingency table, 132–134
Continuity correction, 221–226
    definition, 223
Continuous
    data, 15, 285
    variable, 15, 285
Convenience sampling, 6
Correlation, 85–90
    and cause-and-effect, 89
    definition, 88
    coefficient, 90–96
Critical value
    from a standard normal distribution, 275–278
    from a T-distribution, 286–288

D

Data, 2–4
    set, 3–4
Decile, 70
Decision rule, 307–313
    in tests about means, 330
    in tests about proportions, 315
    in the modern approach, 328
Degrees of freedom, 286–289
    in confidence intervals, 287
    in hypothesis tests, 331
Descriptive statistics, 2
    in boxplots, 31
    on the TI-83 Plus, 16–18
Deviation, 54
    absolute deviation, 58–59
    in computing variance, 58
    mean deviation, 58–59
    standard deviation, 59–60
Discrete data, 15, 35

Discrete variable, 15, 35
    compared to continuous variable, 208–209
    related to continuity correction, 222–226
Dot plot, 21, 33

E
Empirical Rule, 71
Error bound, 252, 267–277
    definition, 267
    in confidence intervals about means, 284, 287
    in confidence intervals about proportions, 273–277
Estimator, 242
    unbiased, 242–243, 245–246, 255–256
Event, 119–123
    independent events, 137, 139
Expected value, 189–192
Experiment, 23, 120
Experimental probability, 122–125

F
Factorial notation, 161–163, 165, 178
Fitted line, 96–100
    least squares line, 100–106
    median-median line, 106–111
Five-number summary, 74–76, 78
Frequency, 12, 21–22, 26–29
    frequency table, 12, 21–22, 26–29
    relative frequency, 21–23
Fundamental Counting Principle, 154–160

G
General addition rule, 131
General multiplication rule, 138–140
Geometric probability, 141–145

H
Histogram, 24, 26, 32–35
Hypothesis, 304–305
    alternative hypothesis, 304–306, 314–315
    hypothesis testing, 302–337
    null hypothesis, 304–309, 318–319
    research hypothesis, 304–306

I
Independent events, 137, 139
Inferential statistics, 2, 236, 266, 338
Interquartile range, 74–75
Interval data, 14–15

L
Law of Large Numbers, 124–126
Least squares line, 100–106
Level of
    confidence, 252–253, 267–276, 292–293
    significance, 311, 313
Line graph, 5, 35–36
Line plot, 33–35
Lower class limits, 27–30
Lower fence, 76
Lower quartile, 74, 76
Lower tail area, 269–270, 275–276
Lower-tailed test, 315–316, 331, 336–337

M
Margin of error, 252, 267–277
    definition, 267
    in confidence intervals about means, 284, 287
    in confidence intervals about proportions, 273–277
Mean, 50–55
    deviation, 58–59
    of a binomial random variable, 204–206
    of a data set, 50–55
    of a random variable, 189–192
    of a sample, 50–55
    of the distribution of sample means, 245–246, 250
    of the distribution of sample proportions, 255–258
Median, 50, 52–54, 64, 66, 74
Median-median line, 106–111
Misleading graph, 37–43
Mode, 49, 52–54, 64, 66
Modern approach, 323–327
Multiplication of choices principle, 154–160
Mutually exclusive events, 127–128, 130–132

N
Negative correlation, 88–96

Nominal data, 12
Null hypothesis, 304–309, 318–319
    in tests about means, 329
    in tests about proportions, 315
Numerical data, 11–15

O

Observed significance level, 319–327
Ordinal data, 14–15
Outcomes, 120–130
Outliers, 54, 76–79

P

P-value, 319–327
pdf, 203
Pearson's correlation coefficient, 90–96
Percent, 15, 22–23
Percentile, 70
Permutations, 160–168
Population, 2–4
    mean, 237
    parameters, 3–4, 7, 51
    proportion, 254–255
    standard deviation, 60, 64, 291, 329–330
Positive correlation, 88–96
Probability, 119–145
Probability density function (pdf), 203
    for a binomial random variable, 203
Probability distribution, 186–189

Q

Qualitative data, 11
Quantitative data, 11–15
Quartile, 64, 70, 74–76

R

Random
    sample, 4–6, 8–9, 234–242
    variable, 183–189
Range, 34, 54, 57, 66
Ratio data, 14–15
Raw data, 3–4
Rejection region, 312–316, 318–319
Relative frequency, 21–23
    tables, 25–30

S

S_{xx}, S_{xy}, and S_{yy}, 90–94
    computing formulas, 93
    definition, 90
    and Pearson's correlation coefficient, 93
    and the least squares line, 103
Sample, 2–4
Sample mean, 50–55
    compared to population mean, 237–240
    distribution of sample means, 242–253
Sample proportion, 253–260
Sample space, 121–123, 128, 130–136
Sample standard deviation, 59–60, 63–66
Sample statistics, 3–4, 51, 242–243, 260–261
Scatter plot, 85–90
Significance level, 311, 313
    observed significance level, 319–327
Simple random sampling, 4–6, 8–9, 234–242
Slope, 86–88, 97–98
Standard deviation
    for a data set, 59–60, 63–66
    of a binomial random variable, 204–206
    of a random variable, 196–199
    of the distribution of sample means, 248, 250
    of the distribution of sample proportions, 259
Standard error, 281–282, 292
Standard normal curve, 207–213
Statistically significant, 313
Statistics
    descriptive statistics, 2, 16–18, 31
    inferential statistics, 2, 236, 266, 338
    sample statistics, 3–4, 51, 242–243, 260–261
Stem-and-leaf plot, 23–24, 30–32
Stratified sampling, 7–8
Systematic sampling, 6–8

T

T-distribution, 285–290
Tail heavy, 286
Test statistic, 310–314, 321–323
    in tests about means, 328–330
    in tests about proportions, 315
Theoretical probability, 122–125
    compared to experimental probability, 187, 200–202

Total tail area
    in finding confidence intervals, 275–276, 283
    in testing hypotheses, 307
Two-tailed test, 314–317
Type 1 error, 307–309
Type 2 error, 307–309

U

Unbiased estimator
    for a population mean, 242–246, 250
    for a population proportion, 255–256, 258
Univariate data, 23, 85
Upper fence, 76
Upper class limits, 27–30
Upper quartile, 74, 76
Upper tail area, 269–270, 275–276
Upper-tailed test, 314, 320–321, 331

V

Variable, 3–4
    categorical and numerical, 11–12
    discrete and continuous, 15
    random, 183–189
Variance
    of a binomial random variable, 204–205
    of a data set or a sample, 57–59
    of a random variable, 193–196
    of the distribution of sample means, 246–248, 250
    of the distribution of sample proportions, 257–259
Venn diagram, 132, 136

W

Weighted average or weighted mean, 55–56
    related to mean of a random variable, 190
    related to median-median line, 110
    related to variance of a random variable, 194–195
With replacement, 163, 167–168
Without replacement, 138, 149, 161, 163, 167, 171

Y

y-intercept, 98–100
    of least squares line, 103–105
    of median-median line, 109–110

Z

z-score, 70–73, 196–197